Korbinian Stöckl
Love in Contemporary British Drama

CDE Studies

Edited by
Martin Middeke

Volume 31

Korbinian Stöckl

Love in Contemporary British Drama

―

Traditions and Transformations of a Cultural Emotion

DE GRUYTER

ISBN 978-3-11-111123-0
e-ISBN (PDF) 978-3-11-071470-8
e-ISBN (EPUB) 978-3-11-071476-0
ISSN 2194-9069

Library of Congress Control Number: 2020948028

Bibliographic information published by the Deutsche Nationalbibliothek
The Deutsche Nationalbibliothek lists this publication in the Deutsche Nationalbibliografie;
detailed bibliographic data are available on the Internet at http://dnb.dnb.de.

© 2022 Walter de Gruyter GmbH, Berlin/Boston
This volume is text- and page-identical with the hardback published in 2021.
Printing and binding: CPI books GmbH, Leck

www.degruyter.com

Contents

Acknowledgements —— IX

1   Introduction —— 1

Part I **The Making of Romantic/Erotic Love: Traditions and Transformations**

2   **Philosophy of Love: Selections** —— 17
2.1   Love as Compensation —— 18
2.1.1   Plato's *Eros* —— 22
2.1.2   Plato's Romantic Legacy —— 31
2.1.3   Nostalgic Love: Psychoanalysis and the Desire of the Displaced Self —— 36
2.2   The Precariousness of Love —— 46
2.2.1   The Unwanted Guest: Alcibiades and the Desire of the Other —— 47
2.2.2   The Endangered Self: Love's Precariousness in Modern Philosophy —— 54

3   **Romantic Love in Sociology** —— 65
3.1   Modern Functions of Romantic Love: A Very Brief History —— 66
3.2   (Post-) Modern Attitudes to Love —— 72
3.2.1   Love and the Culture of Capitalism —— 72
3.2.2   The Problem of Authenticity: Idealist and Realist Narratives —— 74
3.2.3   Individualism and Romantic Rationality —— 76
3.2.4   Postmodern Precariousness of Love: Choice, Irony, Recognition —— 80

Part II **Analysis: Love in Contemporary British Drama**

4   **"Why isn't love enough?" Commitment in Patrick Marber's *Closer*** —— 93
4.1   Style and Genre: A Modern Comedy of Manners? —— 93
4.2   Synopsis —— 95

4.3		*Closer* in Context: The Crisis of Love at the Turn of the Millennium —— 97
4.3.1		Isolation, Egoism, Self-realisation —— 98
4.3.2		Postmodern Incredulity towards Love —— 103
4.4		Timeless Issues: Compensation and Precariousness —— 116
4.5		Conclusion —— 122
5		"if you're not with me I feel less like a person": Sex, Drugs, and the Myth of Self-Sufficiency in Mark Ravenhill's *Shopping and Fucking* —— 124
5.1		Synopsis —— 125
5.2		"Why are there so many sad people in this world?" The Play's Society —— 127
5.3		Addicted to Love —— 130
5.4		Gary's Desire —— 135
5.5		Conclusion —— 141
6		**Autopsies of Love: Sarah Kane's Erotic Plays —— 142**
6.1		The Love Experience —— 142
6.2		The Aesthetics of Kane's Dramatic Plays —— 145
6.3		"A spear in my side, burning": Precarious Desire in *Phaedra's Love* —— 154
6.3.1		"No one burns me": Hippolytus' Refusal of Love —— 155
6.3.2		"Don't imagine you can cure him": The Need for Compensation —— 158
6.3.3		"If there could have been more moments like this": The Joy of Death —— 160
6.4		*Cleansed:* Fragments from the Laboratory of Love —— 164
6.4.1		A Panopticon of Love —— 167
6.4.2		Heaven or Hell: *Cleansed*'s Ambivalent Eschatology of Love —— 170
6.5		Conclusion —— 180
7		**"Not saying I don't want things though": Emotional and Material Desires in Dennis Kelly's *Love and Money* —— 183**
7.1		Synopsis —— 184
7.2		Everybody Needs Somebody —— 189
7.3		I Would Do Anything for Love … But I Won't Do That —— 192
7.4		All You Need Is Love —— 196
7.5		Conclusion —— 201

8	"Love at first sight and the lost city of Atlantis": Penelope Skinner's *Eigengrau*, Or 'A Fairy Tale of Blind Love' —— 203	
8.1	Synopsis —— 204	
8.2	Urban Loneliness and the Desire for Meaning —— 212	
8.3	"I don't want you to need me": Love's Precariousness and the Desire for Control —— 214	
8.4	Conclusion —— 220	
9	"We don't need ties": Rebellious Love in Mike Bartlett's *Love, Love, Love* —— 222	
9.1	Synopsis —— 222	
9.2	Act One: Breaking Free —— 228	
9.3	Act Two: Trapped —— 231	
9.4	Act Three: The Lost Generation —— 237	
9.5	Conclusion —— 241	
10	"this poetical … *shit*": Coming to Terms with Love in debbie tucker green's *a profoundly affectionate, passionate devotion to someone (-noun)* —— 243	
10.1	Synopsis —— 244	
10.2	'Understanding' and 'Devotion' —— 251	
10.2.1	Love You More Than I Can Say: Epistemological and Terminological Limits of Understanding —— 252	
10.2.2	Devotion: I Want You to Want Me —— 260	
10.3	Conclusion —— 265	
11	Coda —— 267	

Works Cited —— 274

Index of subjects —— 286

Index of authors —— 293

Acknowledgements

I wish to thank all the people who have contributed to this book in one way or another. The present monograph is a slightly revised version of my PhD thesis that I was happily allowed to write under the excellent supervision of Martin Middeke who, besides providing expertise advice and having me in his team at the University of Augsburg, found a felicitous balance between little coercion and much freedom that enabled me to produce an independent study without ever feeling lost. The theatre trips to London and New York were inspiring experiences and helped maintain and intensify the fascination with contemporary drama which was bound to arise at his Chair of English Literature. In this context, I would also like to thank the many members of the Oberseminar who have for years engaged in thorough discussions of contemporary plays, and particularly those who were prepared to extend discussions long into the night: Felix Aping, Christian Attinger, Anja Hartl, Georg Hauzenberger, Timo Müller, and Christina Schönberger-Stepien. My heartfelt appreciation also goes to Mary Wölfle for her meticulous proofreading and to Nicole Held and Carolin Steinke for their editorial assistance. I am grateful to my wonderful colleagues at the University of Augsburg, Eva Ries, Martin Riedelsheimer, and David Kerler, for creating an atmosphere that has always been warm and cooperative, and to the unforgotten and sadly missed Christoph Henke, whose impassioned way of teaching English literature planted the seed of what became the idea of writing a PhD thesis many years ago.

Special regards are due to the German Society of Contemporary Theatre and Drama in English for their inspiring conferences, the publication of many eminent studies on which I could rely for my research, and for distinguishing my thesis with the CDE Award.

I would also like to express my gratitude to my friends and family at home, who always give me a warm welcome when I need it, and especially to my parents Sylvest and Elisabeth, whose unceasing support during the last years has been an invaluable source of confidence.

Most of all, I am indebted to Theresa, the unreferenced primary source for this project, who knows that the language of love is so citational that the only way of speaking innocently of it is to quote self-consciously:

> And once the amorous subject creates or puts together any kind of work at all, he is seized with a desire to dedicate it. What he makes he immediately, and even in advance, wants to give to his beloved, for whom he has worked, or will work.
> Roland Barthes, *A Lover's Discourse* (78)

1 Introduction

To ask about love in contemporary British drama opens up two trajectories that cannot be followed in isolation. From the start, the objective of this study has been to arrive at a better understanding of a selection of plays by enhancing close, text-based interpretations with an analysis of culturally dominant discourses of love. Equally from the start, it has transpired that this sort of discourse analysis would bring the project within the scope of cultural studies, splitting up, as it were, the focus of interest. Even though this book is still primarily intended as a work of literary criticism reaching for a more informed comprehension of specific texts, the enquiry into the formative discourses that shape and influence these texts inevitably places the cultural and social significance as well as the historical evolution of these discourses within the range of vision. As a result, the literary trajectory has been widened by a cultural one to include the contention that an approach that seeks to work out both the historicity and the peculiarities of contemporary conceptions of love is capable of shedding light on one of the most crucial factors that shape and define the time, culture, and society we are part of. In tracing the origins and carving out the distinctive features of contemporary love, I argue, we may arrive at a deeper understanding of some of the notions, beliefs, and emotional dispositions which determine what Raymond Williams has called the "structure of feeling" (Williams 22–24; see also Sharma/Tygstrup 1–2) of our own day and age.

If this claim is to hold, a few prerequisites are necessary. The first, and least debatable, is the assumption that love is still important today, which is testified by the plethora of books, articles, and websites that offer sociological, philosophical, psychological, and biological approaches to love to a broad readership.[1] As philosopher Simon May observes in *Love: A History* (2011), "[a]cademic books, chat shows, pop lyrics, internet dating sites, self-help manuals – all buzz with curiosity about the conditions for successful love" (xi) in a culture that, as he argues, has substituted love for religion and other grand narratives "as the one democratic, even universal, form of salvation open to us moderns" (1). The *New York Times* column "Modern Love," which advertises to publish "weekly essays that explore the joys and tribulations of love" since 2004, is just one of countless possible examples for the ubiquity of love as a topic of public interest.

---

[1] The most influential strand of non-fictional love literature are products of the self-help and therapy industry, which combine and popularise the findings of psychoanalysis, psychology, neuroscience, and sociology and have been increasing their output and discursive impact ever since the 1920s (cf. Illouz, *Cold Intimacies* 9–10; 43).

At the same time, new insights in the fields of neuroscience and anthropology are popularised through approachable books such as biological anthropologist Helen Fisher's influential *Why We Love: The Nature and Chemistry of Romantic Love* (2004). In the wake of what has come to be called the 'affective turn,' academic interest in love has also increased in the humanities. The Call for Papers for a conference titled "Love etc.," organised by Rita Felski at the University of Southern Denmark in October 2019, echoes the same convictions underlying this book: "Endlessly invoked, celebrated, assailed, abused, and parodied, love has been hailed as the meaning of life and disdained as the ultimate cliché. This conference is inspired by the conviction that love is poised to become a focus of renewed interest in the humanities" (Call for Papers). These examples may suffice to illustrate the unceasing or even increasing significance of love for contemporary authors and readers, even before mentioning the vast field of fictional writing, from cheap novelettes to highbrow literature, which, as Catherine Belsey states, "remains the supreme location of writing about desire" and thus "constitutes the primary material for analysis" (*Desire* 11). Moreover, the examples indicate what is a second prerequisite for my approach: that there are distinctively modern or contemporary concerns about love. As I will be able to conclude after Chapter 3, they add up to the paradoxical situation that lasting love is desired and idealised in a society that is afraid of commitment and has largely ceased to believe in transcendental ideals – an attitude that, following Belsey and Eva Illouz, I will refer to as 'postmodern.' It is precisely this ambivalent status of love as "a source of existential transcendence" in a secularised culture contesting and demystifying all transcendental ideals which convinces Illouz that "[t]o study love is not peripheral but *central* to the study of the core and foundation of modernity" (*Why Love Hurts* 9).

An approach that seeks to uncover what is new and typically contemporary in discourses of love must necessarily also look back at the history of thinking about love, and this view will uncover the peculiarities of the present as well as the lasting influences of the past. The third prerequisite underlying this project is the premise that love has a long and continuous history, during which essential ideas reoccur in time-specific variations until the present day. As the historical overview in Chapter 2 will demonstrate, different concepts and variations of love have always imbricated and influenced each other, making neat compartmentalisations difficult. However, the two terms most suitable to denote the kind of love I am concerned with are *erotic* and *romantic*, which designate a form of love that implies emotional and sexual longing for intimacy with a specific and unique other. The use of these words is principally based on their connotations in modern everyday language, where they mean the love between two usually unrelated persons issuing from bodily attraction and a sense of inner connectedness. But as Chapter 2.1 will show, erotic or romantic love, if it is to be more than mere infatuation or sexual lust, must also include components such as reciprocity and altruism that are usually conceived as integral to other kinds of love like *philia* or *agape*. The intricacy of the matter is made obvious by attempts to analytically disentangle the components of love. Fisher, for example, argues that what is commonly called romantic love – a mating drive that focuses amorous attention exclusively on a single beloved for a period of time for reasons of procreation and species preservation (cf. 126–52) – "is deeply entwined with two other mating drives: lust – the craving for sexual gratification; and attachment – the feeling of calm, security, and union with a long-term partner" (78). What makes the experience of love so unfathomable and its definition almost impossible is the fact that the three drives flow into one another in various constellations. Only a scientifically grounded separation of the three distinct drives, Fisher argues, can bring clarity because "once you begin to envisage lust, romantic love, and attachment as three specific mating drives, each producing many gradations of feeling that endlessly combine and recombine in countless different ways, love takes on tangibility" (94). The tripartite classification allows her to view different styles of love as different combinations of these drives. Fisher demonstrates this with reference to psychologist John Alan Lee's Colour Wheel Theory of Love (1973), another systematic approach that distinguishes *eros*, *ludus*, *storge*, *mania*, *agape*, and *pragma*, which are, according to Fisher, explicable as forms of love in which one or more of the three mating drives are present or absent (cf. 95). Considering that in Lee's theory combinations of several types of love such as 'agapic eros' or 'agapic ludus' are possible, it transpires how multifaceted descriptions of love can become. I will not dip deeper into these or similar attempts at categorising love but I have referenced them here to demonstrate that what is often simply called love consists of a likely indefinable number of features which may or may not occur concomitantly. For my purpose, whenever I speak of romantic or erotic love in the following it will be in a broad sense in which an element of sexual passion and attraction towards a specific beloved are necessary features but from which a longing for the security of long-term companionate togetherness and the willingness, even eagerness, to selflessly care for the other are by no means excluded.

* * *

With love's unbroken topicality, its modern peculiarities, and the persistent influence of its history, I have already addressed three premises on which my approach rests. The fourth and most important one is the claim that love is not a radically unique, individual experience but one that is, within a culturally and temporally determinable group, to a considerable extent collective and comparable. On this assumption depends the very possibility to communicate reasonably about love and, more specifically, my approach, which treats the presentations of love in contemporary British drama as artistically individual yet collectively recognisable illustrations of our cultural and social knowledge or, to borrow a term from Roland Barthes, our collective "image repertoire" (6) of love. As regards current debates about affects and emotions in the humanities, this means taking up a middle position, as I will try to explain briefly.

In her oft-quoted essay "Love and Knowledge: Emotion in Feminist Epistemology" (1989), Alison M. Jaggar challenges the view of emotions as standing opposed to knowledge and reason, which had long prevailed in Western thought. She describes how emotions, or 'passions,' were largely considered irrational bodily sensations unrelated to and often in conflict with reason and cognition. In the second half of the twentieth century, cognitivist approaches then successfully promoted the idea of emotions as intentional judgements connected to physiological sensations. One result was the reinstallment of the old dichotomy between bodily sensation and mental cognition within the category of emotions itself so that emotions were thought of as consisting of two components: "an affective or feeling component and a cognition that supposedly interprets or identifies the feelings" (53). In the introduction to the *Palgrave Handbook of Affect Studies and Textual Criticism* (2017), Donald R. Wehrs describes precisely this dichotomy as formative for twentieth-century conceptions of human emotions. In his wording, the legacy of mind-body dualism is expressed by means of distinctive terms for the two components: "In the twentieth century, 'affect' was commonly associated with bodily causality and natural science, 'emotion' with ideas, outlook, social science, and the humanities" (1). Some theories in the emerging field of affect studies contributed to maintaining this dichotomy between pre-intentional, pre-linguistic, purely physical affect on the one hand and emotion, which requires intentional consciousness, on the other (cf. Wehrs 39–40; see also Leys). The turn to affect is hence often a turn away from the impacts of learning, socialisation, and cultural heritage towards the pre-cognitive and innate elements of our emotional/affective apparatus – a focus, to use the terms suggested by Wehrs, on the human mind as a product of "nature" rather than "nurture" (2). My own approach, necessarily, is more indebted to the 'nurture' side. If the sharp distinction between affect as "a dimension of bodily experiences and encounters [...] that remains, significantly, non-semantic and non-representational" and emotions as "a somehow translated, signified and subjectified version of the elusive, pre-discursive affective matter" (Sharma/Tygstrup 7) is maintained, then I will treat love as an emotion because an analysis and interpretation of literary texts is evidently dealing with already verbalised and semanticised material. Moreover, the very claim that contemporary experiences of love are shaped by the long history of thinking, writing, reading, and hearing about love presupposes at least a mild constructionism in the concept of love. For my part, I will refrain from making any hard distinction between affect and emotion in my treatment of what Margaret Wetherell describes as "phenomena that can be read simultaneously as somatic, neural, subjective, historical, social and personal" (11).

This inclusive conception allows to view emotions as an inextricable assemblage not only of bodily and cognitive but also of biological and cultural constituents. Jaggar puts much emphasis on her argument that emotions are not merely "presocial, instinctive responses, determined by our biological constitution" but "socially constructed on several levels" (54). People are socialised in an environment where appropriate responses to situations and appropriate forms of expressing emotions are predetermined. Moreover, the assumption that mature human emotions are intentional in the sense that they are, or at least contain, judgements about specific objects or situations, presupposes a subject in possession of concepts and categories along which such judgments can be made. "For this reason," Jaggar writes, "emotions are simultaneously made possible and limited by the conceptual and linguistic resources of a society" (54). This view accounts for cross-cultural differences in the experience and expression of emotions: despite a "considerable overlap in the emotions identified by many cultures" there are some that are "historically or culturally specific" and others that appear to be universal but yet differ in some respect from one culture to another (cf. 54). But it also accounts for a certain degree of homogeneity regarding the experience and expression of emotions within one cultural group or society. To sum up Jaggar's position, which I will largely adopt,

> mature human emotions are neither instinctive nor biologically determined, although they may have developed out of presocial, instinctive responses. Like everything else that is human, emotions in part are socially constructed; like all social constructs, they are historical products, bearing the mark of the society that constructed them. (60)

In adopting this position, I do *not* necessarily turn against the view put forward by Fisher or social cognitive neuroscientist Matthew Liebermann that the foundations of our emotions have developed for evolutionary reasons (cf. Liebermann 9; 43; 236–37) and that their chemical, endocrinological component is universal among humanity and has hardly changed since the human brain has had its present shape (cf. Fisher 150). Jaggar's concept leaves room for the evolutionary origins of emotions as factors contributing to and probably even determining their physical symptoms while being far from constituting all that is implied in the term 'emotion.' The reason for pointing out this possible congruence is that I reject both a reductionist biologism and a radical constructionism in my treatment of emotions. Mature human emotions are neither entirely universal and unchanging products of evolution nor pure social constructions without any relation to general human characteristics whatsoever, but the products of both 'nature' and 'nurture.' Like Jennifer Harding and E. Deidre Pribram in their introduction to *Emotions: A Cultural Studies Reader* (2009), I "adopt an approach to studying emotion which is both constructionist and culturalist" but I "reject those aspects of social constructionism which focuses on the interiority, depth and individuality of emotion" (12). In other words, I have no intention of suggesting that what individuals actually feel when they experience emotions is socially or discursively constructed. Rather, I would like to "investigate emotions in the contexts of more widely felt, commonly shared spheres" (12). I suppose that besides the individually felt interiority of emotional experience (which, due to the biological/chemical constituents of emotions, might not even be so individual after all), there is that part of an emotion that is socially constructed and thus shared by members of a social group.

Which brings me back to the approach I have chosen for this study. As Lila Abu-Lughod and Catherine A. Lutz attest, "[i]f emotions are social phenomena, discourse is crucial to understanding how they are so constituted" (106), and discourse will indeed be my object of analysis. I am interested in the discourses of love that are repeated in contemporary British drama – how the characters talk about love and how dramatic situations and plot developments express culturally embedded narratives and conceptions of love – but also in the way romantic love itself can, partly, be seen as a discourse, as a set of pre-scripted action sequences, behaviours, and expectations. As Dominic Pettman puts it in *Love and Other Technologies* (2006), "[h]owever we choose to label this phenomenon – romance, love, lust, affection, passion, intimacy, eroticism, longing, dependence, desire, sexuality, obsession – we are dealing with a *discursive* constellation. It is constituted through discourse, embedded in the different cultural and historical genealogies of that-which-is-called-love" (20). Again, this is not to say that romantic love as an emotion is entirely discursively constructed: I have no doubt that the affective, physiological symptoms of love point towards its origins in evolutionary biology. But I am equally convinced that, as a mature emotion, love can neither be thought nor felt detached from the cultural 'knowledge' that has formed around it and which influences or even determines the expectations, explanations, expressions, and evaluations of what is felt. Pettman, too, refers to Barthes' 'image repertoire' as a term that captures what is meant by the conception of love as discourse. "The image-repertoire," he argues,

> is the discursive legacy of love stories which we have imbibed and internalized since we were children, often through literature but also through television, movies, gossip, fairy tales, magazines, and so on. The image-repertoire can be considered the world data bank of images of love, with which we construct our own imaginative mis-en-scènes. (20)

It is hardly possible, this implies, to talk about or experience love without having recourse to this 'data bank.' As an emotion, which is more than a precognitive affective state, love always goes along with culturally embedded modes of sense-making and behaviour – with a 'code,' as Pettman puts it elsewhere – and this code, to draw the line back to the aspect of love's historicity, is necessarily the product of a long, continuous history. "As with computers," Pettman argues,

> no code is born ex nihilo, or reinvented one hundred percent anew. Each version is designed to be 'backwards compatible,' so that it can function without too many issues. And so, we have the same buggy palimpsests in the contemporary codification of intimacy, never able to make a quantum leap from previous ideological eras. Indeed we feel these contradictory cohabitations very intimately as lovers. For instance, at one moment we are obliged to be chivalrous or courtly, and the next we are in the grip of modern love, or free love, or cynical postmodern love. No wonder we sometimes freeze up or crash! ("Love Materialism" 13)

In *Tales of Love* (1987), Julia Kristeva started out with the claim that today "we lack a code of love; no stable mirrors for the loves of a period, group, or class" (6). But what might be closer to the truth (and what is indeed indicated in the rest of her book) is that what we have is rather an overwhelming plethora of codes that, as Pettman suggests, overtax lovers with a multitude of incommensurable and often conflicting rules and demands. To dig into this palimpsest of love, to excavate the codes of bygone ages and uncover their traces in those of our own time, is the aim of this study.

* * *

The plots and characters of contemporary plays, fictional as they are, lend themselves to this sort of analysis. If the emotion of love is not a solipsistic, monadic experience but, in large parts, a culturally shared phenomenon, then art and literature can find ways for its representation that are collectively recognisable within this culture. Any art form that is mimetic in the sense that it seeks to represent or reflect an element of reality will, if it approaches the topic of love, reproduce the 'image repertoire' or 'codes' that determine its audience's conception of love and thus guarantee its recognisability despite all artistic inflections or distortions. Contemporary British theatre, where the tradition of broadly realistic representational drama has remained stronger than in many of its continental counterparts (cf. Pattie 395; Saunders, "Persistence"), offers a wide range of plays fulfilling this condition. As Middeke, Schnierer and Sierz observe, "since Osborne's *Look Back in Anger* even the most experimental plays of the new British drama have featured recognisable traits of everyday life, language and images" (xv), making British drama a mine for discourse-analytic approaches. Even the 1990s, which are marked as a decade of change and new departures in British theatre (cf. Tönnies 2; Aragay/Monforte/Zozaya ix; Pattie 393–95), have not initiated a sustained rejection of dramatic and mimetic theatre but rather, with the phenomenon of in-yer-face theatre, produced a mostly dramatic, mimetic, and, despite its exaggerative style, substantially realistic alternative to the postdramatic tendencies of the continental avant-garde (cf. Nikcevic 255–64). In the 2000s, as Andrew Haydon points out (critically), the entire concept of 'New Writing' was even understood as implying a naturalistic representation of contemporary society (cf. 72–73). My analysis will focus primarily on this kind of text-based, dramatic, and mainly realistic play which still dominates the British theatrical landscape and which allows me to examine the discourses of love on the level of the characters' speeches and actions.[2]

I assume that what the characters say and do are fictional yet ultimately mimetic manifestations of the discursive fragments or 'figures' (cf. Barthes 3) that constitute what is called love in modern Western societies. In the selected plays, there is no clearly visible level of fictional mediation and no form of narrative introspection into the characters' minds. They remain unknowable, incomprehensible 'others' to each other and to the audience save for what their utterances and behaviours disclose. Consequently, in expressing a character's emotions in a play like this, the playwright has to steer a difficult path between unrealistic clarity and eloquence on the one hand and realistic but ineffective awkwardness and silence on the other. Yet what the playwright always can do is to draw on the image repertoire shared by characters and audience, that is, to make the characters say and do

---

[2] Sarah Kane's *Cleansed* is the exception with regard to realism and Dennis Kelly's *Love and Money* and debbie tucker green's *a profoundly affectionate, passionate devotion to someone (-noun)* are the clearest exceptions with regard to dramatic form in the sense of linear, action-driven plot development, as the plays consist of non-chronological or only minimally connected episodes and transfer significant action into a narratively presented past. These are, however, aesthetic choices motivated either by the wish to create an experiential presentation of emotional states or by the attempt to cover an extended period of time and *not* by a disbelief in the possibility or relevance of individual agency which, according to Hans-Thies Lehmann, underlies many postdramatic works (cf. 39–43). This point will be taken up again in more detail in the Coda.

things that are recognisable as belonging to the discursive field of love. This can take various forms, from the social realism in Marber's *Closer* or Bartlett's *Love, Love, Love* to the highly metaphorical imagery in Kane's *Phaedra's Love* and *Cleansed*, but all the plays discussed in this volume are mimetic in the broad sense that they reproduce, in one way or another, recognisable variations of our shared narratives of love. In contemporary British drama, that is to say, circulating discourses are moulded into a specific art form that necessitates the recognisability of these discourses. It is a site of heightened intertextuality, citing the established 'figures' and 'codes' of love because they are the only means of expression available to the dramatic figures and the most natural way of ensuring effective communication between stage and audience.

Selecting plays from the period between 1996 and 2017 allows for the treatment of both the plays and the discourses they reflect as contemporary. This is not to say that there have been no changes regarding the concept of love, its sociocultural context, and the structure of feeling around it during this time. Divorce rates, for instance, are now significantly lower than at the beginning of the twenty-first century³ and Anna Malinowska and Michael Gratzke observe that despite (or because of?) the neo-liberal demand for flexibility and non-commitment, "people seek to commit more than they may have done in the late 20th century" (4). However, the wide-ranging congruence of sociological research on love from the past three decades, which will be explored in Chapter 3, suggests that these changes have been gradual rather than fundamental and have not essentially altered contemporary conceptions of romantic love. Another reason for going back to the 1990s, but not further, is that the decade marks a watershed in British drama. This refers not only to the rise of in-yer-face theatre and its peculiar aesthetics but also to the general turn of interest away from large-scale political issues towards private affairs and intimate interpersonal relationships (cf. Middeke/Schnierer/Sierz xiii–xiv). Not that the theatre of the 1990s ceased to be political (cf. D'Monte/Saunders); but, as David Pattie argues, the typical 'state-of-the-nation' plays of the previous era gave way to treatments of more private or intimate matters of political relevance such as same-sex relationships, the crisis of masculinity, or identity troubles of marginalised and isolated groups and individuals (cf. 393–95). Even after the "re-emergence of political theatre" (Howe Kritzer 7) in the late 1990s and 2000s and the return of broader political topics such as globalisation, terror, migration, and national identity (cf.

---

[3] According to the Office for National Statistics, in 2017 divorce rates in England and Wales had reached the lowest level since 1973 while the average duration of marriage at the time of divorce was 12.2 years, a high last seen in 1972 (cf. "Divorces in England and Wales: 2017").

Middeke/Schnierer/Sierz xiv; Holdsworth/Luckhurst; Sierz, *Rewriting* 1–12), the popularity of stories about love and intimacy continued unabated, as is indicated, for instance, by the thirty-page chapter about love as a topic in British drama in Sierz' *Rewriting the Nation* (2011). And yet, the sustained presence of love as dramatic theme has not met with academic interest in the historicity and contemporary peculiarities of the discourses that are digested in these plays. Studies dealing with the dramatic representation of intimate relationships instead seem to focus predominantly on their political aspects such as gender- and sexuality-related roles or power structures (e.g. Grassi; Greer), or the difficulties of relationships in an over-sexualised and/or profit-oriented and/or postmodern society (cf. Sierz, *Playwriting* 54–68). These approaches are necessary and justified but they tend to take love for granted when they analyse it within a specific area of tension and their attention seems to be directed at the political context more than at love as an emotion. My approach, too, is interested in the tribulations of love within particularly contemporary areas of tension, but I will sometimes background the political and social context while putting my focus on love itself, on what it *is* and *means* for the characters in the plays and how they experience and express it.

Finally, what qualifies the selected plays for my approach is that they deviate from mainstream popular culture in their presentation of love. As Lauren Berlant argues,

> In the popular culture of romance [...] dramas are always formed in relation to a fantasy that desire, in the form of love, will make life more simple, not crazier. Boy meets girl, boy loses girl, boy gets girl: this generic sequence structures countless narratives both high and low (sometimes with the genders reversed). The fantasy forms that structure popular love discourse constantly express the desire for love to simplify living. (89)

Belsey similarly observes that popular romances "generally centre on a reassuring tale of obstacles finally overcome and love ultimately and eternally requited" (*Desire* 22), and Illouz partly explains the success of the highly commodified and formulaic products of the romance industry with the fact that they provide "guidance" and "a sense of direction" to the reader (*Hard-Core Romance* 26). Nothing could be farther from the depiction of love in the plays to be discussed here. The 'generic sequence' Berlant describes hardly surfaces at all, and even if it does, the stories are anything but 'reassuring' or offering models for imitation. Love simplifies nothing in these plays. In all of them, love is deeply entwined with images of lack and loss, unattainable needs, and an inevitable precariousness, and it is on this distinctive feature that the main thesis of this project is based.

* * *

I have indicated that, despite their otherwise fundamental opposition, biological/anthropological universalism and social/cultural constructionism both imply that it is reasonable to talk about and analyse love as not just a uniquely individual but a collective experience whose bodily symptoms and cognitive conceptualisations are shared to a considerable extent by the members of a group – be it the human species as such or a culturally distinct subgroup. They both imply, in other words, that it is possible to talk about love in a way that others can understand because their own experiences of love are comparable. Certainly, neither can two experiences of love be assumed to be more than only comparable, nor can language be assumed to enable faultless articulation and communication of an emotional sensation. And yet, the common ground of love suggested by both evolutionary and constructionist theories of emotions justifies attempts to investigate a culture's or society's concepts of love by analysing its discourses and to treat recurring narratives as particularly significant.

Maybe the two most persistent narratives are those about love as a form of *compensation* and about love's inherent *precariousness*, both broad and variegated discourses that have been inextricably entwined throughout the history of love, as the human longing for compensation through love has always been in conflict with the various sources of its precariousness. In fact, due to the compensatory function that has for centuries been ascribed to it in different forms, love might be regarded as the most prominent focal point of human precariousness, of which it is both cause and effect. The notion of precariousness as fundamental human condition has received special attention in the wake of Judith Butler's ethical and political writings that take our shared vulnerability as starting point. Her thoughts, and those of Emmanuel Levinas, on which they are largely based, have proved fruitful for analyses of contemporary drama and theatre as is testified by books like Katharina Pewny's *Das Drama des Prekären* (2011) or Mireia Aragay and Martin Middeke's edited volume *Of Precariousness: Vulnerabilities, Responsibilities, Communities in 21st-Century British Drama and Theatre* (2017). Pewny applies a broad conception of precariousness as that which is 'difficult,' 'delicate' or 'uncertain' ("das Schwierige, Heikle, das Ungesicherte," 13) to her analysis of dramatic and, more importantly, postdramatic theatre and puts a central emphasis on the live encounter of performers and audiences in the performance situation as a re-enactment and reproduction of the pervasive uncertainty and sense of vulnerability characterising modern lives (cf. 21, 24). Her focus is less on the thematic than on the performative level of theatre productions and the ways in which they create moments of precariousness. Her argument that any performance is an encounter with precariousness rests not only on the fact that the co-presence of actors and audience always implies some sort of interaction that remains to variable degrees uncontrollable and uncertain, but also on the assumption that the encounter between performer and audience member is an encounter with the other in the Levinasian and Butlerian sense, that is, an encounter with the other's ontological vulnerability (cf. 107, 117). This notion of ontological vulnerability is also the starting point of Aragay and Middeke's *Of Precariousness*, which follows a clearly ethical and political trajectory and applies an understanding of precariousness that combines its denotation of social, political, and economical states of precarity with its more philosophical meaning of a shared human vulnerability and interdependence in order "to underline the ethical and even political significance of (ontological) precariousness" (3). While the present study is decidedly less concerned with performance and the postdramatic and treats larger social and political matters only in passing, it is this notion of ontological precariousness that links it to the two works just mentioned. For Butler, if there is something like a general human condition, it is our shared vulnerability as bodily beings who are from the beginning of life dependent on and exposed to others (cf. 20; 31). "In a way," Butler suggests, "we all live with this particular vulnerability, a vulnerability to the other that is part of bodily life, a vulnerability to a sudden address from elsewhere that we cannot pre-empt" (29). From here, Butler goes on to enquire into the implications of this "general conception of the human" (31) for the ethics and politics of human social life. In contrast to that, in my own study the notion of ontological precariousness will occur primarily as the first origin of human love in many of its theoretical conceptions, which at the same time mark love as a constant source of precariousness. In this model, which keeps on reoccurring in texts and theories from Platonic philosophy to contemporary British drama, love is conceived as a powerful remedy against our fundamental human vulnerability but is at the same time notoriously uncertain, uncontrollable, and dependent on others. Part I is intended to illustrate this interconnection with a series of spotlights on formative philosophical approaches from various centuries (Chapter 2) and a selection of sociological perspectives emphasising the specifically contemporary sources of love's precariousness (Chapter 3). I will sketch the history of love as the history of the function(s) attributed to it in both ancient and recent philosophy and in sociological theory and conclude that this function, despite significant changes, has been and still is predominantly compensatory. At the same time, I will try to show that love has always been conceived of as inherently precarious, but that different historical periods have explained and assessed this precariousness differently.

If Part I hence seeks to establish the historicity and continuity of the concepts of love articulated in contemporary British drama, this might raise questions concerning the channels and exact modes of circulation which have enabled the perseverance and reoccurrence of these discourses. Such questions, however, cannot be answered satisfactorily in a study that uses methods of historical discourse analysis primarily to reach a better understanding of works of literature rather than using literature to exemplify the socio-historical circulation of discourses. Exactly how certain concepts of love are handed down, preserved, and modified throughout the ages, how they find their way from one field into another so that they circulate between philosophy, psychoanalysis, sociology, neuroscience, biology, popular knowledge, and literature, whether as forms of direct reference and conscious adoption or via more obscure channels – these are questions that must be reserved for research that follows precisely this interest. The present study will content itself with observing the longevity of discourses and the prominent role they play in their specifically modern variations in contemporary drama where they, this much I am prepared to claim, make their appearance usually not because the playwrights take recourse to specific texts and theories but because the discourses have become part of popular knowledge and are recognisable to both playwright and audience as culturally dominant concepts of love.

If the exact roads of travel of historical discourses have to remain uncertain, the ways they react with their new, modern contexts are clearly observable: love in contemporary British drama still finds itself enmeshed in the ancient struggle between the need for compensation and love's precariousness. The peculiarities of our time in this conflict are to be found in the concrete deficiencies love is supposed to compensate today and in the specifically modern sources of its precariousness. In various forms, the selected plays voice the expectation that love will make up for the shortcomings of a post-traditional and late capitalist society, where the freedom won in historical processes of liberation is often accompanied by experiences of alienation and lack of orientation. This hope for compensation, however, is counteracted by modern commitment phobia enhanced not only by an abundance of choice but by the tendency to perceive romantic commitment as a threat to autonomy and self-realisation. The traditional concepts of love, in other words, are still at work in contemporary relationships (cf. Rathmayr 300), but they are transformed and adapted according to changed socio-historical and cultural circumstances. Awareness of this simultaneity of timelessness and topicality can enhance the interpretation of contemporary plays, which reflect how historical continuity interlaces with the peculiarities of modern society in discourses and concepts of love.

Part I **The Making of Romantic/Erotic Love: Traditions and Transformations**

2 Philosophy of Love: Selections

It is a commonplace persuasion often repeated in sociological and historical accounts that romantic love is not a universal but a historically and culturally specific phenomenon, a genuinely European 'invention' of the Middle Ages (cf. de Rougemont, esp. 170–208; Jaggar 54) or even the (early) modern period when, as is sometimes suggested, the strong sense of individuality necessary for romantic love began to emerge (e.g. Solomon 505–07; cf. also Hendrick/Hendrick 21–22 and Jankowiak/Fischer 149). To some extent, this view cannot be refuted. As I will later illustrate in more detail, there are close links between our contemporary notion of romantic love and the emergence of capitalism and modern individualism. The interrelation between love and identity formation, the broad consensus that love must precede and be the only legitimate foundation for marriage, and the development of love into a rather clearly demarcated social subsystem are bound to socio-historical processes taking place in modernity. But while the function love fulfils in a given society as well as its assessment and interpretation unquestionably change from one generation to the next, it is less likely that all constituents of the emotion change to the same degree. This is not to deny that socio-historical circumstances have an impact on feelings and that assessments and interpretations of love influence how it is experienced. But to conclude that the emotion of romantic love is entirely a product of modernity is, I think, unjustified.

To be more precise, I share the position of philosopher Simon May, who argues "that the emotion of love is universal but that the way this emotion gets *interpreted* varies greatly from one society and epoch to another" (11). This position is supported by the results of Elaine Hatfield and Richard L. Rapson's empirical study (1987), which strongly suggests that romantic love is experienced universally, with only minor deviations between groups of different genders, ages, and cultures, while major differences occur in the way these emotions are evaluated, lived out, or suppressed, depending on the cultural and historical context and its dominating ideology. Their results have been strongly corroborated by the research of William Jankowiak and Edward Fischer (1992), whose analysis of 166 distinct societies was "able to document the occurrence of romantic love in 88.5 per cent of the sampled cultures," leading to the suggestion that "romantic love constitutes a human universal, or at the least a near-universal" which, however, is bound to produce "cross-cultural variation in the styles of romantic expression" (154). With reference to such findings, social psychologists Susan Hendrick and Clyde Hendrick conclude "that the physical experience of love or falling in love may be a possibility for most of the species in most eras. What passionate love means and how it is channeled, however, is shaped by culture" (24).

It is this cultural shaping that will be the point of interest in this chapter. What I want to uncover are traditions in the Western philosophy of love that eventually lead into our modern concepts and are still discernible in contemporary discourses. For this purpose, I will rely often on Irving Singer's monumental three-volume study of the philosophical history of love (*The Nature of Love*) and I adopt his view that modern concepts of love cannot be comprehended without recourse to their discursive heritage, as the initial statement to his short and synoptic *Philosophy of Love* makes clear:

> To someone doing the kind of research I did, it was apparent that many elements of nineteenth-century Romantic love derived from sources in ancient Greek philosophy and literature, in Hellenistic fables, in the burgeoning of Christianity, in the reaction against Christianity during the Renaissance, and then in a diversity of seventeenth- and early-eighteenth-century modes of thought. You can't really separate this continuum in two periods, the first of which was prior to any ideas about Romantic love and the other consisting in the thinking of the last two hundred years with its great focus on it. The claim that Romantic love is an *invention* of the latter period is therefore of limited value, and, on the face of it, mistaken. (2)

The following survey into love's philosophical history will exemplify the continuity of ideas Singer indicates by focussing on two motives which keep reoccurring with surprising persistency. Throughout Western thinking, the notion of love as *compensation* and warnings about its *precariousness* have largely determined the discursive field, both as separate motifs giving rise to various theories and narratives and as closely interrelated and conflicting principles.

2.1 Love as Compensation

Why do people fall in love? If purely biological approaches are left aside, answers to this question have always implied ethical evaluations of love as well as philosophical assumptions about its nature and origin. In *Upheavals of Thought*, Martha Nussbaum explains why, in the course of history, "philosophers have not often been friends of erotic love" (463). In contrast to other forms of love such as *philia* or *agape*, the nature of *eros* as an amalgamation of neediness, possessiveness, partiality, and excessiveness has made it appear unsuitable as a contributor to ethical life. Conceived as "a particular kind of awareness of an object, as tremendously wonderful and salient, and as deeply needed by the self" (477), erotic love, unlike its ethically superior relatives, was considered ultimately self-seeking and conditioned upon the flourishing of the loving subject.

The ancient distinction between forms of love of different ethical value finds a modern equivalent in the philosophy of Harry G. Frankfurt. In *The Reasons of Love* (2004), Frankfurt argues that true love is unconditional. Love is defined as "disinterested concern" for a particular other, a pure altruism typically absent from "relationships that are primarily romantic or sexual" (43). The notion of 'disinterested concern' implies not only that the lover expects no reward from the beloved but also that, unlike in *philia*, there is no obvious, objective value in the beloved to which love is an adequate response (cf. 11–12). The worth and value that any object of love surely has for its lover is not the main reason for love but rather its main result: "The lover does invariably and necessarily perceive the beloved as valuable, but the value he sees it to possess is a value that derives from and that depends upon his love" (38). Consequently, Frankfurt seeks not to answer the question what it is exactly *in* or *about* another person that elicits love but the question why people love in the first place or, to be more precise, why they love 'disinterestedly,' seemingly without any good reason. His answer is that it is "essential to our being creatures of the kind that human beings are" (17), that we have a "generic" and "fundamental interest in loving as such" (51). This interest is grounded in our desire to experience life as meaningful, which can be satisfied by committing to final ends like those embodied by an object of love.

> By providing us with final ends, which we value for their own sakes and to which our commitment is not merely voluntary, love saves us both from being inconclusively arbitrary and from squandering our lives in vacuous activity that is fundamentally pointless because, having no definitive goal, it aims at nothing that we really want. Love makes it possible, in other words, for us to engage wholeheartedly in activity that is meaningful. (90)

Regarding the concrete choice of objects, Frankfurt implies, love has its own logic that cannot and need not be fully understood by lovers, sociologists, or philosophers. "The heart has its reasons, which reason does not know," as Pascal had once put it.[4] What can be reasonably understood, however, is that within and through this incomprehensible logic of love human lives and actions are enabled to acquire meaning which they would lack otherwise.[5]

---

[4] Number 277 of Blaise Pascal's *Pensées*, at p. 98 in the used edition.
[5] Cf. also Singer: "Our entire existence is characterized by the creation of meaning. Implicit in the need to be loved is the need to have a meaningful relationship with persons that matter to us" (*Pursuit* 4).

There is a logical contradiction in Frankfurt's theory, which he candidly spells out. If it is in the nature of love to gratify the insurmountable human longing for meaning, can it then really be said to be 'disinterested'? Is the relationship between lover and beloved not ultimately one in which the latter serves as a means to reach a desirable goal (cf. 59–60)? Frankfurt cannot solve the problem convincingly. Although he emphasises that only love that seeks nothing but the well-being of the other is love of a final end in itself and hence capable of generating meaning (cf. 60), he inevitably evokes suspicion about the selflessness of love when he writes that "[t]he most rudimentary form of self-love […] consists in nothing more than the desire of a person to love" (90). In the end, the conception of "the essential nature of love as a mode of disinterested concern" (43) seems untenable in the strict sense. Even Frankfurt's ideal of love, which he strictly distinguishes from "infatuation, lust, obsession, possessiveness, and dependency" (43), cannot be construed as completely unconditional and selfless but appears to be essentially striving and hence *erotic*.

The possibility of unconditional and selfless love has been questioned by philosophers before. The religiously inspired writings of C. S. Lewis, Josef Pieper, and Robert Spaemann, for instance, regard it as a natural fact that there is an element of need-gratification in all kinds of human (as opposed to divine) love, which stems from the existential insufficiency of the solitary human being. Any denigration of the need to be loved amounts to a denial of human nature and a form of human hubris that claims godlike self-sufficiency (cf. Pieper 62–63). Moreover, in addition to the need to be loved they postulate an equally natural human need for bestowing love which results in an undeniable joy of giving. In Lewis' terminology, besides the genuinely human "Need-love" humans are also capable of "Gift-love" (1), albeit not in its divine purity but infused with the permanent element of human neediness (cf. 21) so that even a mother's affection for her child "is a Need-love but what it needs is to give. It is a Gift-love, but it needs to be needed" (40). Spaemann emphasises the significance of 'Need-love,' the wish to be loved, for the sincerity and absoluteness of love. A true lover, he says, cannot assume the attitude of a completely self-sufficient and undesiring entity as this would deprive the other of the opportunity to bestow love and enjoy the pleasure of giving. The feeling that our love is wanted and needed is no less important than the feeling of being loved, and wanting 'the good' for another person means enabling him or her to experience both. The logical consequence of these assumptions is that human love, no matter how disinterested it may appear to be, is ultimately conditional.

This argument is brought forward most unambiguously by Simon May. May echoes Lewis and Pieper when he criticises the notion of unconditional love as an illusion that has been modelled after religious conceptions of God's love for humans (cf. 2). In the age-old struggle between desiring *eros* and selfless *agape*, May claims, a majority of writers have valued the latter as morally superior, disregarding its unfitness for the human race, and "our age not only sustains this centuries-old ideology" (235) but has even reinforced the view of "genuine love as unconditional, enduring and selfless" (236), despite all evidence to the contrary. According to May, human love is not always and not necessarily enduring and, above all, never fully unconditional and selfless. All human love satisfies a vital need and this function of love is the prerequisite for its coming into being. All love is first of all *erotic* in that it seeks to gratify this one essential need, and only if this condition is fulfilled, forms of love that seek nothing for the self can ensue (cf. 247).

May's one and only condition for love is related to the notion, discussed more broadly in Chapter 3, that love has taken over from religion the task of providing meaning and security. The promotion of love to the only acceptable metaphysical phenomenon in an otherwise irreligious culture, May suggests, was stimulated by a need that itself underlies all religious needs – a need for what he variously calls "groundedness, rootedness, at-homeness" (6) in the world we live in. "Love," he argues,

> is the rapture we feel for people and things that inspire in us the hope of an indestructible grounding for our life. [...] If we all have a need to love, it is because we all need to feel at home in the world: to root our life in the here and now; to give our existence solidity and validity; to deepen the sensation of being; to enable us to experience the reality of our life as indestructible (even if we also accept that our life is temporary and will end in death). This is the feeling that I call 'ontological rootedness' [...]. (6)

This feeling, then, is May's condition *sine qua non:* "far from being unconditional, love is inescapably conditional on this promise of ontological rootedness" (7). Echoing twentieth-century existentialism, he describes the human condition as a "relation with an uncontrollable and alien world into which birth has thrown us" (10), a relation that is characterised precisely by a lack of 'rootedness' or 'at-homeness' and by a sense of vulnerability. For May, "love is born in extreme vulnerability and seeks to overcome that condition through a correspondingly extreme invulnerability" (10). It is the feeling we have for someone who alleviates our vulnerability, our ontological precariousness, to use Butler's term, our inescapable insufficiency and natural human condition as beings of want and lack. As I will demonstrate, this thesis is firmly rooted in a long and immensely influential tradition in which love has, above all, a compensatory function.

2.1.1 Plato's *Eros*

Love as compensation is arguably the most pervading image of love in Western philosophy. Irving Singer starts the second volume of *The Nature of Love* with a distinction between an idealist and a realist tradition in Western thinking, generally opposed to each other in their most central beliefs. However, Singer points out, "[t]here is one point on which realist and idealist accounts of love tend to agree. They usually begin with the loneliness of man" (*Courtly and Romantic* 4). From ancient myths to modern psychoanalysis, feelings of loneliness and separateness play a crucial role in explanations for love. The idealist tradition, according to Singer, usually holds that humans can overcome their loneliness and separation in an act of 'merging,' which is conceived of as the obtaining of or the return to a state of (primordial) wholeness or oneness. And while the realist tradition casts doubts upon the image of merging as impossible or even undesirable (due to the loss of individuality it implies) and as a rule disregards the idea of utopian wholeness as an idealist chimera, it very often agrees with the idealist tradition in viewing an experience of lack and insufficiency as the first cause of love (cf. 4–11). Jean-Luc Nancy summarises this common denominator in the thinking about love (which he himself considers a misconception he seeks to replace) as follows:

> Since the *Symposium* – or, if you prefer, since before Plato, in Heraclitus or Empedocles, in Pythagoras or Parmenides – the general schema of a philosophy of love is at work, and it has not ceased to operate even now, determining philosophy as it understands and construes itself, as well as love as we understand and as we make it. If it were necessary to take the risk of grasping this schema in a formula, one might try this: love is the extreme movement, beyond the self, of a being reaching completion. ("Shattered Love" 86)

It is no coincidence that Nancy mentions Plato's *Symposium* as the starting point of this "general schema" of love, even though he also mentions Plato's rather obscure and often half-legendary pre-Socratic predecessors. For many, Plato provides the first securely handed down philosophical treatment of erotic love – and at the same time one of the most influential and lasting theories of love in Western cultural history. As Singer puts it: "In the philosophy of love [...] I am convinced that every discussion must *start* with Plato. Courtly love, Romantic love, and major emphases in religious love all take root in him" (*Plato to Luther* 47). Plato's influence on later philosophies and theories of love challenges any conception of romantic love as a modern invention. Even though romantic love as understood and lived in modernity cannot be thought without modern individualism and modern social structures, many of its core elements are derived from Platonic idealism.

The *Symposium* is the central document of Platonic love philosophy (cf. Singer, *Plato to Luther* 48) and, according to May, "the first extended discussion of love in Western philosophy and, together with the commands from Deuteronomy and Leviticus, the most influential text on love in the Western world" (40). As a sequence of seven speeches in praise of the god Eros, all differing from and sometimes contradicting each other, it is a vivid portrait of love as discursive field in which various discourses struggle for a definition of the phenomenon. It is common usage to see Socrates both as Plato's spokesman and as the winner in this competition of speakers (cf. Hähnel/Schlitte/Torkler 37), as he first points out errors in the previous speeches and then elaborates on a theory of love which contains the fundamental tenets of Platonism, outshines the other speeches in terms of philosophical complexity, and remains undisputed. On the other hand, this victory does not make the other speeches unnecessary as they bear testimony to the swarming multitude of theories about love in Plato's days.[6] Moreover, Socrates has neither the last word in the *Symposium*, nor is he the speaker who makes use of the most striking and most permanent imagery. His portrait of love as an ultimately bodiless, spiritual, and intellectual striving which is eventually equated with philosophy itself is framed by the speeches of Aristophanes and Alcibiades, the former creating one of the most powerful mythological images of desire, the latter giving expression to the precariousness and pain inherent in love, and both acknowledging the intensity of passionate desire and its bodily dimension that is neglected in Socrates' account. The Socratic theory of *eros* is certainly that which exhibited most direct influence on later thinking about love, especially after its appropriation through Christian theology in the form of Neoplatonism and its modulation and 'humanisation' during the Middle Ages in the system of courtly love. But it is limited because of what it excludes, as the following analysis will aim to unfold. In order to cover all that the *Symposium* has to say about compensation and precariousness I will therefore look into the speeches of Aristophanes, Socrates, and, at the beginning of the next chapter, Alcibiades.

* * *

---

[6] The seven speeches in the *Symposium* are only those which Plato's fictional narrator Appolodorus remembers as most noteworthy from the account he had heard from one Aristodemus who, although present at the legendary symposium when it took place years ago, had also forgotten some (parts of the) speeches (cf. Benardete/Plato 178a; 223d). What Plato indicates with this narrative structure and the hint at the text's incompleteness is that the discursive field of love was/is even more extensive and varied than presented in the *Symposium*.

The setting for the *Symposium* is the house of the tragedian Agathon, one day after his victory in a competition of playwrights. Weary from the heavy drinking of the previous day's celebration, the guests of the banquet agree on giving speeches in praise of Eros rather than indulging in drinking again. After the contributions of Phaedrus, who praises Eros as a guide towards virtue and bravery, Pausanias, who distinguishes between purely sexual Pandemian and the more praiseworthy Uranian Eros, and the physician Eryximachos, who extends the duality of Eros to an all-pervasive cosmic principle of a struggle between harmony and discord, the *Symposium* reaches a first climax with the speech of Aristophanes.

Aristophanes' vision of Eros is doubtlessly the most powerful image in the *Symposium* and narrativises a particular idea or experience of love that is so intuitively convincing that it reoccurs over and over in philosophical and popular notions of love until the present day (cf. May 43). Its compelling persuasiveness has prompted Allan Bloom to conclude that "Plato makes Aristophanes the expositor of the truest and most satisfying account of Eros that we find in the *Symposium*. There has probably never been a speech or poem about love that so captures what men and women actually feel when they embrace each other" (478). In this mythical account, Eros appears as the god or force that tries to alleviate a tormenting feeling of incompleteness that afflicts all mankind and dates back to primordial times. Originally, Aristophanes explains, humans were spherical beings, with four arms, four legs, two faces, "and two sets of genitals, and all the rest that one might conjecture from this" (Benardete/Plato 190a). Moreover, the human race consisted of three genders – male, female, and a combination of the two, the androgynous (cf. 189e). Proud and wanton because of their strength and perfection, they began to rebel against the gods so that Zeus decided to punish and weaken them by splitting them in halves (cf. 190b–c). The operation had an unforeseen effect: so intense was the pain of separation and incompleteness that henceforth each human being would do nothing but search for the lost half, "and throwing their arms around one another and entangling themselves with one another in their desire to grow together, they began to die off due to hunger and the rest of their inactivity, because they were unwilling to do anything apart from one another" (191a–b). Zeus reacted by removing their genitals from the back to the front,[7] enabling sexual intercourse during the embrace. Thus, if two halves of the formerly androgynous gender met and embraced, they could

---

[7] The original, spherical humans had worn their genitals on the outside but had not copulated with one another but "gave birth [...] in the earth, like the cicadas" (191c). After their reshaping, the genitals had to be moved to that side of the cut which was the new front.

now procreate and secure the survival of the human race, something they had previously forgotten in their desire to grow together. And if two halves of the same sex met, Aristophanes explains, "there might at least be satiety in their being together; and they might pause and turn to work and attend to the rest of their livelihood" (191c). Sexual intercourse, in this myth, is a kind of outlet valve for the tremendous desire to become one which, without the possibility of sex, would paralyse the entire human race. It is not, as Aristophanes makes clear, what two lovers ultimately want from each other. Even though the ancient nature of humanity and the true reason for love have long been forgotten, "no one would be of the opinion that it was sexual intercourse that was wanted" (192c). To some extent, what makes the Aristophanes myth so appealing is his depiction of human beings as lovers who are convinced that love is more than sexual desire while they are unable to tell what it actually is – a picture that in turn serves as an explanation for the confusion of the language of love. As Aristophanes puts it, "the soul of each plainly wants something else. What it is, it is incapable of saying, but it divines what it wants and speaks in riddles" (192c–d). In an interesting passage, Aristophanes imagines Hephaestus offering lovers to weld them together forever and concludes:

> We know that there would not even be one who, if he heard this, would refuse, and it would be self-evident that he wants nothing else than this; and he would quite simply believe he had heard what he had been desiring all along: in conjunction and fusion with the beloved, to become one from two. The cause of this is that this was our ancient nature and we were wholes. So love is the name for the desire and pursuit of the whole. (192e)

This passage not only contains what can be considered a definition of love for an entire strand of philosophy ("the desire and pursuit of the whole") but also, subtly, points to the improbability of ultimate satisfaction or fulfilment in love. Apart from the problem that humans are only distant descendants of their complete, spherical ancestors and, therefore, have no original halves with whom they could desire reunion, there appears the more fundamental problem that neither with an original half nor with an adequate substitute actual merging seems possible as merging is only a hypothetical possibility that would require divine help from the blacksmith of the gods. Aristophanes seems not to believe in this possibility and ends his talk with the advice to be content with the second-best solution:

> [O]ur race would be happy if we were to bring our love to a consummate end, and each of us were to get his own favorite on his return to his ancient nature. And if this is the best, it must necessarily be the case that, in present circumstances, that which is closest to it is the best; and that is to get a favorite whose nature is to one's taste. (193c)

Eros must be praised for his effort to soothe the pain inflicted by human incompleteness, but it is beyond his power to heal and restore the original condition. A "favorite whose nature is to one's taste" is all that can be hoped for in this world. Objects of love can only ever be substitutes for what is really desired, and communion with them can never attain to the perfect union that would end the fundamental separateness and incompleteness of humans.[8] This is the tragic message of the comic poet Aristophanes, and its main elements reoccur frequently in the philosophy of love. Bloom even sees it as a hallmark in the history of philosophy:

> The cutting, the wound to human nature, inflicted by the Olympian gods, gave birth immediately to what is most distinctively human: longing, longing for wholeness. Thus, what is perhaps the most important strand of philosophy and literature came into being. Man is essentially an incomplete being, and full awareness of this incompleteness is essential to his humanity and ground for the specifically human quest for completeness or wholeness. (480)

Whether such direct influence of the Aristophanes myth on later philosophy can be assumed or not, the explanation of love as a consequence of the human condition along these lines is strikingly reminiscent of later ideas of Romanticism and psychoanalysis and at least proves that similar ideas have haunted the thinking about love from the beginnings of philosophy. The idea of erotic love as compensation is indeed what could be called a grand narrative up to the present age.

The next speech has nothing of the philosophical and existential depth of the previous one. What the tragic poet Agathon produces is, in contrast to his comic counterpart, hardly more than exuberant but shallow praise, embellished with a display of rhetorical skills in the final litany of fine qualities and gracious gifts for which Eros must be eulogised (cf. 197c–e).[9] Its main function may well be seen in preparing the ground for the centrepiece of the *Symposium*, the speech given by Socrates, which contrasts with this paradigm of empty rhetoric in its plainness of style, its informational content, and its philosophical depth. Socrates starts by engaging Agathon in one of his typical Socratic dialogues during which the tragic poet is forced to contradict himself and admit errors in his own speech. Agathon has to agree that Eros is not love of nothing but of something, that he necessarily desires this something, that Eros, like everyone else, desires only that which he does not already have, and that thus, since Eros is

---

8 On this somewhat darker side of the Aristophanes myth, see also May (44–45).
9 On the distinctly lower quality of Agathon's speech see, for example, Bloom (485–93).

love of beauty, he cannot be beautiful, as Agathon had claimed in his speech (cf. 199e–201c). Agathon is then released and Socrates recounts how he himself was taught all he knows about Eros by the wise Diotima of Mantinea. From her he learned that Eros is the son of Poros, the god of resourcefulness and plenty, and Penia, the goddess of poverty (203b). As the offspring of Resource and Poverty, Eros has an ambiguous character, "always dwelling with neediness" (203d) but at the same time striving and ingenious. Notably, this is already the second account that links love to a fundamental neediness. The decisive difference between the two accounts is spelled out by Diotima in the next passage. While in the Aristophanes myth the object of love is very specific – the lost other half or an adequate substitute – Diotima's *eros* is defined, rather vaguely, as desire for the permanent possession of the good (cf. 206a) which, for mortals, can only be acquired by giving birth. All mature human beings have this desire to produce offspring (cf. 206c), either in the form of sexual procreation (cf. 208e) or, more importantly, in a spiritual way. The entire theory of erotic ascent that follows refers to lovers "who are pregnant in terms of the soul" (208e–209a) and give birth to offspring of a spiritual nature: virtuous speeches, knowledge, great poetry, or systems of laws and entire states.

Pregnancy in soul does not exclude sexuality. In fact, the ascent on the 'ladder of love'[10] begins with sexual attraction, even though the goal is never sexuality or the production of biological children. "He who is to move correctly in this matter must begin while young to go to beautiful bodies. And first of all, [...] he must love one body and there generate beautiful speeches" (210a). The lover then realises that the beauty he recognises in one body is related to the beauty in all other bodies, that it is the same principle that makes every beautiful body beautiful, "that the beauty of all bodies is one and the same" (210b). Next, he sees "that the beauty in souls is more honorable than that in the body," and from there he turns from the beauty in human beings to "the beauty of sciences" (210b–c). The final level of the ascent is "a certain single philosophical science" (210d), which turns out to be nothing less than the Platonic philosophy of eternal forms or ideas. In Diotima's words:

> Whoever has been educated up to this point in erotics, beholding successively and correctly the beautiful things, in now going to the perfect end of erotics shall suddenly glimpse something wonderfully beautiful in its nature – that very thing, Socrates, for whose sake alone all the prior labors were undertaken – something that is, first of all, always being and neither coming to be nor perishing, nor increasing nor passing away; and secondly,

---

10 This term is frequently used for Plato's ascent theory. See, for example, May (49) and Singer (*Plato to Luther* 178). It is also the title of Bloom's chapter on the *Symposium* (429–546).

not beautiful in one respect and ugly in another [...] and not as being somewhere in something else (for example, in an animal, or in earth, or in heaven, or in anything else), but as it is alone by itself and with itself, always being of a single form [...]. (210e–211b)

To catch a glimpse of the idea of beauty is the ultimate goal of both erotics and philosophy – or, as it is justified to conclude, Platonic erotics and philosophy are one and the same. The Platonic ladder of love, "beginning from these beautiful things here, always to proceed on up for the sake of that beauty, using the beautiful things here as steps" (211c) ends with the vision of "the beautiful itself, pure, clean, unmixed, and not infected with human flesh, colors, or a lot of other mortal foolishness" (211e). The ladder has taken us far away from a concept of love as interpersonal desire between two human beings as presented in the Aristophanes myth or as dominant in both popular and philosophical notions of romantic love. And since the final speech in the *Symposium*, that of the unhappy lover Alcibiades, belongs to the next chapter, it is now time to analyse the Platonic image of love from different angles.

* * *

As Singer remarks about the Aristophanes myth, love did not exist in the "primordial Golden Age" (*Plato to Luther* 51), when human beings were still whole. Rather, love comes into existence only after they become "guilty of an original sin" (52) and are punished accordingly by the gods. Aristophanes understands erotic love not as a universal driving force but as a very specific desire for another human being for the purpose of alleviating this punishment and restoring lost wholeness. In Socrates' account, on the other side, *eros* seems to be a virtually all-embracing concept of love. Given the fact that in the Platonic notion of desire "[n]o one desires anything unless he at least *thinks* it will do him good" (Singer, *Plato to Luther* 54), there seems to be no intentional action that is not an attempt to possess the good. Singer follows this train of thought to its end:

> This being so, it would follow that all human activity is motivated by love. Since Plato believes that everything – not just man – strives for the attainment of some good, the entire universe would seem to be continuously in love. And this is precisely what Plato suggests. It is love that makes the world go round; without love nothing could exist. (54)

Platonic *eros*, as presented in the speech of Socrates, is not only never 'disinterested' but striving for the good, it is also a cosmic principle. It is not hard to see that love between particular human beings can only play a minor part in such a concept. It is only a negligible element in the pervasive striving towards the idea of the Good or Beautiful[11] in which all nature partakes. Only the best of men will recognise the true object of love and climb the ladder in conscious search for this idea, while the rest of mankind and nature remains unaware of the final goal of all love (cf. 54). But consciously or not, everything in the imperfect material world is striving towards this perfect idea, "the highest of forms, the pinnacle of being, the ultimate category in terms of which all other realities are to be explained" (59). These differences aside, the speeches of Socrates and Aristophanes are connected by a common feature that is essential for Plato's philosophy of erotic love: *Eros* is *desire of something* (of the lost half or the Good) and is inevitably *part of human nature* (as a result of the split or as an all-encompassing cosmic force). No matter what its final goal is, human erotic love is defined as desire, and desire is defined as "an attempt to eliminate a state of need or want" (86). Whether this elimination takes the form of (re)union or not, love is the result of human insufficiency. As May puts it, "Socrates thinks that love originates in lack, Aristophanes that love originates in loss" (47). But in both cases, I would like to add, it fulfils a compensatory function.[12]

Plato's philosophy cannot be overestimated in its impact on Western theories of love. Of the several influences May observes still today, the most striking might be the conviction that "sexual desire [...] is *not* what the highest love is finally about or where it is really consummated" (51). For most people, erotic love can still not be reduced to sexuality but is "based on 'higher' things we share with our partners, like common values or ideals" and, as May notes, "[a]mazingly, no amount of sexual liberation has put this view to flight" (51). Yet, this modern idealism of love differs considerably from its Platonic forerunner. If today love is considered to be more than sex, the reason is not a belief in the possibility of gaining access to the realm of eternal forms but the belief that the true worth of love can be found not in the body but in the character, personality, or individuality of the beloved. This principle however, which is so deeply entrenched in

---

11 Singer explains the terminological indeterminacy regarding the final object of love as follows: "This supreme object – and there must be only one, since all things make a unity – Plato calls *the Good*. He also calls it *absolute beauty*. To the Greeks, beauty was a function of harmony. It arose from a harmonious relationship between parts that could not cohere unless they were good for one another. From this Plato concludes that what is truly beautiful must be good and what is truly good must be beautiful" (*Plato to Luther* 54).
12 Achim Geisenhanslüke calls this origin of love in lack or loss the "negativity of desire" ("Negativität des Begehrens") and shows in his study that this Platonic idea has been adopted by numerous theories of love through the ages and especially in psychoanalysis (cf. 12).

the modern concept of romantic love, is conspicuously absent from Platonic love as presented in the speech of Socrates.

That the idealisation of love as ascent on a ladder towards philosophical insight comes at the expense of truly interpersonal love has been noticed repeatedly in the critique of Plato's *Symposium* (cf. Singer, *Plato to Luther* 67–70; Vlastos, esp. at 26; Nussbaum, *Upheavals* 498–500; May 51; Neu 323–24). As Singer points out, "Plato scorns all fixations upon individuals" (67). Love and desire for a single other person is, in fact, the very first stage at which the upward journey of Platonic love begins, but it can only be continued if this fixation is replaced by an unrepressed promiscuity that revels in the beauty of all bodies – until sexual desire, the veneration of beauty in human bodies, is overcome as a whole and superseded by the love of beauty in spiritual things. At no point in this journey is another human being the actual object of love (cf. 67–70). If the final goal of Platonic *eros* is always 'the Good,' every object of love is only loved insofar as it contains or reflects a portion of this ultimate and real object of desire. This renders Platonic *eros* not only elementally selfish but also impersonal. For Gregory Vlastos, "Plato's theory is not, and is not meant to be, about personal love for persons – i.e., about the kind of love we can have only for persons and cannot have for things or abstractions. What it is really about is love for place-holders [...]" (26). To May, this element of Platonic love philosophy means a remarkable disdain of human beings as persons.

> It flattens out their individuality to the point where we could just as well swap our beloved for anyone else, providing they embody at least the same degree of beauty. It makes loved ones valuable only as stepping stones to our greater good as lovers [...] and otherwise gives us little or no interest in their lives or in deepening our relationships with them. Thus, for the sake of the lover's own flourishing, it ends up drawing the truest love from the personal to the impersonal, from the individual to the general, and from the human to the – literally – inhuman. (51)

The depersonalisation of love in Socrates' speech is indeed a salient moment in the *Symposium*. After all, with the exception of the disorderly and confused speech of Eryximachos, all other speeches, as a matter of course, treat *eros* as love between human individuals. Why, then, this movement from the human to the inhuman in the most central contribution? In *Upheavals of Thought*, Martha Nussbaum offers a complex answer. First, the idea of Platonic *eros* has to be seen as the first fully elaborated theory in a tradition that seeks to reconcile the tension between the irrationality and uncontrollable force of erotic desire and the wish to integrate it as a value into the fabric of the good life. Plato's ascent theory is an "attempt to reform or educate erotic love, so as to keep its creative force while purifying it of ambivalence and excess, and making it more friendly to general social aims" (469). In the Platonic and the Christian tradition, this reformation of love amounts to a "dissociation of erotic love from genital sexuality" (476) which was considered responsible for love's "partiality or uneven focus, its excessive neediness and dependency, and its connection with anger and revenge" (527). And indeed, the Platonic ascent theory 'cures' these symptoms of love: partiality is replaced by love randomly directed towards any object that contains reflections of the idea of beauty; anger and revenge – the symptoms of jealousy – do not play a role in love that does not seek to possess the other but only delights in his or her share of beauty; and neediness and dependency, which even in the Platonic account belong to the very nature of *eros*, are gradually brought to the lowest possible level. It is this last effect, Nussbaum argues, that has made Plato's theory so particularly attractive and influential because it makes us forget the shame we otherwise feel in view of our vulnerability and insufficiency (cf. 525–26). True, Platonic love originates in human neediness as we desire that which we do not have. But the only thing we really want, the idea of Beauty or the Good, is permanently accessible to everyone and in everything. By directing love from individual human beings, who are mortal, changeable, and beyond control, to an eternal, stable, and omnipresent idea, the Platonic theory makes the lover independent of other people and thereby eliminates one of the most important sources of vulnerability. Platonic love, in short, is not precarious. It can only be so, however, by banishing the human other from the realm of love. This logic suggests that any form of erotic love that does *not* exclude human beings as objects of love but, instead, regards them as the only truly worthy objects of love, as is the case in contemporary romantic love, is necessarily precarious. The final speech in the *Symposium* points to this very conclusion. Before I turn to the question of precariousness, however, I want to follow the trace of Plato's notion of love as compensation in the philosophy of his successors.

2.1.2 Plato's Romantic Legacy

Plato's disdain for interpersonal erotic love informed the Neoplatonism of late antiquity and the scholasticism of the early Middle Ages. Love as desire for perfection or completion remains an influential idea throughout history, but mostly in the form of spiritual or religious love rather than love between two human beings. It is not before Romanticism that interpersonal erotic love and notions of merging into oneness or wholeness become linked to each other again. Before that, not only did the Socratic approach dominate over the Aristophanes myth; the former was also largely purified into an asexual religious ascent in which the desired oneness was entirely spiritual, while sexual attraction and infatuation between human beings were considered as animalistic instincts or forms of madness detrimental to the striving for a worthy goal. In the Neoplatonism of Plotinus and St. Augustine (cf. Singer, *Plato to Luther* 111–21 and 164–78), for example, human beings are marked by a restless striving of the soul to return to its source, but love for other persons plays only a minor part in the ascent and rather tends to impede than facilitate it. What both Plotinus and Augustine share with Plato is the basic conception of human beings as creatures of lack longing for compensation through union with the ultimate object of love. Love is no surplus energy flowing out from one individual to another but a drive originating in insufficiency and neediness. Christian scholastics like Augustine, Bernard of Clairvaux, and Thomas Aquinas agree on this origin of human (but not divine) love in deficiency and also on its direction, that is, its quest for compensation (cf. 194).

Not surprisingly, Neoplatonic scholasticism tended to follow the path suggested in the speech of Socrates which transforms erotic love into an entirely spiritual enterprise and in which truly interpersonal love eventually drops out. Romanticism, with its symbiosis of the bodily and the ethereal, the sensual and the spiritual, opposes any such lopsided conception of erotic love and liberates the bodily and interpersonal aspects of eroticism from moral and religious scepticism. In this, the Romantics could draw upon the rich cultural heritage of 'courtly love,' the movement which had succeeded to elevate sensual, passionate love between humans from madness or sin to a culturally accepted value.[13] For Simon May, after the occurrence of courtly love "Western love is left with a revolutionary thought: a single human being might be worthy of the sort of love that was formerly reserved for God" (129). This revolutionary thought remains intact even after thinking about love becomes dominated again by Christian Neoplatonism in the Italian Renaissance and it even survives the highly religious period of confessionalisation in Europe. What makes this permanent humanisation of love possible is the increasing conception of the material world as "a site for God's self-manifestation" (134) which allows for a veneration of God in earthly things and beings. From the fifteenth century onwards, "the human, the physical, and the natural are increasingly experienced as places of divine revelation" (135), so that love of these things amounts to love of God. True, as May concedes, the el-

---

[13] For a detailed analysis of the development of different phases and varieties of courtly love see Singer, *Courtly and Romantic* 19–126. For a similar assessment of the significance of courtly love see May 119–20.

ement of sexuality was again significantly reduced in this symbiosis of religious and natural love (cf. 137). But the more the religious aspect of the symbiosis changed from orthodox Christianity into an amorphous pantheism or spirituality, the easier sexuality could re-enter the stage.[14] At the end of a long development, love of nature and love of humans as natural beings was sanctified as a relationship to some sort of universal origin or all-encompassing principle, and every aspect of this relationship, including sexuality, could, at least theoretically, be justified as one of its natural components. This is the world of Romantic love, which inherits from the courtly tradition the overvaluation of a human being as the object worthy of the highest kind of love and combines it with a profound metaphysics unknown to the medieval troubadours. As a result, the idea of love as compensation, which had played only a minor role in courtly love, is revived and elevated to unprecedented importance in Romanticism.

* * *

While human love had been celebrated as splendid and worthy in itself in the courtly tradition, it was not before Romanticism that it acquired the metaphysical and spiritual import that allowed it to replace religious faith as the dominant expression for the human desire for oneness and compensation. Romanticism combines the esteem for earthly, interpersonal, human love of the troubadours with the spiritualism of the Platonic and Neoplatonic tradition and its desire for union. In the majority of cases in which the concept of love as compensation becomes graspable it is expressed as a desire for a union, fusion, or merging that is hoped to result in a state of oneness. "Most Romantics," Singer writes, "thought that love enabled us to know and appropriate the universe by means of endless yearning for oneness with another person, or with humanity, or with the cosmos as a whole" (*Courtly and Romantic* 286). Conceived of in this way, love becomes the path to recognition and appreciation of the ultimate "oneness in all being" (287) – a core belief that most variants of Romanticism shared, be they traditionally Christian, pantheist, or atheist. "In the Romantics as a whole, love is a metaphysical craving for unity, for oneness that eliminates all sense of separation between man and his environment, between one person and another, and within each individual" (288). The compensatory function of love is clearly indicated in its direction from separateness to oneness. A feeling

---

14 May exemplifies this development with Spinoza's pantheism and its implications on his concept of love (cf. 143–51). On the influence of Spinoza on concepts of love in Romanticism, see also Singer, *Courtly and Romantic* 291.

of division or incompleteness characterises the Romantic lover, just as it had characterised the split halves in Aristophanes' or Eros itself in Socrates' speech. Singer quotes Coleridge, whose definition of love indeed reads as if it had been taken out of the *Symposium:* "Love is a desire of the whole being to be united to some thing, or some being, felt necessary to its completeness, by the most perfect means that nature permits, and reason dictates" (qtd. in Singer 288).[15] All differences that can doubtlessly be found between Romanticism and Platonism, or between different variants of Romantic love, cannot obscure their striking agreement on this central assumption of man's neediness and insufficiency and the resulting concept of love as the desire to replace this feeling of incompleteness with a feeling of oneness.

For a series of German Romantic philosophers, for instance, love was capable of establishing a sense of oneness with a "cosmic spirituality that underlay the apparent materialism of nature" (Singer, *Courtly and Romantic* 383). Fichte, Schlegel, Schleiermacher, Schelling, and Hegel all believed in the existence of a guiding or driving spirit behind material reality, conceived of as the "totality of being" or simply "the Absolute" (383), with which a union in or through love was thought possible. Hegel's early fragment on love, as Singer summarises it, describes love as "a process that occurs in three stages: unity, separation of opposites, reunion" (399) – a process in which a compensatory function is clearly inscribed. Hegel starts with the claim that "genuine love excludes all oppositions" (Hegel 117), adding that "[i]n love the separate does still remain, but as something united and no longer as something separate" (118). This view on the explosive question what 'genuine love' means for individuality and autonomy is indeed highly reminiscent of his notion of *Aufhebung*, according to which the synthesis does not only abolish the opposition between thesis and antithesis but also retains (parts of) them on the higher level of the new union. The lovers form "a living whole" and have independence and peculiar individualities only insofar as either of them may die and thus destroy this living whole – just like plants, made up of "salt and other minerals," may rot if one of these components can no longer fulfil its function (cf. 118). It is obvious, however, that the living whole, the union of the lovers, is the higher form of being, just as the plant is a higher form of existence than its components. Accordingly, genuine love requires unreserved dissolution of the separated, demarcated self into this new union of lovers – and Hegel sees that not all lovers might be willing to commit so completely to love. "There is a sort of antagonism between com-

---

15 Samuel Taylor Coleridge. *Shakespearean Criticism*. Vol. 2. Edited by Thomas Middleton Raysor, Dent, 1960, p. 106.

plete surrender or the only possible cancellation of opposition (i.e., its cancellation in complete union) and a still subsisting independence" (118). But true love demands this complete surrender, and true lovers will necessarily feel this demand. Failure to give in fully to loving oneness will eventually cause a feeling of shame: "A pure heart is not ashamed of love; but it is ashamed if its love is incomplete" (119). But why is it that what true lovers desire is a state of complete union in which "consciousness of a separate self disappears, and all distinction between lovers is annulled" (119)? The answer that can be inferred from Hegel's fragment points forward to twentieth century psychoanalysis or other systems of thoughts such as that of Georges Bataille or Erich Fromm, which will be discussed further below. In a few brief sentences, Hegel delineates the process of individuation and the development of the child into an individual as a process of gradual separation.

> Everything which gives the newly begotten child a manifold life and a specific existence, it must draw into itself, set over against itself, and unify with itself. The seed breaks free from its original unity, turns ever more and more to opposition, and begins to develop. Each stage of its development is a separation, and its aim in each is to regain for itself the full riches of life. (119–20)

The development of the human individual is thus a process of alienation that, on the one hand, furnishes the individual with the 'riches of life' but, on the other hand, leads him or her away from a state of unity. Eventually, the child will reach maturity and seek union with another individual as a way of restoring the lost primal unity – and the product of this union will be another child who will repeat the process of "unity, separated opposites, reunion" (120).

Hegel's theory of love is a Romantic argument against the alienation and isolation that come along with individuation. Individuality, in this version of Romanticism, becomes a stumbling block and hindrance to the blissful state of true union. As May shows, it is then only a short way from a "craving for a realm beyond individuality" (171), which characterises much of Romanticism, to a view in which life itself, as inescapably bound up with individuality and separation, needs to be overcome to enable union. But for Hegel, death is not the ultimate goal of love. Rather, love overcomes a form of death which possesses the individual who is unaware of the true status of being. The separated individual is dead to his true mode of existence, while "the oneness lovers feel is indeed their true reality. Through it they become self-aware for the first time and consequently transcend the metaphysical alienation that prevented them from being in touch with their ultimate being" (Singer, *Courtly and Romantic* 403). The 'living whole' into which the lovers merge is not the same as the Absolute, but it comes much closer to the true mode of human existence as non-separated and non-alienated from the rest of the world.

From Spinoza and other sources, Romanticism had inherited powerful notions of pantheism according to which all nature was not only an emanation of God or divinity, but actually infused or even identical to it. Consequently, the "Romantic preoccupation with oneness" (Singer, *Courtly and Romantic* 288) was not restricted to interhuman relations but could also include other forms of love as possible gateways to a fusion with the Absolute (cf. 291). However, the most preferable object to merge with was clearly a human person of the opposite sex. The widespread view that interpersonal love was an act of mutual complementation, in which one partner makes up for the natural deficiencies of the other, defined human beings as the ideal objects of love and the most suitable means to experience a state of wholeness. This may also explain "the pervasive eroticism, however subtle, that belongs to all Romantic theory" (295). With the idea of the fusion of two compatible halves and "its emotional power and blurring of a separate consciousness, sex is the clearest form of merging that most people can understand" (295). But just like in the Aristophanes myth, sex is not the ultimate goal but only a side effect or sometimes a means of this desire. Like Aristophanes, most nineteenth-century Romantics envisage the goal of love as "oneness with an alter ego, one's other self, a man or woman who would make up one's deficiencies, respond to one's deepest inclinations, and serve as possibly the only person with whom one could communicate fully" (Singer, *The Modern World* 4). As Singer demonstrates in his third volume on the history of love, the twentieth century loses faith in this benign notion of love as the union of two complementary halves (cf. 4). Instead, concepts of love gain ground which emphasise the precariousness and, sometimes, the impossibility of love. But as I would like to show, with the notion of love's compensatory function at least one essential feature of Romantic love has survived and, through its adoption by psychoanalysis, has found its way into contemporary notions of love where it remains influential long after Romantic and older beliefs in the metaphysical and transcendental power of love have waned.

2.1.3 Nostalgic Love: Psychoanalysis and the Desire of the Displaced Self

Love as compensation recurs in various but comparable shapes in the psychoanalytic tradition, and the influence of Sigmund Freud on contemporary discourses of love is on a par with that of Platonism and Romanticism, irrespective of the scientific accuracy of his theories. In his later works, especially, which "took on a philosophical cast that he himself might have scorned in his earlier years" (Singer, *The Modern World* 97), the image of love as compensation falls into place, even though there are aspects of it in earlier writings, too. One example for the latter is his description of the development of sexual goals or objects in his *Three Essays on Sexual Theory* (1905), where he argues that any sexual relationship in adulthood is a revival of the child's first experience of sexual pleasure and satisfaction in breastfeeding (cf. 200). Every erotic love relationship is a re-enactment of "this first and most important of all sexual relationships," compensating for the loss of this primal love object and helping to "re-establish lost happiness" (200). In this scheme, love originates as a combination of sexual pleasure, first experienced as stimulation of the erogenous zone of the lips, and a more or less subconscious feeling of gratitude towards the person who answers the existential needs of the helpless child.[16] Love between adults is structured by the secret desire to repeat this combination of sexual and emotional pleasure and to re-experience the absolute wholeness and satisfaction once felt at the mother's breast.

His later elaboration on this state of 'lost happiness' in *Civilization and its Discontents* (1930) offers the most salient (and most Romantic) image of love as compensation in Freud's work. Inspired by a friend's comparison of religious sentiments to "a feeling which he would like to call a sensation of 'eternity,' a feeling as of something limitless, unbounded – as it were, 'oceanic'" (11), Freud sets out to explain this 'oceanic' feeling and to trace its origin in psychological development. He conceives of it as "a feeling of an indissoluble bond, of being one with the external world as a whole" (12), which may occur when the ego experiences a dissolution of the boundaries that separate it from the external world. While usually the "ego appears to us as something autonomous and unitary, marked off distinctly from everything else" (12–13), Freud knows of one state – "admittedly an unusual state, but not one that can be stigmatized as pathological" (13) – where the ego does not experience itself as sharply demarcated from the external world: "At the height of being in love the boundary between ego and object threatens to melt away. Against all evidence of his senses,

---

[16] In "Concerning the Most Universal Debasement in the Erotic Life" (1912) Freud specifies this twofold nature of love as the coexistence of an *"affectionate"* and a *"sensual"* (i.e. sexual) current, which are originally directed at the same object – the mother or a surrogate caregiver – but are separated during puberty. Ideally, affection and sensuality can later be reunited in a relationship where an extra-familial object of sexual desire is also the object of affection, but this reunion is by no means certain. Men especially, Freud states, often suffer from "psychical impotence," the incapability of directing affectionate love and sexual desire to the same person. "Where they love they do not desire," he famously concludes, "and where they desire they cannot love" (253).

a man who is in love declares that 'I' and 'you' are one, and is prepared to behave as if it were a fact" (13). If this symbiotic feeling arises, Freud further assumes, it is basically a revival and re-experience of the primal ego-feeling of undisturbed oneness which characterises infantile self-experience and of which subconscious memories are preserved in the reservoir of mental states long after the ego has learned to conceive of itself as a distinct, separate entity (cf. 13–19). Love, in other words, is capable of providing an 'oceanic' feeling because, for some time and to a certain degree, it dissolves the boundaries of the ego and admits the return of the primal ego-feeling of absolute oneness. The immense pleasure (rather spiritual than sexual in this case) that love evokes is the result of a temporary and partial return to a state of total unity, a fleeting homecoming of the displaced self.

It is plain to see why Singer puts this aspect of Freud's theory in close relation to Romantic philosophies (cf. *The Modern World* 132). The idea of "love as a sense of oneness with the universe" (132) recalls the Romantic obsession with images of merging with the Absolute and there is a pronounced similarity with Hegel's picture of love as remedy against the isolation and alienation brought about by individuation. But *Civilization and its Discontents* mentions yet another aspect of love as compensation, which seems to be independent from the oceanic feeling and its Romantic connotations. Quite simply, Freud presents love as a compensation for life's hardships. "Life, as we find it, is too hard for us," he writes; "it brings us too many pains, disappointments and impossible tasks" (22), which is why humanity has developed "palliative measures" (22) that are supposed to make life bearable. Freud provides an admittedly incomplete list of such measures: chemical intoxication (cf. 25); methods of "killing off" (26) or at least taming the drives which make us strive for unattainable goals; the strategy of "sublimation," in which libidinal impulses are diverted from their original external and hence precarious objects and are invested instead in internal processes of "psychical and intellectual work" (26); or ways of deriving pleasure from illusions, like "the enjoyment of works of art" (27) or the "mass-delusions" (28) of religion. What all these methods have in common is that they try to free the individual from his or her precarious relationship to the external world – either by moulding it through imagination or intoxicants or by replacing external sources of pleasure with internal ones. The last method Freud mentions, however, differs from the previous precisely because "it does not turn away from the external world; on the contrary, it clings to the objects belonging to that world and obtains happiness from an emotional relationship to them" (29). "And perhaps," Freud goes on, "it does in fact come nearer to this goal than any other method. I am, of course, speaking of the way of life which makes love the centre of everything, which looks for all satisfaction in loving and being loved" (29). Such praise of love as the provider of happiness must not obscure Freud's ambiguous and ultimately pessimistic attitude towards interpersonal, 'romantic' love. But although he seems convinced that love is in the end incapable of fulfilling its task, there can be little doubt about wherein this task consists in his system. Whether adult love is presented as modelled on the primal relationship between mother and child, as the reunification of the affectionate and the sensual current, as the revival of the oceanic feeling of oneness with the universe, or as a 'palliative measure' against the adversities of life: in all these cases love has a fundamentally compensatory function.

Many representatives of psychoanalysis have followed Freud in this basic assumption. Carl Jung, while objecting to the attribution of adult love to infantile sexuality, agrees with the conception of love as a "return to that original condition of unconscious oneness" which is both "a return to childhood" or even "a return to the mother's womb" and "a genuine and incontestable experience of the Divine, whose transcendent force obliterates and consumes everything individual" (181). As in Hegel and Freud, lovers are longing for "unity and undividedness" (184) driven by vague memories of a state before the development of individuality and the demarcation of the individual against the surrounding world. Similarly, Margaret Mahler argues that "the psychological birth of the individual" (215) requires a *"separation-individuation process"* (215) during which the infant learns that its existence is essentially independent from that of the mother. Distant memories of the primal symbiotic relationship with the mother and an unconscious wish to return to it are the powerful remnants of this development and determine the erotic life of the human individual. The paradigm of 'object relations theory' underlying Mahler's ideas, according to which drives and sexuality are less significant for infantile self- and personality-development than the child's relations to and interaction with its most salient 'objects,' that is, with the mother and other caretakers (cf. Rivkin and Ryan 122), also shapes Martha Nussbaum's theory in *Upheavals of Thought*. She follows the thesis "that the emotions of later life make their first appearances in infancy, as cognitive relations to objects important for one's well-being, and [...] that this history informs the later experience of emotion in various ways" (179). For a comparatively long period, human infants depend on caretakers, agencies who compensate for the child's helplessness and provide "what the world of nature does not supply by itself – comfort, nourishment, protection" (182). The child experiences the provision of these things as "the restoration of a blissful or undisturbed condition" (183) or as a transformation from a condition of need and helplessness to a condition of plenty and security. The longing for the object which can bring about this transformation and the "kind of rudimentary love and gratitude" (190) that comes with it prepare the foundation and structure for what later is to become mature love. Sexuality, in this view, enters the stage later but then repeats and expresses the very needs that characterise early object relations. "In that sense," Nussbaum writes, "sexual love [...] is a species of a more general category of erotic love and desire that has its origins in the child's longings to control the comings and goings of its mother, seen as the most important and marvellous creature in the world" (460). Unlike in Freudian psychoanalysis, love in adulthood is not a re-enactment of infantile sexuality, but mature sexuality is a re-enactment of infantile love. But this difference is of less concern here than the agreement of both schools on the broader function of love. No matter whether life's first love relationship – that with the mother or caretaker – is sexual or not, it is always regarded as a relationship that compensates for the fundamental insufficiency and helplessness of the child, that is experienced as a state of blissful oneness, and that ends with the individual's recognition of separate existence, leaving behind traces of this benign state of wholeness in a person's emotional memory that structure later love relationships.

The work of Jacques Lacan offers a similar explanation of desire. Catherine Belsey, whose reading of Lacan I will follow, discerns the child's entry into the symbolic order as the moment of loss and separation that turns the human being into a "dissatisfied" and "desiring" being (*Poststructuralism* 57). With entry into the symbolic order the organism turns into a subject through the acquisition of language, which helps the infant to master its surroundings and formulate demands but is also perceived as "irretrievably Other" (58). It is not only 'Other' in the sense that it does not come *from* but *to* the child, but it is also a system, the fundamental principle of which is difference. Through language, the child learns to order and structure the world on the basis of difference, while realising its own affliction by this difference. As Belsey points out, this is an event of loss:

> Something is lost here – experienced, perhaps, as a residue of the continuity with our organic existence, or as wishes that don't quite fit the signifiers that are supposed to define them. Lacan calls what is lost the real. The real is not reality, which is what culture tells us about. On the contrary, the real is that organic being outside signification, which we can't know, because it has no signifiers in the world of names the subject inhabits. (58)

Entering the symbolic order means loss of the real, and this loss is the origin of desire (cf. also P. Barry 109; Rivkin and Ryan 123–25). The 'real' is a rather protean concept in Lacan's system and assumes a variety of meanings in the long succession of his seminars (cf. Johnston). Rivkin and Ryan describe it as "an impossible wholeness of self, plenitude of desire satisfaction (*jouissance*), and continuity of signifier and signified or word and object" (123), a left-over of "the child's original symbiotic relationship with the mother" which caused "a false narcissistic sense of unity" and which is lost with entry into the symbolic and the acquisition of language (124).[17] The effect of entering the symbolic is thus "a combined linguistic/psychological separation of the child both from its initial object, the mother, and from the undifferentiated matter of natural existence" (124). All future desire is an unconscious striving for the retrieval of that which has been lost, while the conscious objects of human desire are only substitutes for the real object. Thus, "all desire is inherently metaphoric" (124) and since "no metaphor can embody what we ultimately desire when we desire anything, we are condemned to slide along a chain of signifiers each of which is a metonymy, a part standing in for the whole we (always) miss" (125). As Belsey puts it, despite the unknowability and unattainability of the real object of desire, "most of us find a succession of love-objects, and fasten our desire onto them, as if they could make us whole again, heal the rift between the subject and the lost real" (*Poststructuralism* 59). This distinction between the real but unnameable object of desire and the concrete objects of our material reality which are substituted for it is also expressed with Lacan's famous *objet petit a*. In Seminar XI he describes it as an object "which is in fact simply the presence of a hollow, a void, which can be occupied [...] by any object, and whose agency we know only in the form of the lost object" (180). The *objet petit a* is a formula expressing the idea that desire is caused by a missing or lost object while indicating at the same time that this 'object' is not an identifiable, graspable, accessible object of material reality but part of the inaccessible Lacanian real. Every object of our reality can only inadequately substitute for this focal point of desire.

The motif of compensation returns in a less prominent statement in Seminar XI, where Lacan explains that entering into the Otherness of the symbolic order is not the only – and maybe not even the most crucial – experience of loss that structures human desire.

> Two lacks overlap here. The first emerges from the central defect around which the dialectic of the advent of the subject to his own being in the relation to the Other turns – by the fact that the subject depends on the signifier and that the signifier is first of all in the field of the Other. This lack takes up the other lack, which is the real, earlier lack, to be situated at the advent of the living being, that is to say, at sexed reproduction. The real lack is what the living being loses, that part of himself *qua* living being, in reproducing himself through the way of sex. This lack is real because it relates to something real, namely, that the living being, by being subject to sex, has fallen under the blow of individual death. Aristophanes'

---

[17] Adrian Johnston marks this conception of the real as belonging to Lacan's earlier works of the 1950s in which "Lacan tends to speak of the Real as an absolute fullness, a pure plenum devoid of the negativities of absences, antagonisms, gaps, lacks, splits, etc. Portrayed thusly, the Symbolic is primarily responsible for injecting such negativities into the Real."

> myth pictures the pursuit of the complement for us in a moving, and misleading, way, by articulating that it is the other, one's sexual other half, that the living being seeks in love. To this mythical representation of the mystery of love, analytic experience substitutes the search by the subject, not of the sexual complement, but of the part of himself, lost forever, that is constituted by the fact that he is only a sexed living being, and that he is no longer immortal. (204–05)

What Lacan describes here is a paradox determining all human life. By being born, by coming to life, immortality is lost. This is more than the simple truism that the gift of life always and necessarily entails the prospect of death. Lacan's cryptic formulation does not merely state that all life is mortal but that there is an actual *loss* of immortality. Moreover, he links this loss to sexual reproduction in a relationship of cause and effect that remains largely enigmatic both in the quoted passage and in the preceding account of his 'myth' of the lost *lamella*, which is supposed to rectify the misleading tale of Aristophanes. Lacan's myth starts at a moment before the birth of the human being. "Whenever the membranes of the egg in which the foetus emerges on its way to becoming a newborn are broken," he asks his readers, "imagine for a moment that something flies off" (197). This thing that "flies off" and is hence lost at a particular moment of sexual reproduction is the *lamella*, a curious concept of an "organ, whose characteristic is not to exist" (197–98) and which is the 'immaterial embodiment' of life as such. "It is the libido *qua* pure life instinct, that is to say, immortal life, or irrepressible life, life that has need of no organ, simplified, indestructible life. It is precisely what is subtracted from the living being by virtue of the fact that it is subject to the cycle of sexed reproduction" (198). In sexual reproduction, at the moment of insemination, human beings lose their 'pure life' or immortality, "[a]nd it is of this that all the forms of the *objet a* that can be enumerated are the representatives, the equivalents" (198).

Lacan's strangely metaphysical speculations and his "equation of sexuality with death" (Freeland 66) may be explicated with the help of Georges Bataille's equally esoteric *Erotism*, originally published as *L'Erotisme* in 1957, seven years before Lacan presented his myth of the *lamella*.[18] For Bataille, too, sexuality and death are closely linked and his theory resembles Lacan's in that it postu-

---

18 As Michel Surya notes in his biography of Bataille, Lacan and Bataille maintained a close and intimate relationship with each other. Lacan stayed informed about the activities of *Acéphale*, for instance, a secret society founded by Bataille complete with secret rituals, nocturnal ceremonies in the woods, and a morbid fascination with mysticism and sacrifice (250–52), and married Bataille's former wife Sylvia. Surya suggests that "[t]he close intellectual and emotional relationship uniting Bataille and Lacan should one day be investigated, and it is a relationship whose effects can more than once be sensed in Lacan's work" (534n34).

lates a fundamental loss which takes place at the moment when life begins: the loss of *continuity*. "Reproduction," Bataille begins, "implies the existence of *discontinuous* beings. Beings which reproduce themselves are distinct from one another [...]. Each being is distinct from all others. [...] He is born alone. He dies alone. Between one being and another, there is a gulf, a discontinuity" (12). In logical consequence, the opposite of discontinuity is to be found in the opposite of reproduction, so that "death is to be identified with continuity" (13). Human beings are torn between the desire to return to the lost state of continuity in which there was no individuality and no separation, and the desire to preserve their lives in the only possible form, that is, as discontinuous lives.[19] "We are discontinuous beings, individuals who perish in isolation in the midst of an incomprehensible adventure, but we yearn for our lost continuity" (15). Eroticism, for Bataille, takes place exactly at the border between these two states. It is the realm in which continuity is approached without fully transgressing the border to death. "Continuity is what we are after, but generally only if that continuity which the death of discontinuous beings can alone establish is not the victor in the long run. What we desire is to bring into a world founded on discontinuity all the continuity such a world can sustain" (18–19). Hence, instead of pursuing actual and total continuity in death, humans seek partial and temporal continuity with other human beings in eroticism. Yet, despite this life-preserving modulation, the uncanny closeness of death and eroticism remains intact. Eroticism aims at a dissolution of the self that is different from death only gradually, not in nature. "Eroticism opens the way to death" (24).

Sexuality and death are connected in yet another way, for which the notion of life as a cycle of becoming and dying is crucial. "The death of the one being," Bataille states, "is correlated with the birth of the other, heralding it and making it possible. Life is always a product of the decomposition of life" (55). Death is necessary for the constant rejuvenation of life on earth, and life, which in contrast to the continuity of death is an immense expenditure and waste of energy, "can only proceed under one condition: that beings given life whose explosive force is exhausted shall make room for fresh beings coming into the cycle with renewed vigour" (59). In the logic of this cycle, sexuality, as the act of the creation of new life, already implies the death of the old. According to Bataille, life thus appears to human beings – and especially to the primitive human beings on the verge of the beginning of cultural history – like "an orgy of annihilation" in which "[s]exuality and death are simply the culminating points of the holiday, nature celebrates, with the inexhaustible multitude of liv-

---

[19] The parallel to Freud's struggle between *Eros* and *Thanatos* is unmistakable.

ing beings" (61). For Bataille, what distinguishes humanity from other forms of life is the conscious attempt, at some point in prehistorical time, to refuse this cycle of nature by imposing primitive laws or taboos that were supposed to counter nature's violent 'holiday' of creation and destruction (cf. 62; 83). And even though this resistance against violence and death has always been only half-hearted and the taboos against murder and sex are frequently transgressed, sexuality never fully loses its forbidden character. "In the human sphere sexual activity [...] is in essence a transgression" (108), Bataille writes, and this is what distinguishes human eroticism from animal sexuality (cf. 108).[20]

Sex therefore stands in a paradoxical relationship to life and death. On the one hand, as the potential origin of new life, it is a movement towards life. On the other hand, as the transgression of a taboo ultimately meant to preserve discontinuous life, it points towards death since in the sexual act mankind's resistance against nature's cycle of reproduction and death is temporarily suspended. Moreover, the sexual act is in itself a temporary dissolution of the separate discontinuity of the two partners. The fusion of bodies allows, if only momentarily and imperfectly, to experience a state in which there is continuity with another being and in which the self is not sharply demarcated (cf. 17; 102). Consequently, "Eroticism [...] is assenting to life up to the point of death" (11). In eroticism, the two fundamental drives of human life, the urge to return to continuity and the desire to preserve one's discontinuous existence, are simultaneously at work. Following this line of reasoning, it is possible to understand why for Lacan the sex drive "is profoundly a death drive and represents in itself the portion

---

[20] Bataille's thoughts seem closely related to Arthur Schopenhauer's in "Metaphysik der Geschlechtsliebe," where love and sex are similarly accused of partaking in an 'orgy of annihilation.' For Schopenhauer, all sexual love is a (subconscious) manifestation of the Will to Life that aims only at the preservation of the species through procreation (cf. 624–27). Love is nature's trick, a delusion that makes lovers assume that they are following their own preferences and interests when all they do is serving the Will (cf. 630–31).In preserving the human species, the Will preserves the indestructible and immortal core of mankind from generation to generation, that which, as the 'thing-in-itself' and detached from the principle of individuation ("als Ding an sich, frei vom *prinicipio individuationis*"), is identical in all individuals (656). This general continuation of life, however, comes at the price of continuing the perennial cycle of individual living and dying which could only be countered with the conscious negation ("Verneinung") of the Will to Life. Love and the sexual act, in contrast, mean consent to the Will and hence to the cycle of individual suffering and death – the very cycle against which, according to Bataille, taboos have been installed. For Schopenhauer, this explains the furtive, surreptitious glances lovers throw each other: "Weil diese Liebenden Verräther sind, welche heimlich danach trachten, die ganze Noth und Plackerei zu perpetuieren, die sonst ein baldiges Ende erreichen würde [...]" (656). As for Bataille, lovers are transgressors or 'traitors' in Schopenhauer's view because instead of negating the will they consent to its indiscriminate cruelty.

of death in the sexed living being" (*Seminar XI* 205). While for Freud the sex drive was essentially a life drive, for Bataille and Lacan it is always also an assent to the mortality inherent in sexual reproduction.

Bataille's whole theory of human self-experience as 'discontinuous' and the experience of eroticism as an act of transgression rests upon assumptions about pre-historical human evolution. Psychoanalyst and social philosopher Erich Fromm follows a similar approach in *The Art of Loving*, where he discerns mankind's evolutionary separation from nature as lying at the heart of erotic desire.

> What is essential in the existence of man is the fact that he has emerged from the animal kingdom, from instinctive adaptation, that he has transcended nature – although he never leaves it; he is part of it – and yet once torn away from nature, he cannot return to it; once thrown out of paradise – a state of original oneness with nature – [m]an can only go forward by developing his reason, by finding a new harmony, a human one, instead of the prehuman harmony that is irretrievably lost. (7)

Displaced from the home which was undivided oneness with nature and made aware of it by the very agent of this separation, the gift of reason, only love can save man from sliding into utter desperation:

> This awareness of himself as a separate entity, the awareness of his own short life span, of the fact that without his will he is born and against his will he dies, that he will die before those whom he loves, or they before him, the awareness of his aloneness and separateness, of his helplessness before the forces of nature and of society, all this makes his separate, disunited existence an unbearable prison. He would become insane could he not liberate himself from this prison and reach out, unite himself in some form or other with men, with the world outside. (8)

The compensatory function of love could not be stated with more clarity. As the most intense form of reaching out and uniting with others it makes up for man's 'disunited existence' which would otherwise be intolerable. "The deepest need of man," Fromm is certain, "is the need to overcome his separateness, to leave the prison of his aloneness" (9), and the only way out "lies in the achievement of interpersonal union, of fusion with another person, in *love*" (17).

* * *

At the beginning of this chapter, I moved from a discussion of the possibility of unconditional love to Simon May's concept of 'ontological rootedness' as the fundamental condition of all forms of love. In his book, May tries to set off his thesis from the long tradition of mythological and philosophical thinking that explained love as a return or movement towards a state of (prior) oneness or wholeness. While in this tradition love's final goal is a form of fusion or merging with the object of love, May emphasises the importance of experiencing the other as "radically distinct" (10). But what I take to be more crucial than this difference is the common ground he shares with said tradition in that he sees love as originating in a state of deprivation. In fact, all the theories of love presented above differ from one another as much as May's theory differs from them. There are, for example, very different objects of love in the speeches of Aristophanes (human beings) and Socrates (the idea of the Good). Romantic theories of love differ with regard to the role sexuality plays in the desired union or fusion, which is variously a union of souls or bodies (or both) with one another, or with nature, or with the totality of being. Psychoanalytic approaches, too, put more or less emphasis on sexuality and suggest quite different objects of loss at the core of desire: the mother, infantile omnipotence, the 'real,' immortality. But all these differences are marginal compared to the overwhelming similarity of the various theories. They all presuppose the neediness and insufficiency of the human being and link love with the desire to overcome this condition. Whether caused by a moment of loss or conceived of as an inevitable feature of human nature, there is a profound lack torturing the human being and love is credited with the task of compensating this lack. It might indeed be possible to use May's terms 'ontological rootedness,' 'groundedness,' or 'at homeness' to describe the ultimate goal of erotic desire in all the theories discussed in this chapter, but this might also generalise too much. It is certainly possible, however, to use the term compensation for the function these theories ascribe to love.

2.2 The Precariousness of Love

Love, supposed to make up for the vulnerability and insufficiency of human life, is itself a dangerous undertaking. As Freud puts it, "we are never so defenceless against suffering as when we love, never so helplessly unhappy as when we have lost our loved object or its love" (*Civilization* 29). To pin one's hopes on love is risky because love brings about another variety of the very state it is supposed to overcome: desired as the remedy against a rather general or 'ontological' vulnerability that is part of the human condition, love introduces a very specific vulnerability that affects the individual as the result of being in love with a particular person; embraced as compensation for a fundamental lack or loss, love comes along with the permanent threat of its own termination. Freud's observation is by no means original. The darker side of love, its inextricable entanglement with the dangers of loss and suffering, has always been part of the discursive field. In fact, the motif of love's perilous nature is so ubiquitous and familiar that there is no need to demonstrate its historical continuity. What I will attempt instead is to give a cursory overview of what has frequently been seen at stake in love and to emphasise the sources of precariousness that can be considered specifically modern. It will turn out that the discourse of love as risk can be divided into narratives stressing the uncontrollability or unattainability of the (actual) *object* of love and narratives focussing on the loving *subject* as threatened with objectification and appropriation through the other and the potential loss of self and individuality within a love relationship.

The term chosen to denote this 'darker' side of love, its never fully absent atmosphere of impending loss and suffering, is 'precarious,' and there are reasons to prefer this term to words such as 'risky' or 'dangerous.' In his *Dictionary of the English Language* (1755), Samuel Johnson provides the following definition of 'precarious': "Dependent; uncertain, because depending on the will of another; held by courtesy; changeable or alienable at the pleasure of another. No word is more unskilfully used than this with its derivatives. It is used for uncertain in all its senses; but it only means uncertain, as dependent on others." This narrow definition describes an inevitable feature of love: if it is supposed to be reciprocal, it is dependent on the will of another. Accordingly, in his *Lover's Discourse* Roland Barthes defines 'dependency' as "[a] figure in which common opinion sees the very condition of the amorous subject, subjugated to the loved object" (82). Love or, more precisely, the wish to be loved in return, throws the loving subject into a state of absolute dependence and vulnerability, entirely at the mercy of the beloved. But I would like to use the term 'precarious' also in the more 'unskilful' manner mentioned by Johnson, where it denotes a kind of uncertainty that does not necessarily depend on the will of the other. As the *Oxford English Dictionary* explains, since the end of the seventeenth century the meaning of the term in common usage was "[d]ependent on chance or circumstance; uncertain; liable to fail; exposed to risk, hazardous; insecure, unstable" (cf. also Gilliver). Even though the independent will of another person is frequently a crucial factor in love's precariousness I do not want to exclude this broader conception from my usage as it additionally covers the threats posed to love by external circumstances as well as all sorts of obstacles emerging from within the loving subject.

2.2.1 The Unwanted Guest: Alcibiades and the Desire of the Other

When Freud talks about love as one of the strategies to cope with the adversities of human life, he unambiguously stresses the precariousness of this option. Seeking happiness in love, Freud says, man "made himself dependent in a most dangerous way on a portion of the external world, namely, his chosen love-object, and exposed himself to extreme suffering if he should lose it through unfaithfulness or death" (*Civilization* 48). The uncontrollability of the love-object is the main problem in human love relationships and therefore, Freud goes on, "the wise men of every age have warned us most emphatically against this way of life" (48). One of the first to do so, again, was Plato. Socrates is commonly seen as the most authoritative voice in Plato's dialogues and, as we have seen, his vision of love is one in which the unstable, ephemeral object of common, earthly, sexual love is eventually replaced by an indestructible, eternal object – the idea of the Good. A love like this is no longer precarious, no longer subject to chance and risk or dependent on another person. Although Nussbaum doubts "whether what is left at the end still contains what was originally valuable and wonderful in love, whether it is still erotic at all, still love at all" (*Upheavals* 469), it is what the *Symposium* seems to suggest as the best form of love. In this traditional reading, the last of the speeches, delivered by Alcibiades right after Socrates' account, functions as an example for the disastrousness of unrefined, unsublimated love that fails to break away from human beings and to ascend the ladder of love towards the realm of ideas.

Alcibiades,[21] not one of the invited guests, invades the symposium "very drunk and shouting loudly" (Benardete/Plato 212e) with the intention of congratulating Agathon on his victory in the dramatic competition. It immediately transpires that some kind of erotic relationship is going on between him and Socrates, who says that "[t]he love I have of this human being has proved quite bothersome. For since the time that I first loved him, it is no longer possible for me to look at or converse with even one beauty; or else in jealousy and envy of me he does amazing things, and abuses me and hardly keeps his hands off me" (213 c–d). Apparently, the usual pattern of pederastic love, in which an older *erastes* pursues a younger *eromenos*, has been reversed: the young and beautiful Alcibiades is jealous of his lover, which puts himself into the role of lover and turns Socrates into the beloved (see also Nussbaum, "The Speech" 279–80 and 300). Alcibiades then complies with the rules of the symposium and agrees to give a speech – except that instead of Eros he will praise Socrates. Like the sculptures of satyrs, he begins, which have a hid-

---

[21] Alcibiades is an historically well-documented political and military figure and the *Symposium* is set at a point in time when he was at the peak of his popularity in Athens. This popularity would have been known to readers of Plato's text just as well as his later downfall. For a detailed analysis of the interrelations between Alcibiades' depiction in the *Symposium* and the historical facts about his person see Seth Benardete, "On Plato's 'Symposium'" (181–82 and 198–99), Bloom (525–26) and Neu (329).

eous appearance but, "if they are split in two and opened up, they show that they have images of gods within" (215b), the notoriously ugly Socrates possesses an inner beauty that is "so divine, golden, altogether beautiful, and amazing that one had to do just about whatever Socrates commanded" (216e). Alcibiades then speaks quite openly about his love for Socrates, which started out as intellectual admiration but soon acquired a sexual dimension. But no matter what he tried, Socrates would not yield to his advances. Neither naked wrestling, nor long intimate conversations, nor a night spent together under one blanket could move Socrates sexually, so that a frustrated Alcibiades proclaims "that though I slept the night through with Socrates I got up without anything more untoward having happened than would have been the case if I had slept with my father or elder brother" (219c–d). What Alcibiades desires, it must be noted however, is not the satisfaction of his own sexual needs. Although he has become the pursuing *erastes* in their unilateral relationship, he offers the sexual service of the *eromenos*, which means that it is not his own sexual gratification but that of Socrates he wants to take care of (cf. 218c–e).[22] Socrates refuses to be gratified, however, leaving Alcibiades in the shameful position of having offered himself without being taken. And yet, Alcibiades cannot bring himself to abandon his desire or even be angry with Socrates, "enslaved by this human being as no one has been by anyone else" (219e), and so he continues praising Socrates for his constancy, courage, modesty, "and many other amazing things" (221c). An ensuing argument between the two is interrupted when "a large group of revellers" (223b) enters the scene and puts an end to the series of speeches, turning the symposium into a mere drinking feast.

Commentary on Plato's *Symposium* has often paid little attention to the speech of Alcibiades and has treated the episode as a kind of comic relief after the serious discussion that preceded it. At best, it has been understood as exemplification of an inferior alternative to the version of love presented by Socrates/Diotima – a form of carnal love practised by those unable to aspire to the spiritual love of ideas (cf. Secomb 18–19; Geisenhanslüke 65, 70). But to disregard the role of Alcibiades means to disregard what he alone, as Martin Koppenfels points out, introduces into the discussion of *eros* in the *Symposium*, namely its dark, destructive, and violent nature (cf. 276). As Linnell Secomb

---

22 Drawing on Kenneth Dover's *Greek Homosexuality* (1978), Nussbaum explains that in Athenian pederasty sexual contact usually did not go any further than the *eromenos* allowing the *erastes* to touch his genitals and perform intercrural intercourse. "The boy may hug him at this point or otherwise positively indicate affection. But two things he will not allow [...]. He will not allow any orifice of his body to be penetrated [...]. And he will not allow the arousal of his own desire to penetrate the other" ("The Speech" 300).

writes, "[s]upplementing Socrates with Alcibiades reveals the multiple possibilities of love and its perplexing articulation of the carnal and the ethereal, of voluptuousness and torment" (19). The Socratic ascent of love may have been Plato's ideal, but the appearance of Alcibiades in the *Symposium* clearly marks it as an ideal which does not correspond to the lived reality of most people, whose erotic love is likely to resemble the passion of Alcibiades. Plato seems to be well aware that the shifting of erotic desire from fellow human beings to the realm of ideas – a move which makes the lover insusceptible to the pain interpersonal love can cause – may be an option for the philosopher but likely not for the ordinary lover.

This significance of Alcibiades as important realist corrective to the ethereal concept of Socratic love is put forward with special emphasis by Martha Nussbaum. She draws attention to the fact that Socrates' theory of love, which denigrates or at least neglects interpersonal love, is followed "with a counterexample, a story of intense passion for a unique individual as eloquent as any in literature" ("The Speech" 282). The entirely risk-free ascent love of Socrates/Diotima satisfies the "deep demand of our natures for self-sufficient love" (294) but it also has an enormous disadvantage: it is "remote from human nature" (292), unlikely to be suitable for someone who is not Socrates. In his praise of Socrates, Alcibiades describes him as almost impenetrable, insensitive to hunger, cold, pain, fatigue, alcohol, or sexual temptation, and capable of immersing himself in contemplation to the point of forgetting the world around him (cf. "The Speech" 295; Plato 219e–221c). These characteristics, Nussbaum argues, are more than "intriguing pieces of biography. [...] They show us what Diotima could only abstractly tell: what a human life starts to look like as one makes the ascent. Socrates is put before us as an example of a man in the process of making himself self-sufficient" (295). Socrates is an exceptional human being whose way of life is not for everyone. The *Symposium* portrays him as a rather detached and unemotional person and Nussbaum doubts whether a life like this is still desirable (cf. 295). The speech of Alcibiades brings down to earth the lofty visions of the Socratic account, confronting the speculations of the philosopher with the profane experience of the unhappy lover. As Bloom puts it, Alcibiades' speech is "the opposite of a sermon given by a dry academic type," a "full-blooded" account "preserving the perspective of normal, healthy human beings" (528).

Appreciating Alcibiades' desire as genuinely human, however, does not yet explain it. If he is attracted not by Socrates' body but by his inner beauty, why does his desire acquire a sexual dimension? And what is the goal of this sexual desire if it is not seeking his own sexual gratification? In his Seminar VIII on 'Transference,' Lacan offers an extraordinary and sometimes idiosyncratic reading of the *Symposium* which puts the accent on Alcibiades and approaches these very questions. Socrates' contribution is only of interest to him because he develops the definition that love/*eros* is always love *of something* into his principle that "desire is in itself identical to lack" (63). From here, Lacan moves to an analysis of the speech of Alcibiades, always with the question in mind whether what the beloved has "bears a relation [...] to what the other, the desiring subject, is lacking" (34) – a question he already answers to the negative at the beginning of the seminar: "What the one is missing is not what is hidden in the other. This is the whole problem of love" (39–40). In other words, what Alcibiades really wants are not Socrates' beautiful qualities hidden inside his ugly shell. It is not something that Socrates could easily give – as would have been possible if the object of desire had been his knowledge, advice, or friendship. What Alcibiades does not know, while Socrates seems to have understood in his rejection to love Alcibiades, is that, as Lacan famously puts it, "love is giving what you don't have" (34). This somewhat obscure formulation makes perfect sense in combination with another equally ambivalent definition that occurs repeatedly in Lacan's writing, namely that "man's desire is the Other's desire [*le désir de l'homme est le désir de l'Autre*]" ("Subversion" 690) or, as he puts it elsewhere, "Desire is desire for desire, the other's desire" ("On Freud's 'Trieb'" 723). Alcibiades' desire to gratify Socrates' sexual needs is, in other words, a desire to be desired. What he really wants is a manifestation of Socrates' need and desire for him (cf. *Seminar VIII* 176). He wants Socrates to give his desire – and since desire is identical to lack, he wants him to give what he does not have.

Lacan's approach to the structure of desire implies that love is bound up with a demand for recognition. What is desired, Lacan specifies, "is that which is desiring in the other. This cannot happen unless the subject himself is situated qua desirable. Which is precisely what he demands when he demands to be loved" (*Seminar VIII* 357). The demand for love is a demand for the confirmation of one's desirability, of being needed. But of course, Lacan's statement that *le désir de l'homme est le désir de l'Autre* also describes another fact about love which Lacan by no means forgets: that desire is desire *for* the other. In the seminar on 'Transference,' Lacan makes explicit that all moral and ethical scruples about the notion of objectifying another human being cannot obscure "the fact that (in this elective, privileged relationship which is the love relationship) the subject with whom, among all others, we have a bond of love, is also the object of our desire" (146). A lover's desire thus seems to aim at two objectives: the confirmation of one's own desirability through the desire of the other (which requires the other qua subject) and the possession of the other as an object of love. Socrates refuses both. He refuses to take Alcibiades as the object of his desire and he refuses to be Alcibiades' object of desire, for which Lacan provides an explanation that ties in with his general assumption that "love is a comical feeling" (33). The comedy of love, as Achim Geisenhanslüke succinctly concludes, is the very fact that no object of love can ever live up to the loving subject's true desire (cf. 78), that, to repeat the quotation from Lacan's seminar, "[w]hat the one is missing is not what is hidden in the other" (39–40). Socrates, whom Lacan construes as an ancient forerunner of psychoanalysis, knows all that (cf. Geisenhanslüke 83). He knows that he does not have what Alcibiades really desires, that every empirical object of desire is only a substitute for the unattainable *objet a*.

★ ★ ★

The speech of Alcibiades exemplifies the kind of precariousness that results from the uncontrollability and uncertain attainability of the beloved and his or her desire. This phenomenon has fuelled love plots in literature from antiquity to the present day. In comedies and tragedies alike, it takes the form of external obstacles which the lovers may or may not be able to overcome, or the form of one-sided, unrequited love which may or may not develop into mutuality.[23] Always, love is portrayed as potentially dangerous, threatening the lovers with uncertainty and dependence. Different attitudes are possible in reaction to this form of precariousness. Lucretius and Ovid, for instance, advise their readers to strictly avoid the riskiness and dependence involved in passionate love, either by separating the necessary act of sexual procreation from its unnecessary appendages love and passion – a separation that Lucretius argues is most effectively achieved through marriage (cf. Singer, *Plato to Luther* 133) –, or as Ovid suggests, by learning to master the game or art of love until it is only the other who passionately desires and depends on the lover but not vice versa (cf. May 76–78). Both strategies are unsatisfactory, of course. The first deprives people of the pleasure passionate love can bring, the second turns love into a deceptive game and struggle for power which rules out permanence and guarantees pain and frustration for one party as much as it prevents it for the other. Passionate love without precariousness, it seems, is hard to imagine.

---

[23] In *Desire: Love Stories in Western Culture* (1994), Catherine Belsey shows that the masses of trivial modern romances follow the well-established formula, telling "of obstacles finally overcome and love ultimately and eternally requited" (22), while "the great love stories, the ones we remember and distinguish now, do not have happy endings" but offer "unfulfilled passion and unhappy love" (38). In both cases, however, the very existence of an obstacle, of precariousness, is indispensable for a story to evolve. Helmut Kuhn, with his brief discussion of the Hellenistic romance novel, reminds us that the pattern of facing and overcoming obstacles in trivial romances is by no means a modern invention (cf. 68).

The concept of 'courtly love,' or to be more precise, its specific variety of *fin' amors*,[24] is an approach to love that, instead of seeking ways to avoid precariousness, makes a virtue out of necessity and celebrates the dangers of love as signs and prerequisites for its intensity. It is a basic principle of *fin' amors* that the longing and desire of the lover will never be completely fulfilled by the lady (cf. May 123–25). This principle, however, is not a reason for despair but the basis for the specific value of this kind of love, which, in accordance with its Neoplatonic heritage, lies in an act of sublimation. The desire of the troubadour is as much sexual as it is spiritual, but its sexual component is as a rule frustrated, enabling the lover to approach the goal of spiritual refinement which manifests itself in the writing of beautiful poetry. The beloved lady necessarily has the status of an unattainable being of perfection with whom sexual intercourse is out of the question.[25] At the same time, however, she is still a particular woman, real enough to arouse real sexual desire and to serve as the person whose reciprocation of love is desired, even if it is known to be impossible. The troubadours thus did not try to avoid the precariousness that results from the unattainability of the object but embraced it as a pathway to greater happiness.

While for Lucretius, Ovid, and the troubadours the precariousness of love lay in the uncontrollability or unattainability of the other, in what Singer calls "Romantic Pessimism" the possibility of fulfilment of love *per se* is doubted, at least for the time of earthly existence. In this variant of Romanticism, the desired total union (with the beloved, with nature, or with the universe) is possible only in death (cf. Singer, *Courtly and Romantic* 432–34). What is actually unattainable is no longer the beloved but the consummation of love in this world. Richard Wagner's *Tristan und Isolde* expresses the idea of 'love-death' (*Liebestod*) in perfection. For them, "death is merely a loss of the illusory sense of individual identity, a return to the true nature of one's being, which is what the lovers have been seeking from the start" (470). It is important to point out, as Singer does, the difference between Wagner's version of the Tristan legend and its medieval forerun-

---

[24] Singer shows that the term 'courtly love' has often been misconstrued and applied without necessary diligence to historically and regionally distinct variants in which, for instance, the role of sexuality or the particularity or ideality of the beloved lady differed considerably (cf. *Courtly and Romantic* 19–21; 34–35).
[25] Apart from idealisation, other factors that could make sexual consummation impossible were differences in social status or the fact that the lady was already married. The sacrament of marriage, as Singer points out, was respected by the troubadours, who flattered both wife and husband with their adulterous desire but would not commit actual adultery (cf. *Courtly and Romantic* 51).

ners. In the medieval versions, the precariousness of love consists in unattainability caused by the strict rules of the hierarchically ordered world inhabited by the lovers. Tristan and Isolde cannot have each other because the social world forbids their union. Wagner's opera, on the other side, presents a love which is unfulfillable in this world, regardless of social boundaries (cf. 481). For May, Wagner's *Tristan* is symptomatic of a recurrent paradox in the history of love philosophy – the "paradox that when lovers strive for the absolute – for a good conceived as eternal and perfect – they end up willing their own destruction" (174). This urge to self-destruction is essentially a desire to overcome the separation that is caused by individuality, a desire which we found expressed, for example, in Hegel's or Bataille's concepts of love and which is followed resolutely to its ultimate consequence in the concept of *Liebestod*.

2.2.2 The Endangered Self: Love's Precariousness in Modern Philosophy

In the concept of *Liebestod*, where the impossibility of the lovers' true union in this world is resolved by death, the form of precariousness which is the result of obstacles preventing love's attainability mingles with the form of precariousness concerning the loss of self in and through love. But while the dissolution of individuality is enthusiastically welcomed in 'love-death,' it turns into one of the main sources of fear in modern discussions of love. Loss of self, individuality, and identity became the bogeyman of love in the twentieth century so that the precariousness of love today seems to consist as much in the possibility of losing the beloved as in the possibility of losing oneself in love. Reasons for this development can be found in the ceasing belief in the metaphysical notion of actual fusion or merging (cf. Singer, *The Modern World* 4) and in the modern emphasis on individualism, which have rendered the idea of the dissolution of separating individuality both unnecessary and undesirable. Apart from these large-scale developments, the precariousness of love understood as a threat to the integrity of the self can be retraced to very specific writings of the late nineteenth and twentieth centuries, of which I will briefly present an exemplary selection.

The fear of self-loss in love is closely linked to issues such as power struggles, egoism, or possessiveness, the very topics that are foregrounded in Friedrich Nietzsche's statements on love. In *The Joyful Wisdom* he writes that love between man and woman

> betrays itself most plainly as the striving after possession: the lover wants the unconditioned, sole possession of the person longed for by him; he wants just as absolute

power over her soul as over her body; he wants to be loved solely, and to dwell and rule in the other soul as what is highest and most to be desired. ("Selections" 143)

The use of pronouns here is no coincidence. For Nietzsche, men and women play entirely different roles in this game. For women, the ideal of love is "complete surrender (not merely devotion) of soul and body, without any motive, without any reservation" (146), while a man's role consists in providing the demigod to whom woman can willingly surrender: "Woman wants to be taken and accepted as a possession, she wishes to be merged in the conceptions of 'possession' and 'possessed'; consequently she wants one who *takes*, who does not offer and give himself away [...]. Woman gives herself, man takes her" (147). The full impact of this position comes into view when we consider what Nietzsche says about the cases in which these roles are reversed:

> granted, however, that there should also be men to whom on their side the demand for complete devotion is not unfamiliar, – well, they are really – not men. A man who loves like a woman becomes thereby a slave; a woman, however, who loves like a woman becomes thereby a *more perfect* woman. (146)

In this conception of love, one partner is relegated to the status of a slave, a mere possession. It is only because female desire is construed as the desire to be possessed that love does not appear precarious. If, however, we accept that love is a desire for possession of the other but refuse the idea that being turned into a possession is what women actually wish for, love becomes a struggle for power in which both partners feel the threat of becoming 'enslaved.'

Nietzsche's view of love as power struggle is taken up by Jean-Paul Sartre, who ties it up with his own ontological theory. For Sartre, every encounter with another human being means an encounter with what he calls 'the look,' the act of being perceived by the other. This act, Singer explains, "must always cause the 'objectification' of a subject" (*The Modern World* 288) and this, in turn, causes shame. "For we are ashamed, Sartre says, to be reduced to the status of a thing" (289), and this is what happens when we experience ourselves as the object of someone else's consciousness. To be objectified by 'the look' means to be thrown into a relationship of possession, a situation the freedom-loving subject seeks to avoid at all cost. At the same time, however, 'the look' is needed for the affirmation of our 'facticity,' our perceivable existence in the world as concrete beings. Relationships with others thus inevitably lead to a paradoxical and confrontational situation, as Singer summarises:

> Because each person wishes to be the foundation for his own being he tries to free himself from ontological possession by the Other. And yet, each of us must also recognize that the Other 'founds' our being to the extent that he is aware of its facticity. That makes us dependent upon one another and, according to Sartre, therein consists the inevitability of conflict. (292)

Ideally, love resolves this conflict. In its ideal form, the lover does not want to enslave the beloved, because this would objectify the other and deprive him/her of the free subjectivity which is needed for the act of affirming or 'founding' the lover's being. But the beloved's free subjectivity can also not be allowed to remain entirely unrestricted, because then the 'look' of this unbounded free subjectivity would still objectify the lover. What the lover really desires is the integrity of the beloved's free subjectivity to be centred on himself – a 'possession' of the other's freedom (cf. 292–95). The lover wants to be that to which the freedom of the other freely submits itself. This would save him or her from becoming merely an object for the other. Instead, he or she would be turned into an end in itself for the other, a being of recognised importance. In short, what the lover wants is to be loved, that is, to be recognised and affirmed.

Sartre's own words betray the closeness of his notion of reciprocal love to Lacan's definition of 'man's desire as the Other's desire': "Each one wants the other to love him but does not take into account the fact that love is to want to be loved and that thus by wanting the other to love him, he only wants the other to want to be loved in turn" (qtd. in Singer, *The Modern World* 297)[26]. Sartre is sceptical regarding the prospects of success when two principally selfish demands clash, as is the case in this model of love. Despite all mutuality, the project of love remains a struggle between two free subjectivities trying to impose limits – or demanding self-imposed limits – on the other's freedom. Sartre's choice of words mirrors this constant struggle and his conviction "that love is a conflict" (Sartre 228): in love, he writes, "it is the Other's freedom as such that we want to get hold of" (229). Even though "the man who wants to be loved does not desire the enslavement of the beloved" and "does not desire to possess the beloved as one possesses a thing," his attitude is still one of possessiveness: "he demands a special type of appropriation. He wants to possess a freedom as freedom. [...] He wants to be loved by a freedom but demands that this freedom as freedom should no longer be free. He wishes that the Other's freedom should determine itself to become love [...] so as to will its own captivity." The lover himself "wants to be 'the whole World' for the beloved" which

---

[26] Jean-Paul Sartre. *Being and Nothingness*. Translated by Hazel E. Barnes, Washington Square Books, 1953, p. 489.

means that to some degree he "consents to be an *object*. But on the other hand, he wants to be the object in which the Other's freedom consents to lose itself [...]." Ultimately, "what he demands is a limiting, a gluing down of the Other's freedom by itself" (229–30). Even though all this is clearly different from Nietzsche's vision of an open struggle for power which aims at and ends with the 'enslavement' of one of the lovers, still the unbridled freedom of the subject is what seems to be necessarily lost or sacrificed in love. In this sense, a kind of self-loss is inevitably inscribed in Sartre's concept of love.

In the wake of Nietzsche and Sartre, struggle for power, either in general terms or more specifically as a battle between the sexes, has come to be seen almost as a natural ingredient of love relationships for a number of writers (cf. Bloom 27). In this context, loss of self is mostly understood as a kind of 'dissolution' of one or both members of a couple, as a process of submissively surrendering one's particular identity and individuality to that of the partner or the relationship. Simone de Beauvoir, for instance, regards inequality between the sexes as the stumbling block preventing genuine love. Like her long-term partner Sartre, she considers the maintenance of freedom the pivotal task in love, but, like Schopenhauer and Nietzsche, she also claims that men and women have divergent ideals of love. In *The Second Sex* (1949) she complains that the ultimate goal of women is "identification with the loved one" (236) – a goal which, due to unequal power relations between the sexes and the submissive attitude of women in love, is tantamount to a dissolution of the female self:

> The woman in love tries to see with his eyes; she reads the books he reads, prefers the pictures and the music he prefers; she is interested only in the landscapes she sees with him, in the ideas that come from him; she adopts his friendships, his enmities, his opinions; when she questions herself, it is his reply she tries to hear; she wants to have in her lungs the air he has already breathed; the fruits and flowers that do not come from his hands have no taste and no fragrance. Her idea of location in space, even, is upset: the center of the world is no longer the place where she is, but that occupied by the lover; all roads lead to his home, and from it. She uses his words, mimics his gestures, acquires his eccentricities, his tics. 'I am Heathcliff,' says Catherine in *Wuthering Heights*; that is the cry of every woman in love; she is another incarnation of her loved one, his reflection, his double: she is *he*. She lets her own world collapse in contingence, for she really lives in his. (236–37)

Unlike Schopenhauer and Nietzsche, de Beauvoir does not consider this submissive attitude in love a natural expression of the female character. The reason why woman "chooses to desire her enslavement so ardently" (234) is not a law of nature but a matter of education and socialisation which construes women as naturally inferior and makes the idea of fusion with the godlike figure of a man, of "amalgamating herself with the sovereign subject" (234), desirable. However, no true love can spring from such preconditions. The woman's overvaluation of the man is doomed to end in disappointment, and men will find it hard to keep up interest in a being who has rid herself of her personality. Both are not in love with the other as an individual person but with a version that is either idolised or de-individualised beyond recognition (cf. 237–39). Only if both partners enter the relationship as full and free beings with a view to maintaining their own and the other's integrity can love be beneficial and mutually enriching. Such genuine love, however, is unlikely to arise in a society where gender inequalities are as deeply entrenched as the differing attitudes towards love that stem from them. As long as male dominated society produces female desire for subordination and self-abdication, de Beauvoir insinuates, love will continue to threaten women with the loss of self.

Many of de Beauvoir's arguments reappear in Shulamith Firestone's *The Dialectic of Sex* (1970). For her, too, genuine love requires equality and facilitates mutual enrichment, and like de Beauvoir she sees little chance for this kind of love in a world where "sexual inequality has remained a constant" (250) and different roles are assigned to men and women in love instead of demanding complete mutuality and equality. What distinguishes Firestone's approach from de Beauvoir's is the emphasis she puts on the necessity of vulnerability in love *per se*. For Firestone, love means "being psychically wide open to another. It is a situation of total emotional vulnerability" (249), and this vulnerability is a necessary condition for the enrichment which mutual love between equals can produce. When she writes that "love demands a mutual vulnerability or it turns destructive" (250) she makes emotional openness, the very potential of getting hurt, a necessary ingredient without which there can be no enrichment.

That love is always necessarily precarious is a conviction shared by three of the most influential figures in contemporary French philosophy: Emmanuel Levinas, Jean-Luc Nancy, and Alain Badiou. Badiou's *In Praise of Love* repeats both Firestone's view of love as inherently precarious and her concept of love as mutual enrichment. The inherent precariousness of love finds expression in Badiou's fierce criticism of what he considers one of the most dangerous threats to love in modern society, the "safety-first concept of 'love'" (6) which promises love free of chance, uncertainty, and the risk of getting hurt. For Badiou, such a notion of risk-free love is a contradiction in terms: "Clearly, inasmuch as love is a pleasure almost everyone is looking for, the thing that gives meaning and intensity to almost everyone's life, I am convinced that love cannot be a gift given on the basis of a complete lack of risk" (7). Like happiness, of which love is one of the highest forms, love must be exceptional to be genuine (cf. "Happiness"). If it does not significantly alter the emotional or psychical condition of the subject but merely represent a mode of contentment with the status quo, then it is not 'the real thing.' Only if love has the power to disrupt ordinary time, if it is able to bring about change, can it assume the exceptionality happiness presupposes. Change and disruption, however, are always risky – and this is precisely what Badiou demands in order to save love from its threats. "Risk and adventure must be re-invented against safety and comfort" (*Praise* 11).

The change of life which love can bring according to Badiou seems to be an elaboration on the 'mutual enrichment' which de Beauvoir and Firestone had in mind. Badiou argues against any ideal of fusion or merging that would turn two lovers into one. Rather, love is the pathway to a certain truth, the answer to a certain question: "what kind of world does one see when one experiences it from the point of view of two and not one? What is the world like when it is experienced, developed and lived from the point of view of difference and not identity" (*Praise* 22)? Difference is indeed the key element in Badiou's concept of love, and he elevates its preservation and experience to the principle of true love. Love is "an existential project: to construct a world from a decentred point of view" (25) in which the difference and separateness of the other is an essential prerequisite because it enables the enrichment of one's own limited perspective. This is a complete inversion of the Romantic programme of love that was so obsessed with the idea of transforming separation and difference into oneness. Here, difference is celebrated as the key to love's truth. And yet, love is no less precarious. The difference of the other is not only a precondition for love, it is also, as it has always been in all concepts of love, a problem, an obstacle that makes love risky, contingent, and uncontrollable. However, the more serious source of danger is to be found elsewhere: "my love's main enemy, the one I must defeat, is not the other. It is myself, the 'myself' that prefers identity to difference, that prefers to impose its world against the world reconstructed through the filter of difference" (60). The main impediment to love is the selfish and narcissistic urge to absorb or incorporate the other, to resist the change of perspective in favour of the status quo. In a sense, then, Badiou's approach asks to accept and even celebrate one source of precariousness in order to eliminate another. The alterity of the other, the reason for love's riskiness and uncontrollability, is declared the essential condition of love so that, in true love, the integrity of both selves is guaranteed.

The factor of difference, which is so important for Badiou's concept of love, may be regarded as the central element in the philosophy of Emmanuel Levinas. Levinas' is not a philosophy of erotic love as his main interest is in the conditions of human ethical interrelation and cohabitation in a broader sense. However, love understood as an ethical posture is of crucial importance in his considerations. As Linnell Secomb writes, "[f]or Levinas the personal, ethical and political relation between the self and the other is central: these relations are founded on or conditioned by love" (58). The kind of love he is mostly concerned with is "the selfless, ethical, love of the neighbour" (58) which makes us take responsibility for the other and even subordinate our needs to those of the other. A fundamental characteristic of Levinas' vision of ethical human interrelations and the operation of neighbourly love is "that in the relation between the self and the other, the alterity of the other is always maintained" (Secomb 61). What Badiou requests for the love couple Levinas has already formulated as prerequisite for any form of human relation: an attitude that accepts and seeks to preserve the other's difference and alterity and makes no attempt at absorbing or assimilating the other (cf. 61).

Although Levinas is mainly preoccupied with the ethical relationship, he is not silent on erotic love which, in his conception, is neither purely selfless nor purely selfish but combines the care and responsibility for the other with the satisfaction of needs and the enjoyment of pleasures (cf. Secomb 64). Above all, however, it is an intimate and exclusive relationship between two persons who, despite all closeness, retain their absolute alterity. As he makes unmistakably clear in "Time and the Other," the "absence of any fusion" (Levinas 51) is a basic element of his concept of love which explicitly refutes all romantic notions of merging or returning to prior oneness. "The pathos of love," he writes, "consists in an insurmountable duality of beings. It is a relationship with what always slips away. The relationship does not *ipso facto* neutralize alterity but preserves it" (49). The precariousness of love is indicated here unmistakably: 'pathos' is unavoidable because love is a relation to something that cannot be possessed, mastered, or controlled, that has to remain 'other' for love to exist. "The pathos of voluptuousness lies in the fact of being two. The other as other is not here an object that becomes ours or becomes us; to the contrary, it withdraws into its mystery" (49). Erotic love is a very special relationship – a relationship that demands intimacy and closeness with a particular person without attempting to possess, absorb, or even understand this person. "It is neither a struggle, nor a fusion, nor a knowledge. One must recognize its exceptional place among relationships. It is a relationship with alterity, with mystery – that is to say, with the future, with what [...] is never there, [...] not with a being that is not there but with the very dimension of alterity" (50). The paradox of this situation is epitomised in the 'caress,' the enactment of tenderness in physical contact that is so typical of the erotic love relationship. For Levinas, the caress clearly seeks more than physical contact, but the impossibility to express what it really wants highlights the difficulty of a love that is a love of alterity. Of course, Levinas says, the caress also involves touching the other's body.

> But what is caressed is not touched, properly speaking. It is not the softness or warmth of the hand given in contact that the caress seeks. The seeking of the caress constitutes its essence by the fact that the caress does not know what it seeks. This 'not knowing,' this fundamental disorder, is the essential. It is like a game with something slipping away, a game absolutely without project or plan, not with what can become ours or us, but with something other, always other, always inaccessible, and always still to come (*à venir*). The caress is the anticipation of this pure future (*avenir*) without content. (51)

The goal of love is a total uncertainty – a "pure future without content." It cannot be otherwise, for Levinas, because of the total alterity of the other which has to be preserved at all cost since any attempt to "possess, grasp, and know the other" (51) is an exertion of power harmful to the ethical relationship. Like Badiou, Levinas rules out one source of precariousness in favour of another. For him, too, love rests upon the unquestioned alterity of the other. In true love, there can be no desire to possess or assimilate – not even to understand or know – the other. Loss of self is precluded in a concept like this. What makes this love precarious is the overwhelming uncertainty which characterises the entire project, from the uncontrollability and inaccessibility of the other to the unknowability of love's goal.

For Levinas, love involves 'pathos' because we have to deny ourselves any form of possession or incorporation of the other. Jean-Luc Nancy discusses precisely this tension between an urge to possess and the demand to guarantee alterity as a defining feature of love in our culture. Love in the modern sense, he argues, follows two principles at the same time: absolute recognition of and intense desire for the other (cf. *Die Liebe, übermorgen* 10). As a result, love is an impossibility in one way or another, violating one principle by being too passionate, emotional, carnal, or selfish, or the other principle by being too ethereal and disinterested and by assuming an arbitrary nature in that it is not reserved for a particular person (cf. 13). But is love really impossible? Are intimate, passionate love of a particular person and a refusal to possess or comprehend this person excluding each other? Is there a possibility to respect the alterity of the other and yet to come to know this other so as to be able to love her as a particular person? For Nancy, it is love itself which makes this possible. Or, in other words, love is the name he gives to that which enables two subjects to share in their alterity. In "Shattered Love," Nancy presents this concept of love in formulations expressive both of its precariousness and of the difficulty of verbalising its ontological significance. Love is not something that two subjects do to each other. "Love arrives, it comes, or else it is not love" (98). Not one subject tries to possess or comprehend the other in a violation of her alterity and subjectivity, but love opens up the subject and violates its autonomy. If I am in love,

Nancy holds, "something of *I* is definitely lost or dissociated in its act of loving" (96). The experience of love is that of a cut or violation of the autonomous self:

> Love represents *I* to itself broken [...]. It presents this to it: he, this subject, was touched, broken into, in his subjectivity, and he *is* from then on, for the time of love, opened by this slice, broken or fractured, even if only slightly. [...T]he break is a break in his self-possession as subject; it is, essentially, an interruption of the process of relating oneself to oneself outside of oneself. From then on, *I* is *constituted broken*. As soon as there is love, the slightest act of love, the slightest spark, there is this ontological fissure that cuts across and that disconnects the elements of the subject-proper – the fibres of its heart. (96)

Love is no healing or completion of a fractured self. Rather to the contrary, it is a breaking up of a self that imagines or even experiences itself as self-sufficient. It "signifies that the immanence of the subject [...] is opened up, broken into – and this is what is called, in all rigor, a transcendence" (97). But love does not open up the self in order to facilitate a fusion of selves. Unlike desire, love does not seek any kind of fulfilment or completion. Love merely "cuts across" (99) and thereby "unveils" (98) the finitude of the subject but does nothing to change it. As Secomb puts it: "In Nancy's formulation, love does not appropriate the other or subsume the other. Rather, love gives. This is to be understood not as a giving of materials, objects and possessions. It involves a giving which is an exposure of the self" (146). The act of violence against the self-enclosed subject is thus not an exertion of power through one of the lovers but the gift of love, which "shatters the atomistic being" (146) and opens it up for the other. Nancy's choice of words bespeaks the intense vulnerability which love brings. Love causes an "ontological fissure" ("Shattered Love" 96), "it cuts, it breaks, and it exposes" (97), and although this does not necessarily entail physical or emotional pain, it is a form of injury. "The crossing breaks the heart: this is not necessarily bloody or tragic; it is beyond an opposition between the tragic and serenity or gaiety. The break is nothing more than a touch, but the touch is not less deep than a wound" (98). The broken heart is, for Nancy, not the result of unhappy or unrequited love but the condition, that is, both the precondition and the situation, of love *per se*. Love breaks and opens the heart for the other and only this unlocking of the atomistic self enables what is commonly called love. Of course, once opened up for the other the heart is exposed to the risk of becoming what is commonly called a broken heart. Love is not to be had without vulnerability.

My last example, Martha Nussbaum's *Upheavals of Thought*, shares the conviction that love always necessarily has to be risky and requires openness and vulnerability. Nussbaum's neo-Stoic concept of love, rooted both in Greek Stoicism and modern psychoanalysis, rests upon "a conception of the self that pictures the self as incomplete and reaching out for something valued" (460). This movement of reaching out is risky because the object to which the self opens up and which it considers necessary for its completeness is beyond control. Yet this element of risk is not a disagreeable side effect of love which would vanish if the object of love could be controlled, but a precondition for the existence of love. Her basis for this assumption is the Stoic concept "that emotions are appraisals or value judgements, which ascribe to things and persons outside the person's own control great importance for that person's own flourishing" (4). Although Nussbaum hesitates to follow the universal claim that all emotions are bound up with precarious objects, she agrees that most emotions involve a "sense of vulnerability and imperfect control" and that, even though some forms of love may be of a different kind, "[e]rotic love notoriously involves the thought of instability in this way" (43). The second important influence on her theory of emotions is the common ground view of the object relations tradition of psychoanalysis that adult emotional life is shaped by the experiences of early childhood: "Early memories shadow later perceptions of objects, adult attachment relations bear the trace of infantile love and hate" (6). The developmental history of love is a history "of longing for protection and comfort, of anger at the separate and uncontrolled existence of the source of comfort" (175). From its very beginning, love is ambivalent. Only with the infant's recognition that the mother is a separate and independent being does the pervasive feeling of omnipotence give way to love towards an object which is endowed with immense value (cf. 209). But at the same time, this recognition of separateness is "a cause for furious anger" (210) and in a further step gives rise to "jealousy – the wish to possess the good object more completely by getting rid of competing influences" (210). In adult love, it is precisely this infantile wish for omnipotence which must be overcome to save love from becoming obsessive or pathological (cf. 212). Adult love has to accept and even cherish the independence of the other and relinquish all attempts at control (cf. 225). That this is not an easy task is corroborated by the many attempts, starting with Plato's *Symposium*, to turn the object of love into something that is not precarious but stable, unchanging, and accessible for everyone who follows the right track. For Nussbaum, representatives of such concepts display a "pathological narcissism" (526), an inability to overcome the infantile wish for omnipotence, "for they long for complete control over the world, and they refuse to abandon that wish in favor of more realistic human wishes for interchange and interdependence" (524). Whatever such philosophical narratives of an ascent of contemplative love describe, it has nothing to do with erotic love, which does not even come into existence if its object is not insecure.

* * *

Precariousness is a constitutive part of human nature. From the moment of birth, human beings are dependent on others, insufficient, vulnerable, in need of care and protection, and a long and influential tradition of philosophy sees erotic love as originating from this condition. But the hope in the power of love to compensate for the precariousness of human existence has always been accompanied by the notion of love's inherent precariousness. In its more 'classical' form, this precariousness results from the uncontrollability of the other. In some theories within the psychoanalytical tradition, this form of precariousness takes a special shape in that the 'real' object of desire is by definition unattainable – irretrievably lost at (or even before) the moment of birth or during the process of individuation – and any object of love is only a surrogate whose deficiency will come to light at some point. With the modern cult of individuality, the threat of self-loss joined the discursive field as another form of precariousness. Diane Enns indicates that while de Beauvoir and Firestone merely warned women not to surrender their identity in unbalanced relationships, the modern self-help industry has extended the request to all lovers in its general "warnings against 'codependency,' a state in which one is considered to have become overinvested in a loved one, dependent and without proper ego boundaries" (35). Rejecting dependence and vulnerability and championing the sovereign autonomy of the self, the outpourings of pop therapy literature suggest emotional detachment, the setting of boundaries, and the cultivation of self-love to protect an inviolate self. The demand is counterproductive, however, as Enns argues:

> What destroys the conditions for love is *in*vulnerability. We cannot love without becoming vulnerable to another, without opening ourselves to the possibility of being wounded by another. Without this opening to the other – an abandoning of the self in the surge of love – we are unable to see ourselves through another's eyes. And if we remain blind in such a way, we can neither give nor receive love. (44)

This premise that there can be no proper love between persons without risk, uncertainty, and vulnerability is the common ground of the approaches presented in this chapter. In all of them, precariousness is intrinsic to love. The preferred remedy against humanity's ontological precariousness is itself utterly precarious. What is supposed to alleviate a fundamental feeling of insecurity and insufficiency produces these very sources of fear.

3 Romantic Love in Sociology

If philosophy, enquiring into the origins or 'nature' of love, tends to ask what love *is*, sociology rather asks what people think it is, or should be, and what function it fulfils in a society or culture. The results are strikingly similar. By sociologists as well as by the people they study, love is attributed with a compensatory function and at the same time perceived as a major source of precariousness. The lines of reasoning, of course, often differ from the philosophical ones. What this chapter is particularly interested in are, first, the socio-historical developments, from the (early) modern period up to the present, which have turned romantic love into a clearly delimited sphere or 'subsystem' endowed with a quasi-religious compensatory function and, second, the interrelations between love's precariousness and a specifically modern and postmodern cultural and social context.

Sociological research collecting what people think about romantic love has yielded results that are surprisingly coherent from one study to another but also inherently paradoxical as they add up to an idealised image of love too optimistic to be seriously accepted by (post-)modern individuals. The features of this image of love, extracted from surveys of fictional literature, psychological counselling and self-help books, magazine articles, newspaper columns, film productions, and empirical interview studies, can be summarised in a list like the one suggested by Leslie Baxter and Chitra Akkoor: "(1) we should follow our heart over other more practical concerns when choosing a partner; (2) love can strike at first sight; (3) there is only one 'true love' for each person; (4) the object of one's love will be perfect; (5) love conquers all; and (6) love will not fade but will last forever" (23–24).[27] The list contains all the major qualities usually connected with romantic love, variously called irrationality, unconditionality, disinterestedness, irresistibility, spontaneity, exclusiveness, uniqueness, infinity, eternality, totality. In addition to these features, love supposedly serves as the primary source of meaning and orientation in modernity. Simon May describes how "love has increasingly filled the vacuum left by the retreat of Christianity" (1) in Western culture, a process also visible in the transference of religious or divine qualities like unconditionality, selflessness, or infinity to human love. He argues that in our post-religious world, love has taken over the task of providing human beings with a shield against feelings of absurdity and existential

---

27 Similar lists can be found in Faulstich (34) and Illouz (*Why Love Hurts*, 159–60). Although Illouz speaks about 'enchanted love' instead of 'romantic love' at this point, the concept she describes equals what she also calls 'romantic love' elsewhere.

anxiety: "Human love, now even more than then, is widely tasked with achieving what once only divine love was thought capable of: to be our ultimate source of meaning and happiness, and of power over suffering and disappointment" (1). Of all the substitutes for God that have been worshipped in the last three hundred years or so, such as Reason, Progress, the Nation, the State, or Communism, only love proved able to provide "the final justification of life's aim and meaning that the Western mind still craves" (4). Love, one might summarise May in this regard, is the last *grand récit* of Western thought.

The elevation of love to the purveyor of meaning was further supported by the modern project of individualisation, which has not only freed people from traditional conventions and compulsory social roles but has also produced individuals who are less integrated into society by means of social ties such as neighbourhood, class, or extended families. While the number of acquaintances is constantly growing, their intensity is diminishing, leaving the individual in need of closeness and intimacy (cf. Beck/Beck-Gernsheim 49–50). The liberation from social constraints which modernisation and individualisation made possible thus also means a loss of stability and security, causing feelings of loneliness and 'inner homelessness' ("innere Heimatlosigkeit," 67), which love is supposed to make up for. Obviously, these qualities and functions add up to an utterly optimistic, idealistic, and arguably unrealistic concept of love which, as I will suggest below, is by no means naively accepted by contemporary lovers but is met with due scepticism and frequently collides with other, incompatible values of the (post-)modern individual. Nevertheless, this concept figures prominently in collective awareness and, whether believed in or not, has a marked impact on collective and individual notions of love. It is thus worth having a closer look at the genesis of this multifarious and inherently contradictory concept, narrative, or 'code.'

3.1 Modern Functions of Romantic Love: A Very Brief History

In sociology, romantic love is frequently regarded as an invention of modern society, inextricably linked to the rise of capitalism and the individual self (cf. Hähnel/Schlitte/Torkler 31–32). Despite the reservations against this idea put forward at the beginning of Chapter 2, its emphasis on the co-development of romantic love and the economic and cultural dimensions of capitalism has some irrefutable explanatory worth as to why love is so overtaxed with conflicting features and functions. It can explain, to be more precise, why it seeks to combine irrationality, meaningfulness, passion, and permanence.

As Werner Faulstich argues, not only did capitalism change the world of work in a way that made the nuclear family a necessity, it was also the foil against which the nuclear family and the marriage relationship were defined. Incorporating romantic notions of love as unconditional, selfless, and eternal, marriage became the conceptual opposite to the cold, callous, and rational principles of capitalism (cf. 31–36).[28] Faulstich delineates how from the eighteenth century onwards the social microstructure of the 'house,' which consisted of an extended family unit usually comprising three generations and including long-term servants and farmhands who all together formed a largely self-sustaining economic community, was gradually replaced by the social microstructure of the nuclear family. As a result of the separation of the formerly united area of work and living into a public sphere of work and a private sphere of home, the family acquired a new function: instead of supplying its members with food and income, the new primary function of the family was the supply of emotional security (cf. 24–26), turning the family into the definite place of love and affection. The new function of the family went hand in hand with a changing attitude towards marriage. Until then, marriage had first of all been a relation of convenience, often arranged by parents and relatives on the premise of financial, economic, or political profit. Ardent, passionate love was generally an extramarital affair, and within circles of nobility the entertainment of courtesans and more or less secret lovers was almost common courtesy (cf. 28). This gradually changed because of the new function of the family as the site of love, a new financial independence resulting from income in the form of wages, and the celebration of romantic love in popular literature. First among the bourgeoisie and then in the other social classes, marriage of convenience was increasingly rejected and love became the only legitimate reason for marriage (cf. 29). This enhancement of love was intensified by the fact that its new function served as compensation for a certain form of female discrimination. The separation of work and family was tantamount to the creation of a male sphere of work defined by achievement motivation and competitiveness and a female private sphere far less prestigious and less productive as a source of identity formation. Raising the prestige of the female role as the provider of intimacy and emotional security worked as partial compensation for this imbalance (cf. 31). Largely excluded from the public sphere, the woman – usually as a wife and mother – would find her identity in fulfilling the function of providing love, a function dia-

---

28 Similarly, Ulrich Beck describes love and the family as the emotional spaces that were constructed as counter-narratives to modern society – as "Nichtmarkt, Nichtkalkulation, Nichtzweckrationalität" (184).

metrically opposed to the very realm from which she was excluded (cf. 32). This idea, Faulstich goes on, still dominates Western mass culture and endows love with a compensatory function which today, however, is decidedly less female (cf. 36). Instead, today love quite generally works as a remedy against the more depressing effects of capitalism. The fundamental opposition to the norms and values of capitalism is of essential importance for the understanding of some of the defining features of romantic love, such as, above all, irrationality and unconditionality. Moreover, the disdain of marriages of convenience, ubiquitous today in Western societies, has likewise to be seen in this light of a resistance to capitalist principles of rational calculation and profit maximisation. Although there is reason to reject the notion that romantic love is entirely a modern invention, it is evident that this particular development has significantly contributed to prevailing notions of love.

Niklas Luhmann, too, describes the emergence of romantic love as a consequence of social and cultural change. Notably, he understands love not as an emotion but as a medium of communication and thus as a source of meaning. Media of communication, according to Luhmann, are our most important means of establishing order and meaning in our all too complex and contingent world (cf. *Übung* 12). While some media of communication like power and money structure our decisions on actions, others like truth, art, and love guide our acceptance of the world as a place that is, despite complexity and contingency, not completely without order and in which not simply everything is possible. More than motivating specific actions, love guides and shapes the way we perceive and experience the world and endow certain persons and events with a special meaning (cf. 14–16). Love is to be seen as a system with its own rules within which things (for example certain objects, gestures, or utterances exchanged between lovers, or their reactions to each other's behaviour) make sense that would appear meaningless or contingent otherwise. Unlike truth, which claims universal validity for all participants in the communication and excludes dissenters from this community as crazy or strange, love claims validity only for the two lovers. The sense and meaning love creates are not universal but highly individual. Love is the foundation of a personalised world of two in which the self-image and world view of the other are the points of reference for one's sense-making process (cf. 18–19). Love thus affirms meaning in a twofold manner: The constitution of an intimate world of lovers gives meaning to the daily life, interaction, and communication in this relationship. In addition, and more importantly, love means an unconditional affirmation of the individual self ("eine unbedingte Bestätigung des eigenen Selbst," 21), an acceptance of one's very personality through the other. Love generates a feeling of being expected and wanted just the way you are – a feeling of making sense as an individual (cf. 21).

The function of love as a medium of communication which reduces complexity and affirms the self becomes ever more important as the complexity of the world is increasing steadily and its contingency is becoming more disconcerting (cf. *Übung* 25–26). According to Luhmann's notion of 'functional differentiation,' society as a system increases in complexity as it generates more and more subsystems such as politics, religion, the law, the arts, or intimate relationships, which take on more and more specific functions. It becomes increasingly difficult and eventually impossible for one system to fulfil the function of another. Consequently, intimate relationships and their medium love are discharged of functions they formerly fulfilled, namely to support the systems of morality, law, politics, and economics. From the end of the medieval period onwards, Luhmann argues, love steadily evolved from a relationship between members of a group into a relationship between two singular, arbitrary individuals, and *amour passion* became its basis and motivation. In accordance with the idea that 'passion' is something one is passively enduring, passionate love became the realm in which human beings are free from social and moral responsibility (cf. 31–33).

The function this passionate romantic love now takes on is twofold. For the loving individuals, it is a powerful remedy against feelings of uncertainty and meaninglessness in a contingent and complex world where other traditional remedies like religion or the integration in extended familial structures have lost effectiveness. On the other hand, although love has been freed from its political, economic, and moral function, it still serves society in one pivotal task: as the only legitimate basis for marriage it replaces economic consideration and parental authority in the selection of partners and thus secures the reproduction of society. As a result, inasmuch as the stability of society is considered to depend on the stability of families and marriages, durability turns into one of love's most vital features and a social necessity (cf. *Passion* 186–87). But what does a concept of love look like that is supposed to free lovers from the complexity of society, liberate their love from any responsibility except towards each other, but still support society by guaranteeing long-term relationships and stable families? Not surprisingly, the concept is a paradoxical one. As the backbone of long-term relationships, romantic love has to be permanent, while at the same time romantic love is passionate love and as such uncontrollable, contingent, and beyond the realm of social responsibility. The concept of love is still heavily determined by the excessiveness it demands ("Gebot des Exzeßes," 83). While everywhere else moderation is to be practised, in passionate love it is regarded a fault. The only limit to this excess is time – but this is a limit, Luhmann warns, that will be reached inevitably and quickly (cf. 93). Because of this finitude of passionate love, it had long been kept separate from the institution of marriage (cf. 95). Conversely, when passionate love became the only legitimate reason for marriage, its permanence had to become at least a theoretical possibility.

Luhmann describes several developments that facilitated this change in the concept of love. One is a new reflexivity of love. The feeling of love itself becomes a central aim in love relationships – love becomes its own object. The lovers love each other not (or not only) because they are good or beautiful, but because they mutually enable each other to experience themselves as lovers (cf. *Passion* 174). In contrast to the image of the amorous conquest, which ends with the possession of the object of love, reflexive love seeks its own maintenance. This focus on durability is supported by the integration of a new notion of individuality into the concept of romantic love. It was not before the eighteenth century, Luhmann explains, that the individual was seen as both a unique and developing personality. Before that, a beloved person was a person possessing a set of desirable objective qualities that could be found in another person just as well. Moreover, since persons and their character were conceived of as unchanging and unalterable, a passionate love that incessantly seeks to reach and conquer new goals and that is said to cease with the possession of what it strives for, could not be directed permanently at one person (cf. 125–26). Only in combination with the notion of a developing and alterable self could *amour passion* be reconciled with the idea of lasting love and marriage. It is the notion of a self that has its own, unique and specific relation to the world (cf. 173) and that is shaped and influenced individually by this world and by its personal environment. A self like this is capable of variously redefining itself and its relation to the world and to others, thus demanding from its partner the ability and willingness to re-adapt continuously – a demand that has to be seen as a gift, as it enables passionate love to repeat itself towards the same person. Instead of putting an end to love, marriage between two such individuals can be the site where love is constantly renewed (cf. 178). Furthermore, only an individual self of this kind can be the object of a romantic love that is characterised by exclusivity and uniqueness (cf. 167). Both eternality and uniqueness, two essential features of the concept of romantic love, are thus connected to the rise of the modern individual. Not that eternality and uniqueness had not existed as ideals of passionate love long before, but then they were usually only attributes of a passion that was kept alive by the impediments and obstacles that prevented its consummation. The attempt to theoretically reconcile passion, uniqueness, and permanence within an intimate relationship presupposes the modern concept of the unstable, volatile self.

Despite this theoretical solution, however, the reconciliation of passion and permanence remains a major unsolved problem in the actual reality of love relationships (cf. Rathmayr 301). The theory of enduring passion due to alterable individuals could never integrate *amour passion* in its unbridled, excessive form in which time is its only limit. Passion within marriage had to be of a calmer kind, and the idea to base marriage solely on *amour passion* in its unrestrained form was largely rejected already in the early nineteenth century (cf. Luhmann, *Passion* 188). It is obvious that the renewal or renegotiation of love within marriage cannot adequately substitute for the excitement and emotional turmoil of freshly inflamed passion. Moreover, the high degree of individualism that is necessary for the prolongation of passion and for the self-perception of romantic love as unique is at the same time seen as a threat to long-term relationships by Luhmann and other sociologists (cf. *Passion* 46). Therefore, despite the innovations in romantic love that try to reconcile passion and marriage, the contradictions between the ideal of durable love and the still influential and powerful ideal of *amour passion* remain intact. They are, in fact, the reason why Luhmann considers marriage and family based on passionate love as high personal and social risks (cf. *Übung* 66). This ties in with the analysis of Beck-Gernsheim, who talks about the 'trap of romantic love' ("die Falle der romantischen Liebe," 115), in which the passion that initiated a relationship cannot be kept alive and the high expectations for love and marriage are frequently disappointed (cf. 124–25). Ulrich Beck adds that the common notion of love today is not only disappointed in relationships but that it also urges partners to end their relationship if their love cannot live up to the standard ideal circulated by mass media (cf. 229). In a vicious circle, love that is not simultaneously impulsive and durable demands its own extinction.

The problematic interrelation between the depiction of love in media and literature and our actual conception and experience of it are frequently addressed in sociological literature (cf. Munteanu 330–35). Karl Otto Hondrich argues that today's dominating ideal of romantic love is the kind of love celebrated by the entertainment industry (cf. 12). Ulrich Beck warns that the construction of a concept of love is guided by its representation in song texts, romance literature, advertising, pornography, and psychoanalysis (cf. 13). In Luhmann's concept of love as a communication code the experience of love as meaningful is only possible due to a semantics of love handed down through culture and literature (cf. *Passion* 47). Typical images, storylines, and a well-known language of love enable us to 'know' love, to practice and experience it, in the first place. Eva Illouz offers an especially intriguing examination of the mutual dependency between love and its representation in media and cultural products. Her work on love, however, comprises much more than this interrelation and deserves a systematic overview.

3.2 (Post-) Modern Attitudes to Love

3.2.1 Love and the Culture of Capitalism

Illouz is dedicated to the question how (post-)modern romantic love differs from earlier variants of this emotion. With Catherine Belsey we may conceive of postmodern love as an inherently paradoxical concept whose contradictory nature stems from the fact that it is both a reaction against and a product of modernity. In line with Beck and Faulstich, Belsey first describes love as that which, by common consent, is beyond the mechanisms of the capitalist market.

> While sex is a commodity, love becomes the condition of a happiness that cannot be bought, the one remaining object of a desire that cannot be sure of purchasing fulfilment. Love thus becomes more precious than before because it is beyond price, and in consequence its metaphysical character is intensified. More than ever, love has come to represent presence, transcendence, immortality, what Jacques Derrida calls *proximity, living speech, certainty* – everything, in short, that the market is unable to provide or fails to guarantee. ("Postmodern Love" 683)

What makes the contemporary concept of romantic love specifically postmodern is the very suspicion against all that is here attributed to love and renders it a kind of *Ersatzreligion* or metanarrative. While love is vested with a quasi-metaphysical power to endow our lives with meaning, we have at the same time adopted postmodernity's "skeptical attitude to metaphysics, a radical questioning of presence, transcendence, certainty, and all absolutes," which is why "the postmodern condition brings with it an incredulity towards true love" (683). This situation of a love that is the object of enormous hopes and serious doubts at the same time is one key topic of Illouz' *Consuming the Romantic Utopia* (1997), in which she demonstrates that the source of doubt is a heightened suspiciousness regarding the interrelation of media representations and personal expectations and experiences of love.

Illouz stresses the necessity to re-examine the concrete interdependence between contemporary notions of love and the economic framework of late capitalism. As she argues, love has become deeply intertwined with consumerism itself through two reciprocal processes which she calls "the romanticization of commodities and the commodification of romance" (26). Illouz delineates how during the twentieth century love was increasingly drawn into the sphere of influence of the growing leisure time industry. In particular, film and advertising industries used images of romantic love as reliable guarantees for commercial success (cf. 28). In advertising, consumer goods were frequently actively associated with love and romanticism, no matter how artificially this association had to be established. Household appliances were associated with the comfort and cosiness of the marital home, "'ego-expressive' products" (34) such as cosmetics, jewellery, or clothing were associated with the excitement and impulsivity of romantic situations. This "romanticization of commodities" was accompanied by a "commodification of romance" because very often the romantic situation used in advertising contained itself a form of consumption – mostly that of leisure time. The romantic couple is shown in situations that can be purchased, such as dinner at a restaurant, a car journey, a romantic picnic, going to the cinema, or going on holiday, so that the romantic situation, to some degree, becomes an object that can be sold and bought (cf. 33–43). Consequently, "[a]s participation in the leisure markets became increasingly associated with romance, the experience of romance became increasingly associated with consumption" (65). The numerous conventions and stereotypical images around the rituals of dating, celebrating anniversaries, or planning and organising romantic moments, usually linked to consumerism and expenditure, bespeak this commodification of romance (cf. 66–76). Naturally, this situation contradicts the notion of romantic love as the counter-narrative to capitalism and complicates love's status in capitalist culture. Above all, the interconnection between consumerism and many of the signs and gestures available for the expression of love, once recognised, seriously questions the possibility of an authentic and individual performance of love.

Illouz' analysis of advertisements in which goods are in some way or other connected with romantic situations produces an image of love that is the very opposite of the routine of work and everyday life. Nowhere does this become more visible than in the motif of the romantic holiday, the epitome of the simultaneity of amorous togetherness and escape from the ordinary, usually connected with images of exotic and unspoiled nature surrounding the affectionate couple (cf. 89). In general, closing off romantic love from everyday life is one of the most important features of the concept and Illouz describes various strategies lovers make use of to bring about this separation (cf. 113–18). As these strategies frequently involve consumption, they not only demarcate romantic love from everyday routine but tend to model it after love's representation in mass media and advertisement. As a result, love is, on the one hand, a remedy against the discontents of capitalism in that it enables the lovers to break out from the routine and alienation of work life, to experience themselves and their lives as if liberated from any social restrictions, while, on the other hand, this break-out takes the form of consumption, of purchasing the romantic situation, thus reaffirming the capitalistic cycle of work discipline and hedonism.

3.2.2 The Problem of Authenticity: Idealist and Realist Narratives

The interplay of romantic love and late capitalism has another crucial implication which adds up to a crisis of authenticity in the experience and expression of love. Sociological accounts sometimes imply that one of the most obstinate stumbling blocks in modern romantic love is the discrepancy between expectation and reality and that unrealistic expectations result from the representation of love in mass media (cf. Beck/Beck-Gernsheim 114–15; 124–25, Hondrich 55–56; 169). However, while unrealistic expectations may well cause disappointment in relationships and marriages, this is anything but a new phenomenon and the connection between media representations and personal notions of love is more complex.

First of all, as Luhmann asserts, as a medium of communication which enables otherwise unlikely communication, love is a pre-existing code, a set of rules according to which emotions can be expressed and which is already elaborated before one enters the game of love (cf. *Passion* 23). Actual authenticity in love is impossible for Luhmann, as its experience and expression depend on culturally transmitted semantics (cf. 47; 88). Since at least the early modern period, this relationship between the encoded expressions of love and their affective content has been conceived of as problematic, and each attempt to express love added further to the complication. In fact, the notion that people's ideas of love are shaped by the romantic stories and images they encounter in literature and other media is old enough to have become a cliché itself. Illouz quotes La Rochefoucauld's famous seventeenth-century aphorism that "many people would not have fallen in love had they not heard of it" and Flaubert's *Madam Bovary* as well-known examples for the widespread suspicion that feelings of love may be far less individual than one would like to think (cf. "The Lost Innocence of Love" 161–62). Her view that "long before the so-called postmodern age, the relationship between romantic fiction and romantic sentiment was actively scrutinized" (163) begs the question whether there is a specific postmodern development in this relationship. Catherine Belsey asks and answers the same question for the production of literature:

> Desire was probably always citational. Certainly the Elizabethan poets knew this when they translated, cited, adapted, and reformulated Petrarch, not to mention Theocritus, Ovid, and Catullus. [...] The vocabulary of desire shifts as alternative traditions come into view and new sources are adapted and reformulated. What is specific to postmodern writing is that it foregrounds the citationality of desire, affirms it, puts it on display. ("Postmodern Love" 691)

As we will see later, this is true for contemporary British plays about love. But it can also be transferred from the writing of literature to lovers' reflections on their own romantic feelings. There, too, the citationality of love is foregrounded and made conscious, leading eventually to an ironic or sceptical attitude towards love in general. Today, Illouz argues, La Rochefoucauld's maxim has to be reformulated to bring it in line with the considerably high level of reflexivity of lovers. As she puts it, "in the postmodern condition, many people doubt they are in love precisely because they have exceedingly heard about it" ("The Lost Innocence of Love" 183).

In *Consuming the Romantic Utopia,* Illouz corroborates this claim with an empirical study for which she interviewed fifty men and women from working to upper middle class in urban areas along the East Coast of the USA. She found that her interview partners had two differing – in fact, mutually exclusive – narrative frames at their disposal which they applied alternatingly when responding to a selection of differing love stories. If they applied the *romantic* narrative frame, they would declare that true love needs to be impulsive, passionate, irrational, and adventurous. If they applied the *realistic* narrative frame, however, they would reject these very assets of romantic love as typical Hollywood clichés and detrimental to serious relationships. They would then dismiss love-at-first-sight as an improbable illusion, warn against basing a relationship upon passionate infatuation, and stress the importance of confirming the "compatibility" of partners before deciding for a long-term commitment (cf. 156–166). Surprisingly, when the participants were asked to recount their own most memorable love experiences, they all told stories that resembled the romantic narrative frame (cf. 166). Moreover, only three of the fifty participants told stories about their current partners. Almost all the stories that were considered romantic and memorable followed the narrative structure of adventurous romance in that they had a clearly determinable beginning, contained some kind of obstacle that prevented marriage or a long-term relationship, and were demarcated from the routine of ordinary life by their ending (cf. 169). In their "tight dramatic structure" (169), which is characteristic of the 'adventure' and the 'extraordinary' and detaches images of romantic love from the ordinariness of day-to-day experiences, these personal favourite stories displayed the same equation of romanticism with an escape from daily routine that is used in advertising. The interview study suggests that the romantic narrative frame is considered unsuitable as a basis for enduring monogamous relationships, and to ground marriage upon infatuation is rejected as short-sighted and imprudent. But at the same time, the features of romantic love – spontaneity, impulsiveness, passion – are still the features that are desired and expected in any kind of erotic relationship. The result is a seemingly irresolvable dilemma: "The contemporary romantic self is marked by its persistent, sysiphian attempt to conjure up the local and fleeting intensity of the love affair within long-term global narratives of love (such as marriage), to reconcile an overarching narrative of enduring love with the fragmentary intensity of affairs" (175–76).

What distinguishes Illouz' results from Beck-Gernsheim's notion of the 'trap of romantic love' is the awareness the participants had both of the discrepancy between expectation and reality and of the inauthenticity of their favoured concept of love. They proved to be fully conscious that the romantic concept they simultaneously celebrated and doubted is essentially a product of literature and mass media, that the love-at-first-sight-story and their own romantic stories imitated the dominant, stereotypical pattern of romantic love which they considered improbable and unrealistic:

> Like postmodern artists and sociologists, the respondents maintain the ironic stance that their representations and experiences are 'simulacra,' imitations of manufactured signs devoid of referents. The romantic self perceives itself ironically, like a pre-scripted actor who repeats the words and gestures of other pre-scripted actors, simply repeating others' repetitions. (178)

This ironic stance and the feeling of a lack of authenticity are central to the treatment of love in contemporary British drama. All the plays analysed below bespeak their authors' uneasiness with a language of love that cannot be realistic and original at the same time and contain attempts to avoid representations of love and its language that seem to be uncritically and unwittingly citational. Strategies vary from omission of typically romantic scenes, deliberate exaggeration or defamiliarisation, to making the characters themselves, like Illouz' respondents, aware of the citationality of their utterances.

3.2.3 Individualism and Romantic Rationality

While the romantic narrative frame is met with irony and scepticism because of its dissemination through mass media, it is by no means the only narrative frame permanently repeated in public discourse. The realistic frame, which the interview partners used when they considered the prospects of success of love relationships, is no less present in media and collective consciousness. While the film industry (just like the postmodern romantic individual) alternately uses both patterns, the realistic one dominates the therapeutic discourse and the blooming self-help literature. When the defining features of the romantic pattern can be said to be passion, impulsiveness, and the opposition to daily routine and the ordinary, the defining features of the realistic pattern are precisely such qualities as may enable love to survive *despite* the ordinariness of everyday life: compatibility, rationality, and communication.

Even if the ideal of unconditional love is still lingering in the background, modern lovers hardly believe that love has no reason. Besides the arguably time-less condition that the other person answer to certain aesthetic, moral, and intellectual standards which make them 'attractive' to the lover, Illouz names a second condition that is distinctively modern and distinguishes the contemporary concept of love from earlier variants. It is the demand for 'compatibility,' the capacity of the other, with all their qualities and characteristics, to be a suitable partner in a long-term relationship. Earlier societies had assumed that love and compatibility would ensue after some time of living together but today, with romantic love as universally accepted prerequisite for marriage, compatibility is an integral part of this prerequisite (cf. *Consuming* 215–17). The compatibility of a potential partner has to be assessed before entering a serious relationship in a process of careful testing and examination, usually conducted during a series of dates. As Illouz notes, the ideal of compatibility has produced a "romantic rationality" (218) that differs from the rationality of arranged marriages in that it is not the parents or relatives who assess the (mostly financial) merits of a future partner but the (potential) lovers themselves, who assess the (mostly personal) qualities of each other. And it is a curious fact which further highlights the paradoxical nature of love in postmodernity that this rational ideal of compatibility is put forward incessantly in popular media culture, regardless of the romantic narrative frame that usually dominates this domain (cf. 218).

What makes compatibility *before* the relationship so significant is fear of change, mutilation, or loss of individuality in love. As I have indicated in the historical overview of love philosophy, this fear is a specifically modern source of love's precariousness, and it is also emphasised in sociological research as symptomatic of a time in which both the male and the female partners insist on their right (or even duty) to follow their individualised, self-determined biographies (cf. Beck/Beck-Gernsheim 191, Luhmann, *Passion* 46, Hondrich 101). The modern relationship, according to Beck and Beck-Gernsheim, is one in which the realisation of the self tends to be more important than the shared identity of the couple (cf. 130). This insistence on self-realisation runs counter to the increased need for closeness and intimacy in a world of loosened familial and traditional bonds, leaving the single individual in the contradictory situation of demanding absolute freedom and romantic commitment at the same time. Eventually, however, if the relationship threatens to impede self-realisation and individual fulfilment, the contemporary concept of love suggests not to change one's personal goals but to end the relationship (cf. 229).

What Beck and Beck-Gernsheim describe here is what Anthony Giddens has called the 'pure relationship,' which he pictures as the kind of attachment gaining ground in present times. The pure relationship is not based on any form of spiritual idealisation whatsoever but 'purely' on mutual gratification. The term, Giddens explains, "refers to a situation where a social relation is entered into for its own sake, for what can be derived by each person from a sustained association with another; and which is continued only in so far as it is thought by both parties to deliver enough satisfactions for each individual to stay within it" (58). Later he specifies: "What holds the pure relationship together is the acceptance on the part of each partner, 'until further notice,' that each gains sufficient benefit from the relation to make its continuance worthwhile" (63). Obviously, the concept of love underlying such relationships is devoid of romantic notions of unconditionality, uniqueness, and eternality. Instead, Giddens speaks of "confluent love" – self-consciously contingent, temporary, and not necessarily monogamous – as the underlying principle of the pure relationship (cf. 61–64). Relationships of this sort are not, as marriage used to be in earlier historical periods, self-affirming guarantors of durability (cf. 137). If modern relationships are supposed to acquire a durability that outlasts the inevitable moments in which mutual gratification is temporarily suspended, Giddens intimates, what is necessary is a kind of self-dedication to the relationship – a commitment – which, however, always bears the potential of disappointment and thus runs counter to the first principle of the pure relationship.

In *Liquid Love*, Zygmunt Bauman examines the same conundrum of modern relationships. The "residents of our liquid modern world" (xi), he claims, are characterised by an inner struggle between a "yearning for the security of togetherness" (viii) and a profound fear of sacrificing freedom and independence through commitment: "They say that their wish, passion, aim or dream is 'to relate.' But are they not in fact mostly concerned with how to prevent their relations from curdling and clotting? Are they indeed after relationships that hold, as they say they are, or do they, more than anything else, desire those relationships to be light and loose [...]" (xi)? At the heart of the problem lies a specific contradiction between expectation and reality. People enter relationships, Bauman says, mainly for one reason: security (cf. 13). But what they get is "perpetual uncertainty" (14), resulting from the very nature of relationships which are uncontrollable to the extent that they involve a second individual. Moreover, according to Bauman, our liquid modern world "abhors everything that is solid and durable" (29): "In lasting commitments, liquid modern reason spies out oppression; in durable engagement, it sees incapacitating dependency" (47). He thus agrees with Giddens in viewing the 'pure relationship' as the "prevailing target/ideal model of human partnership" (45) of our time. Sex, he maintains, free from commitment and the lurking threats of responsibility which had accompanied the act in earlier days, is "the very epitome" (45) of this 'pure relationship,' promising the fleeting but repeatable comfort of togetherness without the demand to confine one's individuality. It is the ideal of ultimate self-realisation, in which the other is welcome and necessary as an antidote to loneliness and insecurity but has no further impact on individual development.

The notion that love has to facilitate self-realisation ties in with Illouz' description of a "therapeutic emotional style" (*Cold Intimacies* 6) which has pervaded the discourse on love ever since psychology and psychoanalysis made their impact on it. In contrast to the comparatively unproblematic concept of the self of the nineteenth century, contemporary selves are enigmatic and problematic for their owners, who are urged to analyse and verbalise as much of their individuality as possible and to clarify and bring to consciousness their needs, desires, and expectations (cf. 28). The new therapeutic emotional style means that emotions have to be recognised, determined, and verbalised in order to make them communicable and practicable (cf. also Beck/Beck-Gernsheim 120–24). The natural consequence of this idea is that subjects have to clarify their emotional constitution before they can think about the adventure of romantic commitment, while also within existing relationships the demand for emotional self-scrutiny as a means to determine the felicity and future viability of the relationship has grown. The resulting trend to verbalise and differentiate emotions amounts to an objectification of emotions which are then conceived of as "definite discrete entities" that can be "detached from the self, observed, manipulated, and controlled" (Illouz, *Cold Intimacies* 33). Intimate relationships become objects of analysis that are "evaluated and quantified according to some metric" (33) proposed by therapists, self-help books, or the lovers themselves. One such 'metric,' and maybe the most universal one, is the satisfaction of needs and desires (including the possibility of self-realisation), which lies at the heart of the ideal of compatibility and which is what communication is above all supposed to ensure. What this points to is that behind the ideal of communication in (and before) romantic relationships lies the resolution of modern lovers to sacrifice as little as possible of their individuality and self-determined biography to a romantic commitment. Communication has to inform the other about one's needs and expectations, and it helps to ascertain the other's compatibility with these needs. Interestingly, Illouz sees this new romantic rationality, which ascertains the compatibility of the other and the chances for self-realisation, especially at work in the early stages of a relationship, while in later stages of long-term relations the lovers develop the capacity for more irrational forms of love. At the beginning, the other is measured against subjective standards of compatibility and love arises to a considerable extent from the recognition that this standard is largely met. This rather rational and self-interested form of love is only gradually replaced by a kind of selfless devotion which is, from a market point of view, irrational (cf. *Consuming* 242).[29]

3.2.4 Postmodern Precariousness of Love: Choice, Irony, Recognition

In *Why Love Hurts* (2012) Illouz carves out reasons for pain and suffering in love which are particularly contemporary or (post-)modern because they arise from distinctly cotemporary circumstances and value systems and are linked to her claim "that something fundamental about the structure of the romantic self has changed in modernity" (6). For a start, she identifies a renewed view of love, no longer "as an experience that overwhelms and bypasses the will, as an irresistible force beyond one's control" but as a matter of choice (18). For Illouz, this development follows logically from the importance the faculty of choice has gained in modern culture.

> Choice is the defining cultural hallmark of modernity because, at least in the economic and political arenas, it embodies the exercise not only of freedom, but also of two faculties that justify the exercise of freedom, namely rationality and autonomy. In this sense, choice is one of the most powerful cultural and institutional vectors shaping modern selfhood. (19)

It comes as no surprise, then, that choice should acquire significance in the sphere of love, which is so highly important for the shaping of modern selfhood. To choose well in matters of love means to assess the compatibility of the other through introspection and communication, which distinguishes modern amorous relationships from those of the past. Formerly, especially in conjugal relations, the partners were "inextricably intertwined and interdependent" but, from a modern viewpoint, "emotionally distant," feeling no compulsion to lay bare all their inner feelings. "In contrast," Illouz maintains, "modern selves expect each other to be emotionally naked and intimate, but independent" (39).

---

[29] As Illouz points out, this is an interesting alternative to the common notion that love starts with irrational infatuation and grows more rational in time (cf. 242). She also indicates that the later, irrational form of selfless love represents what the philosophy of love has called *agape* (cf. 245). Jean-Claude Kaufman enlarges in detail upon the intricacies resulting from the incompatibility of the rational, calculating individual seeking self-realisation and autonomy and his or her desire for commitment, which presupposes irrational *agape* (esp. at 109–12 and 131).

This emotional nakedness, together with the modern determination to remain an independent individual as far as possible, arguably adds a new dimension to the vulnerability that has always accompanied love throughout the ages.

Another aspect of choice is the "pool of potential partners" that has become so much larger than ever before (52). With class distinctions losing significance and with the modern means of travel and telecommunication connecting people all over the world, the scope of people that can potentially fall in love with each other has virtually exploded. In consequence, a new motif was added to the narrative frame of love in Western culture, namely that of the search for Mr. or Mrs. Right in a rather competitive marketplace (cf. 57). According to Illouz, "the *search* for and *choice* of a partner have become an intrinsic segment of the life-cycle, with its own sociological complex forms, rules and strategies" (57). As was the case with individualisation, which liberated modern humans from many of the constraints of religion and tradition but in the process imposed on them ontological insecurity and the obligation to choose a self-determined, individual biography, the possibility of choice in matters of love has removed for modern lovers the obstacles formerly represented by parents and class affiliation but simultaneously forces them to participate in the game of searching, choosing, and competing for a partner, which entails huge risks for the romantic self. As Illouz puts it, "the exercise of freedom can and does generate forms of distress, such as ontological insecurity and meaninglessness" and, accordingly, "sexual and emotional freedom generate their own forms of suffering" (59). For one thing, the "abundance of choice" (91) in the age of Internet and mobility does not make it easier for many people to actually make a choice but rather leads to commitment phobia (cf. 78 – 89). Furthermore, the demand for introspection and self-knowledge that the freedom to choose entails rests on the assumption that lovers are, or should be, a pair of fully self-aware individuals who know exactly what they desire and expect. Accordingly, Illouz summarises,

> finding the best possible mate consists of choosing the person who corresponds to the essentialized self, the set of preferences and needs that define the self. Crucial to this conception of choice is the idea that through introspection, which entails a hyper-cognized process of decision-making, a rational assessment of our own and another's compatibility and qualities can and must be established. According to this model, introspection is supposed to lead to emotional clarity. (92)

However, Illouz also lists a number of arguments that doubt this model and, particularly, the possibility of arriving at an 'essentialised self' or 'emotional clarity' through introspection. The most convincing counterargument argument relies on the fact that in most contemporary psychology, sociology, and philosophy, the self is usually understood as unstable, fragmented, and subject to change. In this conception, self-realisation is not something to be achieved once and for all time, but a constant negotiation between the changing demands of a protean self and the circumstances that have to satisfy them. Naturally, self-realisation of this kind does not go together smoothly with long-term commitment. As Illouz explains,

> [t]he ideal of self-realization is a very powerful institution and cultural force: it is what makes people leave unsatisfying jobs and loveless marriages, attend meditation workshops, take long and expensive vacations, consult a psychologist, and so on. It fundamentally posits the self as a perpetually moving target, as something in need of discovery and accomplishment. [...] The ideal of self-realization disrupts and opposes the idea of the self and of the will as something constant and fixed, and as praiseworthy precisely *because* of its constancy and fixity. To self-realize means not committing to any fixed identity and especially not committing to a single project of the self. (99–100)[30]

Committing to a long-term relationship based on compatibility of the other with an 'essentialised self,' however, would mean exactly this: to conceive of the self as 'a single project,' as a stable entity that will always remain compatible with the self of the other. It also means, as sociologist Jean-Claude Kaufmann points out, to conceive of the other as "someone who will leave everything unchanged," which is a very inappropriate stance since "[m]eeting someone and falling in love with them metamorphoses both their identities. That is what is so irresistibly attractive – and so frightening – about it" (51). In reality, this implies, both selves in a relationship, including their wishes and expectations, are unstable and likely to change, which confronts the couple with two alternatives. The first is to relinquish the claim to incessant and complete self-realisation, to be willing to compromise, to constantly adapt to and re-establish compatibility with the other. Such self-restraint, however, is increasingly seen as an 'illegitimate' limitation of personal freedom in a culture that celebrates emotional and psychological independence and autonomy (cf. Illouz, *Why Love Hurts* 136). Instead, the alternative chosen more and more often, according to Giddens and Bauman, is that of the 'pure relationship,' which also accepts the changeability of the self but does not demand to forego unrestrained self-realisation. Rather, relationships are terminated when they appear to 'fit' the changing self

---

[30] For a concise portrayal of the rise of the ideal of self-realisation, its intertwining with the discourses of therapy, psychology, and self-help, and its identification with psychic health and the concomitant pathologisation of self-denial, see also the chapter "The self-realization narrative" in Illouz' *Cold Intimacies* (43–62).

no longer (cf. also Kaufmann 111).[31] The 'abundance of choice' fuels this tendency twice, as the mass of potential partners not only minimises the risk of ending up alone after a (series of) breakup(s) but also maximises the pressure on individuals within relationships to always keep looking for the most compatible – which means the least constricting – partner.

This pressure may also explain why the romantic image of love at first sight is still so powerful, at least as a distant ideal lingering in the background. Love at first sight, this "overwhelming revelation that imposes itself by its intensity" (Illouz, *Why Love Hurts* 31), renders unnecessary the need to choose on rational grounds since the compatibility of the other is proven by the sheer force of the attraction. As Kaufmann writes, "[a]ll would be well if passion could sweep us off our feet, just as it does in romance fiction. But that is not what is happening: we always have our doubts and we are always in two minds" (111). For the postmodern lover, love at first sight is an unlikely illusion, simultaneously desired and doubted, and ultimately hardly compatible with the ideal of autonomy and self-determination. The pressure to choose well on rational grounds remains intact.

Another distinctly (post-)modern source of love's precariousness results from the situation, already established in *Consuming the Romantic Utopia*, that contemporary romantic desire is structured around two equally powerful narrative frames. In *Why Love Hurts*, Illouz sets out to delineate how this specifically contemporary organisation of desire adds to the vulnerability of the lover. She maintains that there are "at least two cultural structures at work in the emotion of love: one based on the powerful fantasy of erotic self-abandonment and emotional fusion; the other based on rational models of emotional self-regulation and optimal choice" (159). Of these, the former, which she calls 'enchanted love,' is the historically older narrative which now undergoes a process of 'disenchantment' through the collision with the latter. 'Enchanted love,' which is "simultaneously spontaneous and unconditional, overwhelming and eternal, unique and total," the "ideal-type of romantic love," has been a powerful idea for centuries and its "basic components – sacredness, uniqueness, experiential force, irrationality, giving up one's self-interest, lack of autonomy – have remained in the literary models that took over with the spread of literacy and of the romance novel" (161). Modernity has shattered all these believes, has 'disen-

---

31 It is interesting to see how the notion of an unstable, changing self can lead to opposing assumptions concerning the durability of love relationships. As long as duration was an unquestionable ideal, the changing self appeared as a chance for lovers to rekindle their passion for each other again and again (see Luhmann above at 3.1). With the importance of duration diminishing, the changing self instead became the justification for the temporariness of relationships.

chanted' love, however without completely removing its remnants from the collective image repertoire, so that the ideals of 'enchanted love' are still around but met with uncertainty and irony.

Besides the "intensification of technologies of choice, embodied in the Internet" (177), which have exploded the quantity and rationalised the quality of choice, Illouz describes two major developments in the disenchantment of love. First, "the prevalence of scientific models of explanation of love" (*Why Love Hurts* 162) has replaced romantic metaphysical explanations. While biology, neuroscience, and evolutionary psychology have reduced romantic love to brain chemistry at the service of the preservation of the species (cf. 167–68), psychoanalysis particularly attacked notions of self-sacrifice and emotional fusion as threats to the ideal of autonomy and as "the symptom of an incomplete emotional development" (164). Second, the feminist emancipation movement has opened (or at least put the accent on) a particular perspective on (heterosexual) intimate relationships, focussing pre-eminently on power relations (cf. 170–71).[32] The ideal of equivalence and fairness within relationships necessarily made these relationships the "object of scrutiny and control through formal and predictable procedures" (177), which in turn gave rise to the development and implementation of objective categories, measurements, and language for the assessment of love relationships.

The disenchantment of love has considerably transformed the way in which romantic love is conceptualised and experienced. Above all, it has promoted a deeply ambivalent attitude towards love so that today "uncertainty and irony dominate the cultural climate of romantic relationships" (192). Uncertainty results from the increasing regulation of romantic relationships on behalf of the ideal of equality and fairness and the attempted banishment of gender roles and stereotypes from relationships or from the dating process. The goal of political correctness in matters of love coerces modern lovers to keep "the adequacy of one's own conduct" (193) under constant surveillance. In contrast to the playful ambiguity of eroticism based on gender stereotypes and well-established rules of etiquette, this "uncertainty […] is painful and derives from the difficulty of knowing the rules that organize interactions" (193). Eventually, Illouz goes on, uncertainty even "inhibits sexual desire and entails anxiety, because it makes people focus on and interrogate themselves on the rules of interaction, thus making them less able to let themselves feel emotions elicited by the interaction it-

---

[32] "The central concept that has enabled feminism to deconstruct sex and love is the notion of power. In the feminist worldview, power is the invisible, yet highly tangible dimension organizing gender relations, that which must be tracked down and expelled from intimate relationships" (171).

self" (193). Irony, in turn, is the attitude of the modern lover towards notions of 'enchanted love' in the face of their utter improbability. While the principles of 'enchanted love' are still immensely powerful influences in the formation of contemporary concepts of love, they are at the same time met with suspicion and the ironical distance of someone who has looked behind the surface and seen their historical, cultural, and medial constructedness as well as their apparent implausibility in view of scientific explanations. "Modern romantic consciousness," Illouz concludes, "has the rhetorical structure of irony because it is saturated with disenchanted knowledge which prevents full belief and commitment. Irony cannot take seriously a belief central to love, namely its self-proclaimed claim to eternity and totality" (195).

The atmosphere of uncertainty and irony interferes with what is perhaps the most salient function of love for the postmodern subject – the generation of a sense of self-worth. It undermines romantic or 'enchanted' expectations, opposes any sincere belief in the overwhelming and sacred power of love, and consequently prevents the passionate feelings that would accompany this kind of love from arising in the first place. Uncertainty and irony prevent an experience of love which most contemporary individuals are still wishing for even before it can be frustrated in the course of a relationship and its daily routine. They are thus linked with a "change in what makes the self vulnerable, that is, what makes one feel unworthy" (Illouz, *Why Love Hurts* 6). Love, apart from being a desired and joyful emotion, fulfils an essential function for the postmodern individual's generation and maintenance of a feeling of self-worth. It is because self-worth has been made dependent on the experience of love that the postmodern self has become so vulnerable, and this is why, even though suffering has always accompanied love in literature, theory, and reality throughout the ages, "there is something qualitatively new in the *modern* experience of suffering generated by love" (16).

That romantic love, especially the experience of being loved, contributes positively to one's own self-image is not surprising and certainly not new. Nevertheless, Illouz argues that there have been specifically modern developments in the constitution of self-worth so that "ontological security and a sense of worth are now at stake in the romantic and erotic bond" (110). With regard to love's function as a 'recognition' of the self she claims that

> the pressure for self-differentiation and developing a sense of uniqueness has considerably increased with modernity. In other words, whatever subjective validation love may have provided in the past, this validation did not play a *social* role and did not substitute for social recognition [...]. I argue that it is the very structure of recognition that has been transformed in modern romantic relationships, and that this recognition goes deeper and wider than ever before. (112)

Illouz points to the fact that the liberation from class hierarchies and traditional and religious bonds have urged the modern individual to ascertain social self-worth without and outside these institutions and solely in interactions with others instead. Not surprisingly, the type of interaction that is most significant for the constitution of self-worth is that in which the self is exposed, recognised, and validated in its most essential form – the romantic relationship. Being loved romantically, in this respect, means being recognised, confirmed, and valued as a person and for one's unique and individual personality, a self-affirmation that no other relationship can provide.[33] The fact however, that there are no objective parameters along which such recognition and affirmation are afforded "makes love the terrain par excellence of ontological insecurity and uncertainty *at the very same time* that it becomes one of the main sites for the experience of (and the demand for) recognition" (123). The development of the romantic relationship into the realm where self-worth is mainly generated has made it a site of anxiety. The inability to enter or remain within a romantic relationship is increasingly seen not only as bad luck accompanied by painful emotions but as a serious flaw in personality. As Illouz puts it, "[t]he 'fear of rejection' is a danger ever-looming in relationships because it threatens the entire edifice of self-worth" (125). This precariousness of love, in which "one's basic sense of worth and the foundations of one's ontological security" (126) are at stake is a new and specifically contemporary dimension of pain and suffering.

As Illouz rightly observes, pain and suffering have always played a major role in the literature and philosophy of love. However, they have been experienced and interpreted differently throughout the ages and within changing "cultural frameworks" (127). In courtly love, for example, the bitter-sweet pain of love was less feared than desired and was not conceived of as a threat but as a contribution to the development of selfhood. Christianity, too, made a virtue out of suffering, seeing it "as a purification of the soul" and an ultimately "positive and necessary experience" (127). Similarly, Romanticism cultivated images of heroic suffering and self-destruction in love and celebrated the experience of melancholia (cf. 127–28). What the cultural frameworks of courtly love, Christianity, and Romanticism have in common is that they interpret suffering and

---

[33] Similarly, Kaufman argues that "[w]e seek recognition for two reasons: to improve our self-esteem, and to confirm our individual sense of identity" (114). In the competitive, antagonistic society of late capitalism, "in which everyone is trying to outdo everyone else" (112), we experience a lack of recognition for which love is supposed to compensate. "The huge need for love that is now being expressed everywhere often masks a demand for recognition. [...] When we are loved, we feel that we exist. Being loved by a partner makes up for the lack of recognition" (112).

lack of reciprocity not as a threat to the self but rather as a constitutive element in its development, as affirmation and purification of the self. Traditionally, only the cultural framework of medical discourse has rejected such idealisations of suffering, as Illouz illustrates with reference to the conception of love-suffering in the sixteenth and seventeenth centuries as a disease ("lovesickness"), as a kind of bodily and/or mental "disorder" (129) that was to be cured and removed like any other physical ailment (cf. 129–30). But not even medical discourse saw love-suffering as in any way threatening the integrity of the self (cf. 129). This, then, is the specifically modern aspect of love-suffering, which could only develop in a culture where "a well-developed character is expressed through one's capacity to overcome one's experience of suffering or, even better, to avoid it altogether" and where, in consequence, "suffering has become the mark of a flawed self" (130). The special precariousness of contemporary love arises from the fact that the unhappy lover has to endure not only the arguably timeless and universal pain of rejection or otherwise unrequited love, but also the pain of damaged self-worth and ontological insecurity.

There is an interesting parallel here to the historical development of the idea of happiness which, as Darrin McMahon explains, has long been conceived as something that 'befalls' human beings and lies beyond their control and influence. To think about happiness as attainable without divine intervention was tantamount to the hubristic or even heretic conception of man as self-sufficient being (cf. *Happiness* 7–9). Modernity, in contrast, is characterised by the Enlightenment conviction that happiness can be controlled. This gain in autonomy, however, comes at a price. "Arguably," McMahon states, "there is no greater modern assumption than that it lies within our power to find happiness. And arguably there is no greater proof of that than our feeling that we have failed when we are unable to do so" (12). Both love and happiness have acquired a new precariousness in modernity since their absence tends to be blamed on the unhappy individual. Illouz locates the source of this new precariousness in a culture of self-blame set up by therapy discourse and its pathologisation of all kinds of dependency (cf. *Why Love Hurts* 149). The conviction that self-worth is at risk in romantic relationships due to improper forms of emotional dependency has led to a pronounced emphasis on the importance of 'self-love' in much therapeutic discourse (cf. 151). To the same extent that dependence is viewed as pathological, self-love is seen as its natural and psychologically healthy antidote. The ideal is a kind of emotional autonomy, a state of self-sufficiency in which self-worth is entirely at one's own disposal. Illouz regards this dream as an unattainable illusion and rejects the bulk of "pop psychology advice" to simply work on one's self-love to gain emotional independence:

> Such advice – substitute love for self-love – denies the fundamental and essentially social nature of self-value. It demands from actors that they create what they cannot create on their own. The modern obsession and injunction to 'love oneself' is an attempt to solve through autonomy the actual need for recognition, which can be bestowed only by an acknowledgement of one's dependence on others. (151)

To accept dependence on others, as Illouz demands, can of course not alleviate the precariousness of love. It means to accept this very precariousness in the form in which it has arguably always existed – as the possibility of being rejected, left, or otherwise hurt by a beloved and desired person. But it also means to leave it at that and not to aggravate the suffering in unhappy love through the accusation that this suffering is to blame on one's own lack of autonomy and submission to the pathological state of emotional dependency. It means to resist the particularly (post-)modern precariousness of love where the pain that has always accompanied love is reinforced by a culture of self-blame, where the self is not so much threatened by the pain of being rejected than by the pain of feeling pain in a culture that has pathologised emotional suffering.

* * *

What are the results of this survey of sociological literature? First of all, sociology generally appears to think of romantic love as a product of modernity since it developed both as a result of capitalism (the nuclear family as the foundation of both capitalism and romantic love) and as its counterpart (love as the emotional retreat from the cold and heartless rationality of the market). Moreover, romantic love is credited with a pivotal compensatory function as it is seen as filling in for waning structures and beliefs traditionally tasked with the creation of meaning and a sense of belonging, such as religion and rootedness in local communities. The two main functions Luhmann ascribes to romantic love as a medium of communication – the creation of meaning and the affirmation of the self – are specifically modern functions that love acquired in the course of its 'functional differentiation,' which also turned it into a subsystem of almost entirely private and personal concern. The one social function which romantic love retained was to serve as the guarantor of stable marriages and families, which reinforced the ideal of permanence as one of its core elements. Apart from that, romantic love developed into a private sphere in which the lovers are liberated from obligations except those towards each other and where they find shelter from an increasingly hostile capitalist society. However, as Beck and Beck-Gernsheim, Bauman, Giddens, Illouz, and Kaufmann notice, contemporary love is characterised by a tendency to see already obligations towards the partner as unwarranted restrictions of the right to self-realisation. The profit-oriented rationality of capitalism, against which the ideal of romantic love had originally developed, has infiltrated the realm of intimate relationships and operates against the 'irrational' components of romantic love such as unconditionality, self-denial, or the permanent commitment to a single option.

The prototypical notion of romantic love, loaded with idealist qualities such as eternality, unconditionality, passion, and perfectness, is hence regarded with due scepticism – not only by its academic analysts but by 'ordinary lovers' as well. While Belsey claims that contemporary lovers are sceptical about idealist concepts of romantic love because of postmodernity's 'incredulity towards metanarratives,' Illouz emphasises the awareness of postmodern lovers of the constructedness and unreliability of media images of love. They know that the narrative frame of romantic or 'enchanted' love, while it is still the emotionally preferable and thus more or less secretly desired concept of love, is compromised by its commodification in marketing and advertisement and by its indissoluble link to fiction in any form – as literature or film, highbrow or trivial – and the concomitant lack of authenticity. Furthermore, one of its essential characteristics, the clear demarcation from and opposition to everyday routine, makes it an unlikely basis for long-term relationships. The realistic frame, on the other hand, is no less problematic. Not only does it not correspond to the romantic expectations postmodern lovers, despite everything, still have towards themselves, their partners, and their relationships; it is also a narrative frame whose authenticity is likewise questionable due to its omnipresence in media and literature, above all those belonging to therapy culture, psychology, and psychoanalysis in the broadest sense. The realistic frame is no longer (if it ever was) a rational, commonsensical liberation from unrealistic notions of love and hence a protection against frustrated expectations. Rather, it has developed into a narrative of love in its own right, with its own norms and rules that shape contemporary concepts of love. It has not only 'disenchanted' love to a considerable degree but has also introduced compatibility as a new and central ideal, which in turn generates a demand for psychological and emotional introspection and communication and an increasing rationalisation of love – a rationalisation that comes along with feelings of shame and guilt because it violates the ideal of unconditionality and disinterestedness that is part of the still powerful romantic narrative frame (cf. *Consuming* 225–27). Illouz delineates a paradoxical situation in which postmodern lovers take an ironic and sceptical stance towards the very concept of love they actually prefer, making themselves incapable of fully believing in the unconditional, passionate, eternal, romantic love they secretly expect. On top of that, modern individuals have made their sense of self-worth increasingly dependent on the experience of love, that is, on an affirmation of their status as desirable partners and on the capability to enter successful relationships. In that respect, love has become not only more important than ever (because it creates meaning and a feeling of ontological security in an atmosphere of absurdity and insecurity) and more difficult than ever (because of the ironic and sceptical attitude with which it is approached, the difficulty to reconcile it with individualised biographies and 'essentialised selves,' and the pressure to decide well when faced with an 'abundance of choice'), but also more dangerous than ever (because to fail in love means to fail as a human being).

Of course, this is not to say that the precariousness of love is a recent phenomenon. The survey of love philosophy in the previous chapter has shown that precariousness has always been an inherent feature of love, just as much as this chapter has shown that modern love is still tasked with a compensatory function. Little has changed about the age-old problem that the privileged remedy against life's precariousness is itself inherently precarious. A lot has changed, however, with regard to the specific compensatory tasks love has to fulfil and the specific sources of its precariousness, and it is this nexus of an historically continuous paradox and the peculiarities of its contemporary manifestations that I will seek to disclose in my analysis of contemporary British drama. The following interpretations will single out and scrutinise dramatic situations where the fear of vulnerability prevents full-hearted engagement in love; where the dream of compensating an experience of lack and insufficiency with an intimate relationship is foiled by the refusal to accept inter- or codependency; where a lover's hopes for compensation through love are thwarted either by the uncontrollability of the beloved's desire or by detrimental attempts to dry up this very source of precariousness through power and control; where efforts to keep love free from precariousness are revealed as hubristic illusions of self-sufficiency; where total surrender to the desire for wholeness and continuity implies the dissolution of the self; and where desire for self-realisation and fear of self-loss run counter to the desire to commit to a life shared with another person.

Part II **Analysis:
Love in Contemporary British Drama**

4 "Why isn't love enough?" Commitment in Patrick Marber's *Closer*

4.1 Style and Genre: A Modern Comedy of Manners?

Closer is by now what can be called a modern classic. Graham Saunders' introduction to the play in the *Modern Theatre Guides* series delineates the story of immediate and long-lasting success of Marber's second play, from its 1997 premiere at the small Cottesloe auditorium of the National Theatre and the transfer to larger venues in West End and Broadway theatres to its international triumph, the enthusiastic reception by both theatre critics and academic scholars, its celebrated revivals in the new millennium, and its adaptation as a major Hollywood picture starring Natalie Portman, Jude Law, Julia Roberts, and Clive Owen in 2004 (cf. 1–6).[34] The play was met with almost unanimously favourable and often effusive response, with many critics praising Marber's skilful writing (and directing), the play's contemporaneity, and its masterful but discomforting balance of hilarious surface comedy with deep and painful investigations into the abyss of human love (cf. *Theatre Record*, vol. 17, no. 11, 674–79). Nick Curtis, who describes the play as "[c]austically funny and very moving," pointedly observes that "[o]f the many four-letter obscenities in Patrick Marber's thrilling London love story for the Nineties, 'love' is undoubtedly the most brutal" (677). This statement implies two questions that Saunders explicitly asks in his introduction and which I want to address briefly, too, before moving forward to a specific analysis of the play's treatment of love: To which genre can this symbiosis of comedy, love story, psychological scrutiny, and emotional brutality be assigned? And to what extent is it a "London love story for the Nineties," that is, how much is the play bound to a specific time and place?

Unsurprisingly, there is no clear answer to either of the two questions. Regarding genre, most critics have chosen to call it a comedy, while taking into account its bleakness by further categorising it as "tragi-comedy" (G. Brown 674), the "blackest of comedies" (Hagerty 675), or the "darkest of sex comedies" (Billen 678). David Benedict has called it "a Nineties sex comedy" (674) and John Peter "one of the best plays of sexual politics in the language" (679), while Michael Billington, otherwise applauding the play's realistic depiction of the "sex war," laments that the "crazy sexual square dance" the four characters

---

[34] All references to Saunders in this chapter are to his monograph *Patrick Marber's* Closer (2008).

perform has not enough political or wider social implications (678). Michael Coveney finds a very fitting expression when he calls *Closer* a "wonderful new comedy of sex and modern manners" (676), stressing the parallels with the British Comedy of Manners or Restoration Comedy which Saunders also observes (cf. 21), even if in *Closer* the relation between hilarity and seriousness is much less in favour of the former.

The play's verbal explicitness and emotional ferocity have led to its inclusion in Aleks Sierz' study of *In-Yer-Face Theatre*, but as Saunders rightly points out, "it is markedly unlike other work from the period such as Anthony Neilson's *Penetrator* or Sarah Kane's *Cleansed*" in that its violence is exclusively verbal and does not follow the "experiential approach" typical of in-yer-face theatre (49). Indeed, *Closer* shows only a few of Sierz' defining characteristics of in-yer-face theatre, according to which "the language is usually filthy, characters talk about unmentionable subjects, take their clothes off, have sex, humiliate each other, experience unpleasant emotions, become suddenly violent" (*In-Yer-Face* 5). There surely is filthy language, although not as filthy as in many other plays of the decade, and there is humiliation, but it is much more self-inflicted than done to one another. There is no nudity or sex onstage, physical violence is limited to a single slap in the face, and the experience of unpleasant emotions can hardly count as a characteristic feature of any form or genre of drama. Similarly, the emotional audience reactions Sierz mentions in his short analysis of the play are not a defining feature of in-yer-face theatre, and even if the feelings of stifling uneasiness and of being caught red-handed the play produced on many viewers (cf. 189–90) can be compared in kind, if not in degree, with the effect of in-yer-face plays, it is important to note that in *Closer* this effect is achieved through the potential recognisability of situations and discourse fragments – that is, on a cognitive rather than on an experiential level. *Closer* thus seems to fit into the subgenre of "cool" in-yer-face theatre, with which Sierz distinguishes plays that "mediate the disturbing power of extreme emotions by using a number of distancing devices: larger auditoriums, a more naturalistic style or a more traditional structure" (6–7) from "hot" in-yer-face theatre which "uses the aesthetics of extremism" (6). Indeed, *Closer* was soon transferred successfully to larger venues and was and is often praised for its beautiful symmetric structure, for example by Alastair Macaulay, who finds that "[t]he rhythm of *Closer* is terrific" and that "Marber, while writing a play full of f- and c-words and juicily explicit descriptions of sex, paces his material like a classical artist" (675). *Closer* is also decidedly more naturalistic than plays by, say, Kane or Ridley, and it contains numerous elements of comedy, which Sierz sees as "the most effective distancing device" (*In-Yer-Face* 7). However, it remains arguable how useful the category of 'cool in-yer-face theatre' is in the end. After all, if 'distancing devices' are its characteristic feature, it is questionable to what extent such plays can still live up to Sierz' initial definition of in-yer-face theatre as "a theatre of sensation" which is "experiential, not speculative" (4).

With regard to *Closer*'s situatedness in 1990s London, Saunders discusses both Marber's attempt at making the play "more 'universal' in scope" by removing specific references to the nineties in later editions of the playtext and the reception of the play by many critics as "an archetypal metropolitan 1990s drama" (7). Saunders names the Internet and mobile phones as new technologies that were used and presented as such in the play and aligns the 'cool' atmosphere of both the characters' attitudes and relationships and of the entire play to the short cultural phenomenon of 'Cool Britannia' (cf. 7–9). I will resume this question about the time-boundedness of *Closer* after a brief synopsis of the play. As will become clear, the cultural and social context of the nineties undeniably impacts the play's setting and plot, rendering it a minute snapshot of the symptomatic vicissitudes and difficulties of postmodern love as described, for instance, by Eva Illouz. At the same time, the discourses of love in which the characters speak and act are exactly those I have analysed above as dominant ever since the rise of capitalism and individualism in the (early) modern period, adding a more timeless quality to the play's contemporaneity. It comes as no surprise, then, that in his review of the 2015 revival of *Closer* at the Donmar Warehouse, Michael Billington calls the play "alarmingly durable" and "as powerful and pertinent as ever" (*Guardian*): this play, "which in 1997 seemed to reflect the sexual mores of the moment," is for the most part not symptomatic of a single decade but of the modern age in general – and of postmodernity in particular.

4.2 Synopsis

The play opens with the young American Alice sitting alone in a London hospital room with a bloody cut on her leg, examining the contents of a briefcase. The briefcase is Dan's, who has brought her to the casualty ward after she was hit by a taxi. Alice smiles when she discovers Dan's sandwiches and we later learn that she fell in love with him at this very moment, seeing that he had cut off the crusts. The ensuing dialogue is a theatrical masterpiece of comical, teasingly flirtatious stichomythia that combines figure characterisation with the beautiful presentation of a budding romance. Alice, who describes herself as a "waif" (14) without any connections in London, apparently works as a striptease dancer. She is a beguiling mixture of fragility and toughness who claims to "know what men want" (13) and is strangely fascinated with death (before the accident she spent the day watching "the meat being unloaded" at Smithfield

Market and strolling through Postman's Park, which is "a graveyard too," with "a memorial to ordinary people who died saving the lives of others" [9]). Dan is likewise connected to the motif of death, since he is working, as Alice puts it, "in the dying business," earning "a living" by writing obituaries (6) as his "dreams of being a writer" had been flawed by lack of talent (10). Dan has a girlfriend (Ruth), but his attraction to Alice is palpable and they decide to spend the rest of the day together.

The next scene takes up one and a half years later in Anna's photo studio, where Dan is being photographed for the jacket of a novel he is about to publish. As the conversation between Anna and Dan unfolds, Alice has become both his girlfriend and the inspiration for his novel's heroine. After a while, Dan starts flirting heavily with Anna, who lives separated from her husband and seems to be more in tune with Dan, both intellectually and regarding their age, than the younger and somewhat simpler Alice. Dan entreats Anna to meet him again, but she turns him down because he is "taken" (21). Alice comes to pick up Dan and asks Anna to take her photo, too, while Dan agrees to wait in a neighbouring pub. During the shooting, she tells Anna that she overheard her conversation with Dan and his advances to her.

Scene Three, which is unanimously described by critics as the play's greatest comical success, re-introduces Larry, a dermatologist, who has had a brief appearance in the hospital scene examining Alice's leg. Larry and Dan are sitting in separate rooms in front of their computers, logged in to an internet chatroom called "LONDON FUCK," while their conversation is projected onto a screen to make it visible to audiences. Larry is a first timer at the web site and seems to be seriously looking for a sex partner, whereas Dan is visiting the chatroom out of mere boredom. Dan pretends to be Anna, whose refusal he was apparently unable to shrug off, and poses as an insatiable nymphomaniac who orgasms even while chatting with Larry and asks to meet at the aquarium at London Zoo the next day. In the following scene, Larry is waiting for his date at the Aquarium and, coincidentally, sees the real Anna, whom he immediately approaches and talks dirty to for an embarrassingly long time before he realises that he has been pranked. Surprisingly, they hit it off quite well and in Scene Five, four months later, they are a couple, celebrating the opening of an exhibition of some of Anna's photography, Alice's portrait among it. At the art gallery, Dan declares his love to Anna and asks her to leave Larry and marry him, but they are interrupted by Larry.

Scene Six shows the two couples one year later in a dramatically effective and touching 'split scene' in every sense of the word. Dan discloses to Alice that he has been having an affair with Anna for the past year and is going to live with her. On the other side of the stage, Larry comes home from a business trip and, while the audience already knows about Anna's infidelity, confesses to having slept with a prostitute in New York and begs her not to leave him. Anna tells him about her affair with Dan and their plans to move in together, whereupon Larry demands to know all the sexual details about their affair. In the next scene, Larry is seen in a private room of a lapdance club, where Alice, who calls herself Jane, is dancing for him. She pretends not to know him and even after Larry has paid her five hundred pounds for disclosing her real name, she insists that it is "plain ... Jane ... Jones" (63). In Scene Eight, Anna tells Dan that Larry has finally signed the divorce papers but that she has slept with him one more time, as "a mercy fuck ... a sympathy fuck" (70). Two months later, when Dan visits Larry in his surgery, the constellation has already changed again. Anna has gone back to Larry and now Dan wants Anna back. Larry enjoys this reversal of power and reveals that Anna "never sent the divorce papers to her lawyer" (85). He also tells Dan where to find Alice and that she still loves him, but he cannot refrain from hurting him with the information that he has slept with her. In Scene Eleven, Dan and Alice are together again but they soon start fighting when Dan urges her to tell him the complete truth about her and Larry. The last scene brings together Dan, Anna, and Larry in Postman's Park. Larry is in a new relationship and Anna has got herself a stray dog. We learn that Dan and Alice split up again and that Alice returned to New York, where she was killed in a car accident. Larry has found out some time ago that Alice had made up her name – Alice Ayres – which she had adopted from one of the plaques of the memorial in Postman's Park, and Dan reveals that her real name was indeed Jane Jones, as he learned from the police who called him since "[t]here's no one else to identify the body" (105).

4.3 *Closer* in Context: The Crisis of Love at the Turn of the Millennium

As a portrait of a stratum of society and its struggles with love, *Closer* is not, despite some concrete references to space and time, restricted to the London of the 1990s. The way love and relationships are thematised in the play resembles the sociological analyses discussed in Chapter 3 to such extent that the play rather appears like a symptomatic account of the crisis of love confirmed in these studies for post-traditional societies of the late twentieth and early twenty-first centuries in general. In fact, at times the play appears like a dramatic transfiguration of the sociological verdict on postmodern love as eminently volatile, self-seeking, transactional, and terminable. Against the backdrop of the sociological farewell to romantic love at the turn of the millennium, *Closer* feels like a timely expression of a widely shared structure of feeling that regards love as a pleasant myth of bygone times.

4.3.1 Isolation, Egoism, Self-realisation

In the introduction to *Modern British Playwriting: The 1990s*, Aleks Sierz comments on everyday life during the period. His data about family life in the 1990s casts some light on the background of the sociological pessimism concerning love. "The traditional family is dead," he writes, "killed off by cohabitation and divorce," with Britain having "the highest divorce rate in Europe" (2). In the play, cohabitation and separation are presented more or less as daily business and the absence of any form of "traditional family" is indeed conspicuous. As Saunders observes, "friends and family are relegated to off-stage figures" (54) – a figure constellation *Closer* shares with *Shopping and Fucking* and *Eigengrau*. While Saunders ascribes the absence of close non-sexual relations to the play's structure in which "it is purely the key moments of the central characters' relationships to one another that are of importance" (54) and which omits all other details of their lives, I think it is more than a structural side effect that the characters seem to exist in isolation. As Christopher Innes writes, "the way the characters are depicted as almost hermetically sealed off from the world around them is an image of isolation and dissociation that epitomizes the self-absorbed 'Me' generation of the 1990s" (431). Family members are not completely absent from the play, but they never appear on stage, as Saunders rightly states. In fact, parents and families are mentioned quite frequently by the characters – but always in contexts that stress the lack of security or support associated with traditional familial ties. Dan lost his mother when he was still a boy and his relationship to his father seems disturbed. When he dies, Dan asks Anna to accompany him to the funeral although he is still in a relationship with Alice. It is revealing that he wants to use his father's funeral as an opportunity to spark an affair with Anna, and his attempts to convince her bespeak the difficult relationship they must have had. When Anna reminds him of his relationship with Alice, he responds, "She'll survive. I can't be her father any more" (42), indicating that his role as her boyfriend coincides with the function of compensating for the absence of her parents. Even though Alice's story of her parents' death in a car accident turns out to be a lie at the end of the play (cf. 105), she has no contact to them and, as a newcomer to London from New York, lacks any kind of social or familial ties in her life. As Saunders says at one point, "key to Alice's character is her essential isolation" (22). About Larry's family, we only learn that they like Anna very much (cf. Marber 45; 79) and are "upset" after Anna has left him, but his way of coping with the separation – regular visits at the strip club where he "used to wander around like a zombie, blubbing into his ashtray" (79) – indicates that he, too, is in want of emotional support once his only close relationship breaks down. Anna, finally, mentions family and friends once in the play, when they have come to see her exhibition (cf. 44), but the fact that in the previous scene she went to the zoo alone on her birthday (cf. 34) casts some doubt on the closeness of these relations and, as Saunders observes, her "attraction as an artist to projects that include photographing strangers and derelict buildings allude to her own sense of isolation" (29). In all four cases, that is, the absence of friends and families from the stage is accompanied by more or less explicit evidence of the characters' loneliness and isolation, which adds some further significance to the structural concentration on the four characters and their mutual relationships.

The characters' isolation is the result of their advancement in the project of individualisation as described by Beck/Beck-Gernsheim, Luhmann, and Illouz, which has liberated them from traditional and conventional restraints but has also loosened social and familial ties to a point where they are almost non-existent. Moreover, the project of individualisation, with its sacred objective of self-realisation, impedes the characters' attempts to compensate for the loss of security and support through family and friends with romantic relationships. None of them are able to afford the commitment necessary for a durable relationship, which they all clearly subordinate to the goal of self-realisation. The most obvious example is Dan, who uses Alice as a means to his project of self-realisation as a writer. Once he has finished his novel, his need of Alice as source of inspiration vanishes. For his new life plan – "Children, everything. [...] Grow old with me ... die with me ... wear a battered cardigan on the beach in Bournemouth ... marry me" (41) – he thinks to have found a more suitable partner in the more mature, more 'womanly' Anna. His justification for leaving Alice even though he still loves her, as he admits, is revealing: "Because ... I think I'll be happier with her" (51). As a matter of course, he puts his personal happiness before that of a person he claims to love, blatantly legitimating selfishness in matters of love. Dan is, however, not the only character to adopt this attitude. Both Anna and Alice quite openly admit their egoism when it comes to relationships:

Anna: I am sorry. I had a choice and I chose to be selfish. I'm sorry.
Alice: Everyone's selfish. I stole Dan from ... what was her name?
Anna: *thinks.*
Anna: Ruth.
Alice: She went to pieces when he left her. (81)

That putting one's own happiness before that of another, especially before that of a partner in a relationship, is not unanimously accepted as the moral standard is made clear by Dan when he tries to win back Anna from Larry: "If you love her, you'll let her go so she can be ... happy" (84). However, this idea of selfless love that seeks the other's good regardless of one's own – Frankfurt's 'disinterested concern' – remains only that: an idealistic idea, unsuitable to a world of absolute self-realisation. Even Alice, who easily appears to be the most virtuous character and a victim of the selfishness and insincerity of the other three (cf. Baraniecka, *Sublime Drama* 83), is not without blame in this matter. She, too, is willing to sacrifice her relationship with Dan, after they have come together for the second time, to a kind of self-realisation, which in her case consists in the maintenance of her mysterious personality. Until the very end she never tells him her real name, or where she got the scar in the form of a question mark on her leg, or the truth about her parents. Dan is not completely wrong when he says that Alice "and the truth are known strangers" (96), and it is her unwillingness to disclose her personality and her insistence on being in full control of what the other knows about her that eventually lead her to end the relationship: "I don't want to lie and I can't tell you the truth so it's over" (97).

The self-centred behaviour of the characters is not to be understood as expressive of their individual personalities. Rather, the play offers its own critical analysis of an entire society of which the characters are typical representatives. Saunders argues that "*Closer* rejects psychological verisimilitude" and that the description of the dramatis personae (e.g. "*a man from the city*," "*a woman from the country*") is reminiscent of the types and stock characters in Restoration comedy (21). Similarly, Innes holds that by "identifying each only as coming from town/suburbs/city/country," the character description implies that "their backgrounds cover the whole of English society" and that "they are clearly intended to be representative" (431). It is therefore justified to compare the love lives of Marber's characters with what sociologists have written about the love lives of people in Western societies around the turn of the millennium, to treat the play, as it were, as a snapshot of society through the lens of a playwright. In this snapshot, as Innes observes, "infidelity is indicated as habitual behaviour" (431). Anna is already divorced at the beginning of the play, Dan leaves his off-stage girlfriend Ruth for Alice, and the four characters betray one another and swap partners repeatedly. This volatility of relationships evokes the whole spectrum of what Illouz and others have called 'commitment phobia' – and especially the *hedonic* form of commitment phobia, where serious long-term commitment is avoided in favour of a series of exchangeable relationships (cf. Illouz, *Why Love Hurts* 78–87). As Bauman argues, people of our 'liquid modernity' fear nothing more than the limitations imposed on unrestricted self-realisation by commitment to a serious relationship. What they are thus opting for, instead, are "top pocket relationships," a term Bauman finds in the "'relationship' columns of glossy monthlies and weeklies and weekly supplements of serious and less serious dailies" (ix–x) and which denotes relationships the participants "'can bring out when they need them' but push deep down in the pocket when they do not" (x). Anthony Giddens' 'pure relationship,' which remains intact only as long as the partners are convinced that the profit they gain from the attachment makes the effort worthwhile, is another, quite similar form of refusing commitment. In accordance with these sociological views, the characters in the play do not hesitate to end their relationships as soon as they have reason to doubt the compatibility of the current partner or assume they will be happier with a new one.

This attitude is supported by what Illouz describes as an 'abundance of choice,' the availability of large numbers of potential partners, especially in modern urban areas, which puts lovers in constant doubt whether they have made the best possible choice, a looming suspicion that the grass may be greener on the other side of the fence. Put differently, the other is of special interest as long as he or she is merely a placeholder for all one's desires and preferences, but as soon as the other is discovered as a particular individual who fulfils only some desires and frustrates others, interest may shift to someone new. For David Ian Rabey, *Closer* depicts this avidity for the new as "potential tragic compulsions to which the audience might sense their own disturbing proximity and susceptibility" (201). In Rabey's adroit formulation, the play "suggests that it is indefinite possibility, rather than definite qualities, which attract, compulsively; that this attraction sparks the appetite for the sexually definite, compulsively; and that discovery of the definite ultimately separates people, compulsively" (201). Even though there are only four characters in *Closer*, the availability of potential partners is clearly stressed through the references to the characters' previous or new relationships or marriages. Also, internet dating, although still in its early stages, is indicated as a rising market for potential partners. Moreover, that the mere availability of seemingly better choices suffices to terminate relationships is directly stated in the play:

Anna: Dan left you, I didn't force him to go.
Alice: You made yourself available, don't weasel out of it. (79)

If availability of alternatives is the one factor that makes the relationships in *Closer* short-lived, the other is the characters' insistence on unrestricted self-realisation. As I have argued, Dan, Anna, and Alice all sacrifice their relationships at moments when they threaten to interfere with this project. Larry seems to be the notable exception, and what he says to Dan might just as well be directed to Anna or Alice: "You don't know the first thing about love because you don't understand compromise" (86). Indeed, the characters' unwillingness to compromise, to partly abstain from self-fulfilment for the sake of the relationship, seems to be the main obstacle to love. The demand for absolute freedom obviously outweighs the need for closeness and intimacy, and the characters lack the ability to find a middle way.

However, the play is far from naively idealising compromise as the universal solution to the problem of imperfect compatibility. Rather, it foregrounds the difficulty of finding a compromise that does not entail self-sacrifice. The relationship between Larry and Anna, to explicate this with an example, is marked from the beginning by an incompatibility that is felt both by characters and audience. In terms of looks and social standing, Larry seems to be 'not good enough' for Anna, something Larry feels uneasy about on several occasions. When Alice asks whether he is Anna's boyfriend Larry replies "A princess can kiss a toad" (37), as if Alice's question expressed the sheer improbability of Anna's interest in him. Larry feels inferior to Anna due to the contrast between his own humble background and her wealthy and distinguished family, and Anna does her bit to reinforce this feeling. On the night of her exhibition, when Larry meets Anna's parents for the first time, he wears a new pullover she has given him as a present.

Larry: I've never worn cashmere before. Thank you. I'm Cinderella at the ball.
Anna (*charmed*): You're such a pleb. (43)

Larry is concerned about the impression he has made on Anna's family, while Anna seems to find it hard to say anything positive about Larry's family.

Larry: So ... they didn't think I was ... an oik?
Anna: No. You're not ... you're you and you're wonderful.
Larry: Did you like my folks? They loved you.
Anna: Your mother's got such a ... kind face. (44–45)

Larry tries to fight the feeling of inferiority by filling the social gap between himself and Anna with money and a successful career – a plan that appears somewhat justified by the fact that the only description of Anna's ex-husband in the play is that "He made money. In the City" (22). For Larry, however, this adaptation to his partner's lifestyle, this attempt at creating compatibility, amounts to an act of self-renunciation that threatens his personality. His decision to go into private medicine runs counter to his personal ethical conviction and the loss of self-respect this involves becomes clear later when he virtually asks Alice's forgiveness for it: "I'm seeing my first private patient tomorrow. Tell me I'm not a wanker ..." (40). Larry's strange relationship to his bathroom, too, results from the decision to betray his personality in favour of a lifestyle he presumes necessary to maintain a relationship of which he feels unworthy. The expensive "*Elle Decoration* bathroom" he chose for his and Anna's apartment only enforces his inferiority complex: "[...] every time I wash in it I feel dirty. It's cleaner than I am. It's got 'attitude.' The mirror says, 'Who the fuck are you?'" (47). In the last scene, Larry tells Anna a detail about the night they split up: "I never told you this: when I strode into the bathroom ... *that night*... I banged my knee on our castiron tub. The bathroom *ambushed* me. While you were sobbing in the sittingroom, I was hopping around in agony. The mirror was having a field day" (101). In retrospect, Larry might see the bathroom's attack on his knee as a sign that their relationship had always rested on an unsound footing. Led by a feeling of inferiority, Larry has renounced self-realisation to a degree that harms his self-respect. He is right when he accuses Dan of having no idea about compromise but that does not mean that he is himself an expert in the field. What he might think of as compromise is in fact a unilateral betrayal of his very personality in favour of a lifestyle with which he cannot identify but which he deems necessary for his relationship. Thus, instead of simply denouncing the insistence on self-realisation observable in Dan, Alice, and Anna, the play rather intimates that if there is a hope for love it must lie somewhere in between absolute self-realisation and self-renunciation.

4.3.2 Postmodern Incredulity towards Love

The second reason why *Closer* is, in Christopher Innes' words, "aggressively contemporary" (433), is the way it captures the specifically postmodern ambiguous attitude towards love described by Belsey and Illouz, in which traditional concepts of love are simultaneously continued and contested, in which enormous hopes and desires are placed in a romantic narrative which is no longer believed, and in which irony and uncertainty prevent any sincere commitment to romantic relationships.

The first feature of this ambiguous attitude in *Closer* is the mentioning or direct juxtaposition of different concepts of love. There are, for instance, the sporadic allusions to the concept of courtly love, most openly in Scene One, when Alice thanks Dan for rescuing her from distress.

Alice: [...]
    Thank you for scraping me off the road.

Dan: My pleasure.
Alice: You knight.
Dan: *looks at her.*
Dan: You damsel. (7)

While Saunders argues that this borrowing from the medieval tradition "if anything only succeeds in accentuating competitive and painful feelings at the expense of the romantic" (61), I think this short exchange is above all an expression of the characters' awareness of the historicity of love concepts and of the ironic stance they take on love right from the beginning. The motif of courtly love here stands in for an entire series of traditional attributes of love generally termed 'romantic,' but not so much, as Saunders suggests, as a foil against which the dire reality of the characters' love lives is measured, but as a concept of love that persistently remains part of the discursive field and that shapes and influences the characters' expectations and desires, their ironic attitude towards it notwithstanding. In fact, the play frequently juxtaposes notions of love equivalent to the romantic and the realistic narratives of love described by Illouz. In the play, the two opposing concepts struggling for supremacy in the discursive field of love are labelled 'passion' and 'companionship,' but they are divided along the same line as the romantic and the realistic narratives in Illouz' study: spontaneous, impulsive, passionate love, in which the uncontrollable force of the attraction and the feeling of being 'made for each other' proves the couple's compatibility on the one hand, and a calmer form of love, in which compatibility and durability are the result of adaptation and compromise, on the other. The representatives of the two concepts are Dan, who ruthlessly follows his passion and justifies his actions with his own idiosyncratic version of romantic love, and Larry, who trusts in his ability to keep a relationship going by compromising and comprehending the other's personality. Dan adopts from the romantic concept its notion of love as an irresistible force beyond human will or control that connects two kindred spirits, but he does not equally assume its notions of uniqueness and eternality. He is thus able to romanticise his passionate desire for Anna while he is still in a relationship and, according to his own words, still in love with Alice. His romantic attitude surfaces when he spontaneously asks Anna to marry him instead of Larry, even though they barely know each other.

Anna (*laughing*): I don't know you.
Dan: Yes you do. I couldn't feel what I feel unless you felt it too.
Anna: I haven't seen you for a year.
Dan: Yes you have. We've bumped into each other in the street, twice. I manufactured it once, you the other.

4.3 *Closer* in Context: The Crisis of Love at the Turn of the Millennium — 105

Anna: And you just nodded.
Dan: I was scared you didn't feel it too. I felt guilty about Alice. Anna, we're in love, it's not our fault, stop wasting his time. (41)

Both the idea of a spiritual connection and the concept of love as passion – as something that is beyond control and befalls the helpless lovers – are clearly discernible in this declaration of love, and there is no reason to suspect that Dan does not, at least to some extent, believe in what he says. He is trustworthy when, in a later scene, he defends himself against Anna's accusation:

[...] Why did you swear eternal love when all you wanted was a fuck?
Dan: I didn't just want a fuck, I wanted you.
Ana: You wanted excitement. Love bores you.
Dan: No, it disappoints me. (73)

Dan seems to find himself in the 'trap of romantic love' described by Beck-Gernsheim, where exaggerated romantic expectations are disappointed in long-term relationships. Equating love with a passionate union of spiritually connected souls, he is not bored by it but disappointed that his own experiences do not, after a while, live up to his expectations.

His counterpart Larry, on the other hand, repeatedly ridicules the notion of romantic love. His description of his relationship with Anna – "We're in the first flush, it's paradise, all my nasty habits amuse her" (38) – already implies his scepticism towards romantic love and displays his awareness that the romantic bliss of their infatuation is not durable. Later, when Anna is living with Dan, he sarcastically refers to him as her "soul-mate" (77), indicating his disbelief in the romantic idea of kindred spirits. Against such metaphysical romanticism, he puts "[c]linical observation" (87) as the only method of getting to know the other. He considers himself a "clinical observer of the human carnival" (44) and thinks to know what both Anna and Alice need and want. When Dan comes to reclaim Anna in Scene Ten, he demonstrates the superiority of his idea of love over Dan's: "To a towering romantic hero like you I don't doubt I'm somewhat common but I am, nevertheless, what she has chosen" (85). And when Dan argues that "She's come back to you because she can't bear your suffering. You don't know who she is, you love her like a dog loves its owner" he drily replies, "And the owner loves the dog for so doing. Companionship will always triumph over 'passion'" (84).

It is quite characteristic of the play that both concepts of love fail in the end – neither Dan nor Larry stays with Anna for a longer period of time. But their failure is not necessarily an argument for the impossibility of love. Rather, it is their exaggeration or misinterpretation of the concepts, their stock-character-like inflexible fixation with opposing extreme positions, which terminates their love relationships. In Dan's case, his habit of following his passions unreservedly is inseparable from his total commitment to the project of self-realisation. Not to follow his heart, not to sacrifice one relationship for another which promises more happiness and self-fulfilment, would be a betrayal of this project. His romanticism is flawed by egoism and hedonic commitment phobia, and intimate relationships are thus as threatening to him as they are comforting. Larry, on the other hand, overestimates his abilities as a 'clinical observer': neither does he really understand what Anna needs nor is he right about Alice's scar, which he wrongly diagnoses as an act of self-mutilation due to some mental disorder following the death of her parents while it was actually caused by a bicycle incident, her parents being still alive (cf. 105). Moreover, what he sees as companionship and compromise is rather too close to self-abandonment. Where Dan shows not enough commitment, he shows too much, changing his career and lifestyle to please Anna. There is a kernel of truth in Dan's judgement that Larry loves Anna "like a dog loves his owner," and also in Larry's reply that this is why she loves him back. Of the two, Anna is clearly the stronger character, almost intimidating to Larry, who admires her and takes on a subordinate role in their relationship. Anna, on her part, enjoys the security this power imbalance affords her. In conversation with Dan she says about Larry, "I love him. He's a good man. He won't leave me" (41). It is not the thrill of passion that binds her to him but the feeling of security caused by a relationship she seems able to control. Anna's preference for security instead of passion is further indicated by her returning to Larry after her affair with the far less 'controllable' Dan, and by the news Larry breaks to Dan that "she never sent the divorce papers to her lawyer" (85) but always kept her husband Larry as a kind of emotional surety. When, ironically, Larry eventually leaves *her* for a younger woman, she fills this gap, quite consistently, with a real dog, a stray mongrel she "found [...] in the street, no collar ... nothing" (101), a companion, as it were, who enjoys a subordinate position and whose fidelity seems absolutely sure. Besides, that Larry seems indeed to have a general inclination to assume a 'dog-like' position in his relationships is at least implied when we learn towards the end that he gave up private medicine for his new girlfriend Polly (cf. 102) just as he had entered it for Anna in the first place, or when Alice tells Anna that Larry is "mad about" the scar on her leg and "licks it like a dog" (82).

Next to the juxtaposition of passion and companionship, the play provides a whole series of other concepts or discussions concerning love. Many of them are brought forward directly by the characters, who thus serve as representatives of the postmodern individual torn between multiple and often conflicting narratives of love. The concept of love as passion, for example, is not only addressed in opposition to companionship but also in opposition to the notion of love as a conscious and wilful decision. This notion, strongly reminiscent of Roland Barthes' "I never fall in love unless I have wanted to" (190), is held up most decidedly by Alice. While Dan, as we have seen, tries to convince Anna that their being in love with each other is not their 'fault' because love is an uncontrollable passion, and while Anna legitimises their affair with the same argument of having 'fallen' in love towards Alice (cf. 80), Alice is adamant that the human will is not completely without a say in matters of love:

Alice (*laughing*): That's the most stupid expression in the world. 'I fell in love.' As if you had no choice. There's a moment, there's always a moment; I can do this, I can give in to this or I can resist it. I don't know when your moment was but I bet there was one.
Anna: Yes, there was.
Alice: You didn't fall in love, you gave in to temptation. Don't lie to me.
Anna: You fell in love with him, too.
Alice: No, I chose him. I looked in his briefcase and I found this ... sandwich ... and I thought, I will give all my love to this stupid, boring, charming man who cuts off his crusts. I didn't *fall* in love, I chose to. (80–81)

Alice's trust in the role of the will is an important corrective to the romantic concept of passionate love – the idea of *amour passion* which is free from social and moral responsibility. True, the way intimate relations are depicted in *Closer* – childless and detached from families – makes them a realm in which the actions of the characters harm no one but themselves. But Alice's accusation is a strong reminder that this realm is not therefore a moral or ethical vacuum. By rejecting the idea that love suspends the will and thus any responsibility, she holds Anna responsible for the pain she suffered when Dan left her, just as she admits her responsibility for the suffering of Dan's previous girlfriend Ruth (cf. 81). That her 'decision' to love Dan was based on false assumptions – in the last scene Dan reveals that he had cut off the crusts "only that day ... because the bread broke in my hands" (105) – does not invalidate the argument. Nor does the fact that making love dependent on whether someone cuts off the crusts of his sandwich or not is anything but sound reasoning. What is important is Alice's acknowledgement that, no matter how overwhelming the infatuation, love always contains an act of will, a choice. Whether this actually is the "message" of the play, as Georgina Brown argues in her review (cf. 674), is questionable; but it is certainly true that Alice's argument serves as a counter narrative to the more popular concept of uncontrollable passionate love which is beyond moral concerns. For that reason, it is only partly true when Sierz says that

> [i]n the end, the key to *Closer* is its account of the irrationality of desire. Every time the characters swap partners, no reasons are given. There are no rationalizations for infidelity, no elaborate explanations. [...] Much of *Closer*'s power comes from the way it shows chaotic emotions without burdening itself with explanations. What disturbs is its reminder that, for all our reasonableness, we are at the mercy of unreasonable passions. (*In-Yer-Face* 194)

This analysis ignores that the play does at least indicate reasons for the characters' infidelities: Dan and Anna leave their partners because they feel culturally and intellectually closer to one another than to the waif-like stripper Alice or the 'plebeian' self-made doctor Larry respectively. Their explanations – "I think I'll be happier with her" (51) and "He understands me" (53) – may not be very elaborate but they show that Dan's and Anna's decision to swap partners is accompanied by reflection. Above all, however, Sierz' conclusion completely disregards Alice's notion of love as decision and adopts Dan's crooked argument of love as uncontrollable passion, which he uses to legitimise his infidelity. That 'we are at the mercy of unreasonable passions' is not the message of the play but Dan's (and also Anna's) argument to justify their infidelity and deny any moral responsibility. Alice's concept of love as choice or decision is at least a caveat against such a narrative that views love as completely separate from ethics or morality. Dan may subscribe to an eclectic version of romantic love in which passionate feelings prove the rightness of love and liberate from responsibility while other traditional romantic features such as uniqueness and eternity of love have no place. But Alice's counter narrative reminds us that Dan's concept of passionate love is not without alternative and that it may well be a pretext for avoiding responsibility.

Another concept directly addressed by the characters is that of love as the need to be needed. The first character who brings it up is Alice, when she explains to Dan that "Men want a girl who looks like a boy. They want to protect her but she must be a survivor. And she must come ... like a train ... but with elegance" (14). Alice, again, refers to the tradition of courtly love here, an allusion that becomes even clearer in one of her following statements: "I'm a waif. I appeal to your manly instincts? [...] You want to protect me from the ravages of the world?" (14). She evokes the traditional image of the noble courtier serving his lady, proving to her and to himself that he is useful, that he is needed. Alice, the fragile, almost ephemeral waif, embodies the need for protection which, she assumes, all men desire. In a Lacanian sense, Alice is thus able to 'give what she does not have,' that is, her desire, which is the very thing that man/Dan desires. Her claim that desirable women "must come ... like a train" amounts to the same argument. The lover, in this logic, will assume that the intensive orgasm results from the satisfaction of an immense desire, that this desire was for him, and that

4.3 *Closer* in Context: The Crisis of Love at the Turn of the Millennium — 109

only he could fulfil it. It is partly for this reason that, for the two male characters, the sexual details of their partner's affair are so important. Larry needs to know whether Dan is a better lover than he and whether Anna reached orgasm with him, and her confession that she did (twice), that she enjoyed it, and especially her brutally honest "IT TASTES LIKE YOU BUT SWEETER" (57) are devastating for him since they reveal that her sexual desire was not fixed upon him and that he is replaceable as the object of her desire. When later Anna tells Dan about her 'mercy fuck' with Larry, he too wants to know whether she had an orgasm, which she denies. The following dialogue discloses the significance this topic has for Dan:

Dan: Did you fake it?
Anna: Yes.
Dan: Why?
Anna: To make him think I enjoyed it, why do you think?
Dan: If you were just his slag, why did you give him the pleasure of thinking you'd enjoyed it?
Anna: I don't know, I just did.
Dan: You fake it with me?
Anna: Yes, yes I do. I fake one in three, all right?
Dan: Really?
Anna: I haven't counted.
Dan (*hard*): Tell me the truth.
Anna: Occasionally ... I have faked it. It's not important. You don't *make* me come. I come, you're ... in the area ... providing ... valiant assistance. (70–71)

The conversation leaves Dan utterly insecure. Not only can he not be sure whether Anna really faked it with Larry or rather "had a whale of a time" (73), as he suspects, he also has reason to doubt his own central position in Anna's system of desire: she still seems to have feelings for his rival and his own image as the sole object of her sexual desire crumbles with the information that he does not 'make her come.' From a theoretical perspective, however, the most interesting passage of this scene is Anna's justification for her 'sympathy fuck' with Larry: "If Alice came to you ... desperate, in tears, with all that love still between you; and she needed you to want her so that she could get over you, you would do it" (70). This is the play's most explicit reference to the need to be needed, to the significance of the feeling of being wanted for a sense of self-worth and desirability. It implies that sex with Anna was for Larry not a matter of lust or desire for the other's body but a matter of being desired, and it implies that this pattern of desire is not restricted to Larry but rather universal.

The motif of the need to be needed in the form of desire for the other's sexual desire is taken up at least two more times in the play. In the famous chatroom scene, for instance, Larry unveils his "sex-ex fantasy," a sexual fantasy with his exgirlfriends fighting over him, and Dan, posing as Anna, arouses Larry with his 'self-description' as a desperately needy nymphomaniac (26–27). Both sexual fantasies have at their core the (male) desire to be women's object of sexual desire. Later, in the strip club scene, Larry and Alice display a typically postmodern awareness of this structure of desire, simultaneously doubting and accepting it. Larry asks whether Alice is turned on by stripping.

Alice: Sometimes
Larry: Liar. You're telling me it turns you on because you think that's what I want to hear. You think I'm turned on by it turning you on.
Alice: The thought of me creaming myself when I strip for strangers doesn't turn you on?
Larry: Put like that ... yes. (59)

Again, Larry admits that what he desires (i.e. what turns him on) is her desire (i.e. her being turned on). And while he is already turned on by the idea of her being aroused by stripping for strangers – again the image of the needy nymphomaniac – he still seeks to 'personalise' her desire, to make himself its specific object. He is flattered by the idea that in flirting with him she is "breaking all the rules" of the club, which she is not, and when she asks him a question (instead of only replying) he thinks he has found "a chink in [her] armour" (61) since this might indicate interest in him as a person and not as a customer. His desperate attempt to wring her real name from her follows the same objective of a personal relation, and his proposal "Come home with me, Alice. [...] Let me look after you" (64) and his "Put me out of my misery, do you ... desire me? (65) evince his longing for an affirmation of being needed and desired. Alice's answer that she does not desire him is the second time in the play Larry is confronted with his own lack of desirability, which is further emphasised by the fact that his reply "Thank you. Thank you sincerely for your honesty" (65) is an almost exact repetition of his reaction to Anna's confession that she finds Dan sexually preferable (cf. 57).

That the need to be needed is especially precarious because its object – the other's desire – is so totally beyond control is a lesson frequently repeated in the play. Maybe its clearest expression is to be found in the relationship between Dan and Alice, on a less sexual and more emotional level. While their budding romance at first seems to support Alice's theory that what Dan wants is someone to protect and care for, he soon reveals to Anna that he "can't be her father any more" (42), and when he leaves Alice for Anna he claims that "it's because ... she

[Anna] doesn't need me" (48).³⁵ Dan seems to be weary of being needed – or at least of being needed by Alice. At the same time, after the failure of his novel and the death of his father, he feels a need for emotional support which he denies Alice to provide. When he refuses her to accompany him to his father's funeral, Alice quite clearly recognises this as a refusal of love: "I want to be there for you. Why are you ashamed of me? / [...] I love you, you fucker, why won't you let me?" (36). Despite his emotional neediness, he refuses to 'give' it to her. Instead, he asks Anna to come with him to the funeral, and it might also be possible to speculate that he hopes for more understanding as a "Failed novelist" (40) from the intellectual Anna than from Alice. In any case, the other's need or desire is presented as an utterly precarious object of desire, changeable, unpredictable, and uncontrollable.

Another example for the play's postmodern treatment of love is the way Dan and Alice narrativise their relationship. In Scene Eleven, long after their first break-up and shortly before their second, we see Dan and Alice in an apparently routine game of question and answer, recounting the story of their first encounter.

Dan: What was in my sandwiches?
Alice: Tuna
Dan: What colour was my apple?
Alice: Green.
Dan: It was red.
Alice: It was green, I ate it, I know.
Dan: What were your first words to me?
Alice: 'Hallo, stranger.' Where had I been?
Dan: Dancing then Smithfield then the buried river.
Alice: The what?
Dan: You went to Blackfriars Bridge to see where the Fleet river comes out, the ... swimming pig ... all that.
Alice: You've lost the plot, Grandad. (92)

The scene exemplifies how, in a thoroughly Hayden Whitean sense, narrativisation borders on fictionalisation due to subjectivity, selectivity, and failing memory. Dan has difficulties getting all the details right (the apple *was* green) and he mixes up Alice's story about watching the meat being unloaded at Smithfield Market with a story Anna told him during their photo session about the Fleet

---

35 Ironically, Anna tells Larry that she is leaving him for Dan because "I need him" (53), foregrounding misconceptions about the other's needs and desires as another major theme of the play.

river, which had been built over ('buried') in the eighteenth century and from which "one day this big fat boar swam out into the Thames" after it had "escaped from Smithfield" (16). Alice, for her part, remembers the details of their first conversation meticulously but is consciously selective with regard to the downside of their relationship, first blocking out and then diminishing the period they both spent with different partners:

Alice:	[...] Hey, when we get on the plane we'll have been together four years. Happy anniversary, Buster.
Dan:	What about ... the gap?
Alice (*correcting him*):	Trial separation. (92)

Similarly, she is, or purports to be, oblivious to the presence of Larry during the first moments of their love story:

Dan: [...] Do you remember the doctor?
Alice: No ... what doctor?
Dan: There was a doctor. He gave you a cigarette.
Alice: No. (93)

However, whether the details are right or not, the main goal they try to achieve by narrativising their memory is the creation of a shared love story, a proper romance, that is sharply demarcated from the ordinariness of the everyday and that serves as a kind of retrospective reassurance that their love is quite exceptional. In a fashion closely resembling Illouz' description of lovers' construction of temporal, spatial, artifactual, and emotional 'boundaries' in order to mark and delimit the special character of their relationship (cf. *Consuming the Romantic Utopia* 113–18), Dan and Alice set off the beginning of their love from the ordinary by emphasising the spatial and temporal extraordinariness of their encounter (its adventurous nature, its 'accidentality,' its suddenness) and by elevating ordinary objects that are invested with special meaning because they are somehow related to their love (the tuna sandwich, the apple, their first words to each other). This process of mythologising the beginning of their love culminates in Dan's "I saw this face ... this vision ... and then you stepped into the road. It was the moment of my life" (95). The whole scene smacks of postmodern irony and scepticism because it so consciously depicts the personal romantic love story as a form of (wilful) self-deception, a retrospective and selective embellishment of the facts into a romance that stands out in sharp relief against the 'reality' the audience has just witnessed (and is going to witness further in only a few moments when Dan and Alice break up for the second time).

4.3 *Closer* in Context: The Crisis of Love at the Turn of the Millennium

The scene is thus more than slightly reminiscent of Roland Barthes' assessment of such personal love stories:

> There is a deception in amorous time (this deception is called: the love story). I believe (along with everyone else) that the amorous phenomenon is an 'episode' endowed with a beginning (love at first sight) and an end [...]. Yet the initial scene during which I was ravished is merely reconstituted: it is after the fact. I reconstruct a traumatic image which I experience in the present but which I conjugate (which I speak) in the past. [...] Love at first sight is always spoken in the past tense: it might be called an anterior immediacy. [...] [W]hen I 'review' the scene of the rape, I retrospectively create a stroke of luck: this scene has all the magnificence of an accident: I cannot get over having had this good fortune: to meet what matches my desire; or to have taken this huge risk: instantly to submit to an unknown image [...]. (193–94)

The retrospective idealisation of the first encounter, according to Barthes, is what creates the sense of having 'fallen' in love – the 'rape' as he calls it here to emphasise the alleged passivity of the 'falling' individual – and it aggrandises the event so that it may serve as an explanation for this improbability called love. Both Dan and Alice need the story of this magnificent accident: Dan, because without it he lacks the security that meeting Alice was really a 'stroke of luck' and that she really 'matches his desire,' and Alice because she needs an explanation why she has 'taken this huge risk' of submitting to the image of the man who cuts the crusts of his bread. Even though Alice claims not to believe in the concept of falling in love, she believes, or wants to believe, that her decision to love Dan was justified, and she deduces this justification from their idealised romantic love story. In the first scene, they had both parodied the language of courtly love, one of the paradigms of romantic stories, displaying their awareness of and ironic distance from this tradition; now they both create and cherish their own romance. This is precisely the ambiguous attitude between scepticism and adherence with regard to the romantic narrative that Illouz had discovered in her interview study.

Closely connected to this ambiguity is the last example for *Closer*'s embeddedness in a postmodern *zeitgeist:* the characters' treatment of romantic language. Nick Curtis writes that "[t]he language of love is the oldest, most devalued dramatic currency in the book. But Marber's play [...] has a depth and ruthless contemporary edge that makes you look at even the most cliched expression of affection with a fresh, sometimes startled eye" (677). If this is true it is for an important reason Curtis is not mentioning: it is because the characters themselves problematise the clichéd language of love, bemoaning either the impossibility of authentic (self-)expression or the emptiness of the hackneyed old phrases. However, the use of language is inevitable. As Barthes describes it, "[w]hether he seeks to prove his love, or to discover if the other loves him, the amorous subject has no system of sure signs at his disposal" (214). Lacking unmistakable signs to prove love, the lover thus "falls back, paradoxically, on the omnipotence of language" (215). But then, saying 'I love you' is anything but a clear, unmistakable message, as Barthes elaborates elsewhere. "Once the first avowal has been made," he writes, the expression "has no meaning whatever; it merely repeats in an enigmatic mode – so blank does it appear – the old message" (147). Moreover, the phrase is devalued because it has lost all authenticity through its omnipresence in culture and media: "every other night, on TV, someone says: *I love you*" (151). The characters in *Closer* are postmodern lovers in that they are fully aware of this dilemma: that they ultimately need language to assure each other of their love but that this language is old, worn out, inauthentic, citational. There are numerous examples of the characters' reflexivity concerning the citationality of their language. In the same scene where Dan and Alice parody courtly love ("You knight"; "You damsel"), Dan tells her about the moment when she first became aware of him: "You were lying on the ground, you focused on me, you said, 'Hallo, stranger'" (7). Alice's reaction to hearing her first words to Dan is "What a slut," immediately recognising how much the phrase sounds like taken from a saccharine dime novel or romcom. Later, Dan cuts the line "She has one address in her address book; ours ... under 'H' for home" from his novel because it is "[t]oo sentimental" (19). Even though the line most probably describes Alice accurately, Dan is afraid that the truth may sound too much like a sentimental cliché to be credible or authentic. He faces the same problem even more pressingly when he tries to declare his love to Anna:

Dan: [...] I cannot live without you.
Anna: You can, you do.
Dan: This is not me, I don't do this. Don't you see? All the language is old, there are no new words ... I love you, I 'fucking' love you. I need you. I can't think, I can't work, I can't breathe. We are going to die. Please. Save me. [...] (42)

Dan despairs over his incapability to express his love authentically. Inserting a 'fucking' in the ordinary phrase does not make it an authentic expression of his individual feelings – what he lacks is a language of his own, which is, of course, an impossibility. He thus simply strings together a series of well-known phrases, marking them, as it were, as quotations from the pre-existing discourses of love, as old words he is forced to use because there are no new ones. His awareness of the citationality of the language of love, especially in front of the well-educated Anna, makes Dan a truly postmodern lover in the sense of Umberto Eco's illustrative description of the postmodern condition. In the postscript to *The Name of the Rose*, Eco writes:

4.3 *Closer* in Context: The Crisis of Love at the Turn of the Millennium — 115

> I think of the post-modern attitude as that of a man who loves a very cultivated woman and knows he cannot say to her, 'I love you madly,' because he knows that she knows (and that she knows that he knows) that these words have already been written by Barbara Cartland. Still, there is a solution. He can say, 'As Barbara Cartland would put it, I love you madly.' At this point, having avoided false innocence, having said clearly that it is no longer possible to speak innocently, he will nevertheless have said what he wanted to say to the woman: that he loves her, but he loves her in an age of lost innocence. (17)[36]

Though Dan does not reference a specific author, he resembles Eco's postmodern lover in deliberately signalling the loss of innocence in language which, unlike in the playful reference to courtly love in the flirtation with Alice, is troublesome now that he longs for an authentic and credible expression of his feelings for Anna. The best he can do to underscore the sincerity of his message is to draw attention to the unavoidable insincerity of the code and to prevent even the semblance of 'false innocence.'

The citationality of language is foregrounded again when Larry is left by Anna and warns her not to fall back on clichéd phrases to justify her decision ("Don't say it, don't fucking say 'You're too good for me.' I am but don't fucking say it") but at the same moment makes use of equally platitudinous formulations adopted from pop psychology ("You're making the mistake of your life. You're leaving me because you think you don't deserve happiness" [55]). Finally, Alice expresses her disappointment about the emptiness of the language of love and particularly Dan's "I love you": "Show me. Where is this 'love'? I can't see it, I can't touch it, I can't feel it. I can hear it, I can hear some *words* but I can't do anything with your easy words" (98). Words cannot take the place of 'signs' or 'proofs' for love for Alice, and this is not only because Dan has already demonstrated the worthlessness of his words and promises when he left her for Anna, but also because Alice, too, senses the lack of authenticity inherent in a language of love that is necessarily borrowed. The scene of their final break-up recalls the question of authenticity one last time:

Alice *spits in his face. He grabs her by the throat, one hand.*
Go on, hit me. That's what you want. Hit me, you fucker.
Silence.
Dan *hits her.*
Silence.
She stares at him- He looks away.
Alice: Do you have a single original thought in your head? (99)

---

[36] Illouz cites parts of this quotation for her description of the "postmodern romantic condition" in *Consuming the Romantic Utopia* (cf. 179). The German edition *Der Konsum der Romantik* contains the full passage as cited here in German translation (cf. 183).

Dan could not write a fictional novel without using Alice as his inspiration, and now he cannot end his relationship without an act of violence that is not only inspired by Alice but that is also, despite the fact that it happens frequently in reality, a worn-out cliché. Fiction and reality, cliché and authenticity, media images and private lives – in *Closer* these spheres are shown to flow into one another, leaving characters and audiences in doubt as to which degree an authentic experience of love is possible.

4.4 Timeless Issues: Compensation and Precariousness

While the existence of competing concepts of love and the (resulting) incredulity towards love are addressed more or less directly in the play, the two substantial narratives of compensation and precariousness are subtly woven into the play on a less superficial and visible level. With regard to love's compensatory function, I have already described the characters' loneliness and isolation, and it seems quite natural to read their quests for intimate relationships as manifestations of a desire to compensate for feelings of desolation and emptiness. Dan's unresolved trauma after the death of his mother (and later also father), Alice's sadness about life (cf. 37), Larry's performance as a "sad fruit machine spewing out money" (65) in the strip-club, Anna's lone visit to the zoo on her birthday, her photography of "sad strangers" (37) and "[d]erelict buildings" (79), the prospect of her becoming "a sad person" (105) living alone with her dog, and many other little details contribute to the profound atmosphere of sadness permeating the entire play, which suggest that the characters' loneliness is a source of emotional pain they seek to overcome.

Moreover, for Dan and Larry love works also as compensation for a lack of recognition that results in a diminished feeling of self-worth. Larry, the self-made man with the inferiority complex, who resents that his professional career does not win him recognition in polite society and who at the same time despises himself for striving after this recognition, solves the dilemma through being loved by Anna. Being chosen by another person and loved for the person he is boosts his self-worth and gives him the emotional and ontological security that comes along with this form of recognition. The effect is downright palpable in the scene in which Dan asks Larry to give Anna back: Larry is beaming with self-confidence, looking down on Dan as the loser in their competition. Dan, for his part, is grappling with his sense of self-worth due to his unsatisfactory occupation as obituarist and, later, his status as "[f]ailed novelist" who "needed praise" (49) but was denied the longed-for recognition. For him, too, it is Anna who provides the necessary recognition, loving him for what he wants to be loved for – a romantic intellectual.

But the compensatory function of love as a provider of recognition and self-worth has, of course, its precarious downside. If the characters make their sense of self-worth dependent on their intimate relationships as the realm where they are recognised and loved not for the social roles they may put on but for their 'essentialised' selves, then rejection in this realm is an immediate attack on the very foundation of any feeling of self-worth. As Illouz has elaborated, failure in romantic affairs and relationships is then perceived as a withering assessment of the core of one's personality and individuality as not worthy of love and recognition. The play's concentration on the four characters enforces the impression that, in the different couples they form, they are to one another their only sources of affirmation. Anna and Larry are professionally successful as artist and medical practitioner, but their feeling of self-worth depends on recognition in intimate relationships and is severely damaged when they are left by their partners. Larry turns into "Happy Larry," a pathetic, grief-stricken regular at the strip club, and Anna, towards the end of the play when Larry has left her for "someone younger," is on the verge of becoming a sad loner that "give[s] [her] love to a dog" (102). They both cannot transform professional success into a source of self-worth without being recognised 'essentially' in a close relationship. For Dan and Alice, who both lack professional success in the first place, recognition in romantic relationships is arguably even more important. As I have suggested above, one reason for Dan to leave Alice although he still loves her might be the desire to be recognised for what he takes to be his true self – a kind of romantic intellectual – and not only (or no longer) for his role as a protector figure. Alice, finally, represents the wish to have one's 'essentialised' self recognised in its purest form. Moreover, as I will argue, her case sheds light on the dangers and problems that inhere in the very process of making the self an object of acceptance or rejection. Preceding the threat of rejection, the project of opening towards the other necessarily breaks the protective cover of the self.

The precariousness of opening up is embedded in the broader theme of romantic and sexual relationships as power struggles, which many critics have identified in their reviews as one of the play's central issues. For Michael Billington, the play is a "candid, scathing, very modern view of the sex war" in which the partners inevitably "remain out of synch" due to the "extraordinary physical and emotional gulf between men and women" (678). Jane Edwardes, too, recognises the "gulf between men and women" as a focal theme of the play (678), and Sierz, who analyses *Closer* along with three other plays under the heading "Sex Wars" in his *In-Yer-Face Theatre*, holds that it "emphasizes the differences between the sexes" and that "[i]n almost every skirmish, men come off slightly worse than women" (192). Likewise, for Saunders a central concern of the play are the fierce "struggles for power" in its "pessimistic depiction of relationships between men and women being essentially antagonistic" (5). None of these critics, however, specifies what these differences between the sexes exactly are that the play allegedly brings to the fore. There are, of course, differences between the male fantasies about women and the frustrated female expectations concerning men the play thematises – but it is reasonable to see them not so much as examples for the gulf separating men and women but rather as variations of the same, shared phenomenon: the tendency to appropriate and impose an ideal image onto the other. This, indeed, is a major theme in the play, which thus showcases the precariousness of love in both its forms, not only as the risk of losing the beloved but also as the risk of losing the self and falling victim to possession.

In Chapter 2.2.2, I have discussed the notion of love as a potential threat to the integrity and sovereignty of the self in some detail. Regarding *Closer*, it is above all the objectifying effect of 'the look' as described by Sartre that finds expression throughout the play. In the perception through the other, there is always the looming threat of being reduced to the state of an object that can be possessed or absorbed, and while in the ideal love relationship the urge to possess or absorb the other is suppressed in order to preserve his or her free subjectivity (which is needed for the desired recognition and affirmation of one's own self), in reality love for Sartre always remains a struggle between two free subjectivities trying to appropriate the other. Like Sartre, de Beauvoir, Firestone, Levinas, and Badiou all warn in quite similar fashion against the tendency in human (love) relationships to absorb, appropriate, or impose a restrictive interpretive frame onto the other. The play now translates these abstract notions of violation of the other's free subjectivity, which for the quoted thinkers are often located more on an ontological than on an ontical level, into concrete images or stories.

Anna's photography of Alice, for instance, both appropriates Alice for Anna's professional career and imposes an interpretive frame on her. In *Sublime Drama: British Theatre of the 1990s* Elżbieta Baraniecka analyses *Closer* with reference to the Kantian concepts of the sublime on the one hand and the agreeable and the beautiful on the other. For Baraniecka, Anna's portrait of Alice is an example for "how the other is often subsumed and domesticated by the consumerist and cool strategies of the agreeable, which are often presented under the label of the beautiful" (17) – strategies that make the object of perception familiar, understandable, and consumable at the cost of extracting it from the realm of the sublime where it is characterised by its inexplicability, indeterminacy, and mystery (cf. 78–81). The portrait and the whole exhibition of "sad strangers photographed beautifully" are not only "a big fat lie" because "the people in the photos are sad and alone but the pictures make the world seem beautiful" (Marber 37), they also exemplify how the indeterminacy of persons is destroyed by the objectifying 'look' of artist and visitors who frame the portraits as 'beautiful' and as 'art.' As Baraniecka argues,

> [w]hat Alice […] is visibly hurt by is the feeling of being appropriated by the others' 'gaze,' first by Anna's taking the picture of her, and then by the visitors at the exhibition. Her existence is stolen from her and given a beautiful form, and through this form, it is made agreeable and consumable. Commodified, her life and her appearance lose their indeterminacy and sublimity for us. What should remain ungraspable is grasped and presented in a pleasurable form. (79–80)

Such metaphors for the objectification and appropriation of the other abound in the play. Most prominently, Dan literally frames Alice between the covers of his novel or, as Robert Butler puts it in a phrase that evokes an image of violent appropriation, "he has cannibalised her life for his first book" (677). Later, in the strip club, it is Alice again who is objectified through a 'male gaze.' This time it is Larry who bluntly tells her that, as a stripper, "you think you haven't given us anything of yourselves. […] / But you do give us something of yourselves: you give us imagery … and we do with it what we will" (Marber 66). In both cases, the image of Alice is turned into an object ready for possession and usage at will. But while she is little bothered by being used as an object of sexual fantasies of anonymous strangers, Dan's limitation of her self through confinement to the interpretive frame of a novel is problematic. "Unlike the men's limiting gaze at the lap dance club," Baraniecka writes, "Dan's appropriation of Alice's image hurts her deeply. She wanted from Dan love and recognition of her existence in its indeterminacy, for what it was. Instead she was treated yet again as a service provider, equipping Dan with an image he could use to his advantage" (92).

It would be misleading, however, to see in Alice nothing but the innocent victim of Dan, Anna, and Larry, the character with "the least unscrupulous ego," as John Peter suggests (679), or the only character who "remains honest, and […] pays for it," as Curtis puts it (677). True, Alice is the only character in the play who does not betray one of her partners; but apart from that, she shows no more commitment to her relationships than the others. Her principle "It's the only way to leave, 'I don't love you any more, goodbye'" (13), which she applied to a former partner and applies to Dan later in the play, is as cold as it is honest. This and her reaction when Dan breaks up with her – "I'm the one who leaves. I'm supposed to leave you. I'm the one who leaves" (51) – shows that she has no scruples leaving someone who still loves her if she no longer feels the same. There is no reason why Dan's decision to leave Alice for Anna should be morally inferior to Alice's 'decision' to stop loving Dan. Alice is also not only a victim of the limiting gaze of others but views others with the same objectifying 'look' that imposes fixed interpretations on them. Her claim to "know what men want" (13) is one example, her framing of Dan in the opening scene is another. As Baraniecka writes, "[s]earching through Dan's briefcase, she creates an opinion about him by choosing to believe what she herself needs or wants to believe. The idea or the image of Dan cutting off his crusts must reflect Alice's most intimate needs or longings, which she projects on Dan and therefore hopes to satisfy through the relationship with him" (82). Even though we can only speculate about the precise nature of these needs or longings, Baraniecka's assumption is convincing that "[t]he idea of Dan's cutting off his bread crusts is an extremely warm image, associated with the security of home and loving parents. And this seems to be what Alice is in desperate need of" (82). Alice thus imposes on Dan and on men in general an image as providers of security and protection, and this image is as much rooted in (justified) assumptions about male desires and fantasies to assume this role as it is a projection of Alice's own needs and desires. When Dan complains that he "can't be her father any more" (42) he indicates that he indeed perceives this imposition of a fixed image as a limitation of his free subjectivity.

For all this warning against a threatening loss of self through the other's attempts at objectification, limitation, absorption, and possession, the play does not argue for the strict maintenance of absolute indeterminacy and alterity. In two crucial scenes, Larry and Dan demand to know the complete truth from Anna and Alice about their sexual relationship with the other man. In the first case, it is certainly reasonable to see this as Larry's attempt at gaining power over Anna. The more he knows (about) her the more he can feel in control over her – or at least his image of her. As Sierz writes: "It is also a way of possessing her even as he loses her. Knowing all there is to know is less about understanding than about winning" (*In-Yer-Face* 193). In the second case, however, when Dan wants Alice to tell him what he already knows from Larry, it is at least as much about power and control as about something else: a minimal degree of honesty and openness necessary for and in love. Before questioning her about Larry, Dan wants to know – once more – how Alice got her scar and wants to have a look at her passport, but Alice persists in sealing off her past even from those who are closest to her. Baraniecka interprets her refusal to tell anyone her real name and disclose her past even within a relationship as her strategy to block out all information that could be used to construct a defining narrative around her. It expresses "the unrealistic demand to be loved and seen exclusively in terms of her absolute presence in the here and now without a past or future which would determine her identity" (92). Following this reading, Alice keeps up her enigmatic identity in order to provide less material for an objectifying or limiting interpretation. In offering only her here and now she hopes to be recognised only for her immediate presence in all its unexplained indeterminacy. As Baraniecka argues, only when being loved in this way, Alice believes, can she hope to avoid abandonment which is the inevitable result of a love that, instead of recognising indeterminacy, is focused on the definite and slowly becomes bored or frustrated with it (cf. 94).

Alice's hope for a love that does not strive to know and understand the other – a love that is love precisely because it embraces the other's indeterminacy – is reminiscent of Levinas' vision of a "relationship with [...] the very dimension of alterity" (Levinas 50), a relationship that refrains from any attempt to "possess, grasp, and know the other" (51). What Alice demands is that her lovers adopt exactly this attitude of loving her as an absolute alterity. This demand is, however, impossible. For Baraniecka, Alice "sets conditions for a relationship which are too idealistic, absolute and therefore impossible to fulfil. For, becoming closer to one another necessitates giving to the other not only your presence but also your past and future" (92). Knowing each other – and allowing the other to know oneself – are prerequisites of closeness and intimacy, even if a lover should resist the urge to impose a definition or interpretation on the beloved.

> The relationship between Dan and Alice fails because neither of them is willing to admit the impossibility of their demands. Dan is incapable of recognizing the irreducibility of Alice's otherness and creating a more hospitable space for her. Alice, in turn, appears to be so terrified by the possibility of being imprisoned in a frozen image, used and discarded that she does not want to recognize the fact that intimacy requires taking the risk of opening yourself to the other. (94–95)

In this apt analysis of the relationship between Dan and Alice, Baraniecka already provides an answer to her logical question whether it is "possible to be intimate with the other without destroying its otherness by trying to get to know it better" (17). After all, neither Dan's demand for knowledge nor Alice's demand for indeterminacy seem to lead to successful relationships. The answer lies in the phrase "the risk of opening yourself to the other." What is harmful to their love is not Alice's demand that Dan should refrain from defining and interpreting her but her refusal to open herself voluntarily. Her attempt to keep full control over her self and others' image thereof is an attempt at taking the risk out of love – the risk which for Firestone, Badiou, Nancy, and Nussbaum is its necessary condition. If love is supposed to be passionate, intimate love for a concrete person and not an abstract idea, then the object of love must be known – not defined, not limited, but known. But knowing the other without merely speculating or imposing ready-made images is only possible if the other takes the risk of opening up. If there is love, Nancy says, the self is no longer detached, self-possessed, and sovereign but it is opened, exposed, and vulnerable. Love is and has to be, as Firestone puts it, "a situation of total emotional vulnerability" (249). Only if the self is opened and exposed to the other is there a chance for true intimacy because only then the other can love this particular and genuine self and not an abstraction, idealisation, construction, or imposed image. Absolute knowledge of the other is impossible, and any attempt at absolute knowledge can be no more than an objectifying, limiting, imposed definition. But love is not about absolute knowledge or a definition of the other. It is about being allowed into the sphere where the other is vulnerable, where absolute indeterminacy turns into partial knowledge, and where control over one's image is replaced by exposure. Baraniecka is thus right when she writes that "Alice's desire to keep her real identity absolutely indeterminate is [...] just as ineffective and just as inhospitable as Dan's claim to know her absolutely" (94). In her attempt to control the image of her self and to be recognised and loved only for her immediate presence, she avoids the riskiness and vulnerability love requires and she denies Dan the openness and exposure that is needed for true intimacy. It is the tragic irony of the play that Alice's behaviour of hiding behind an armour of enigmatic indeterminacy is understandable and justified through the behaviour of the men she encounters. Both Dan and Larry display a disposition of imposing ready-made images on her – Dan with his novel and Larry with his 'clinical observation' and (wrong) diagnosis. Alice's strategy of protecting herself against this limiting gaze by preserving absolute indeterminacy is understandable, but it is also counterproductive as it prevents true intimacy and leaves the men with only two options: to accept her absolute indeterminacy or to construct and impose their own images on her. Not surprisingly, they opt for the second.

4.5 Conclusion

Closer's treatment of love combines an unsparing analysis of the specifically contemporary and postmodern vicissitudes of sexual and affectionate relations with reflections about more universal concepts of love. The play offers an acute and thoroughly believable microcosmic detail of urban love at the end of the twentieth century that ties in with sociological descriptions of intimate relationships published around the same time (Beck/Beck-Gernsheim, 1990; Giddens 1992; Illouz 1997, 2007, and 2012; Bauman 2003; Kaufmann 2009). Alice, Anna, Dan, and Larry represent and display many of the features and attitudes frequently ascribed to postmodern and contemporary lovers, from their lack of embeddedness in robust familial, traditional, or religious structures, their apparent commitment phobia, and their insistence on happiness and self-realisation, to their ambivalent stance on romantic love between adherence and incredulity, their awareness of the problem of authenticity in communicating loving feelings, and the demand for recognition of an 'essentialised' self. If Alice's question "Why isn't love enough?" (51) when Dan, although still in love with her, ends their relationship is the central question of the play, the answer might be found in the sum of these features of the postmodern romantic condition. That the four characters sometimes appear like type-figures in a modern comedy of manners does thus not distract from the play's realism. Rather, in their sometimes exaggerated adoption of these attitudes they resemble the representative types construed in the necessarily reductive and schematic portrayals of sociological literature which, for the sake of comprehensive social analysis, neglects the peculiarities of individual psychology.

At the same time, the play allows for interpretations that dig beyond the surface of the sociological descriptions and show that behind the specifically modern difficulties of the characters lurk the same old concepts of love: their quest for love is fuelled by the sharply felt necessity to compensate for an experience of insufficiency manifest in their saddening isolation and desperate want for recognition of their selves through a partner in an intimate relationship, while the inevitable precariousness of love, which amounts to a threat of self-erasure for those who make their self-worth entirely dependent on this recognition, strikes with double force at the moment of its attempted suspension. Alice's project of remaining enigmatic and indeterminable is an (over-)dramatised version of making an 'essentialised' self the sole object of the other's love while trying to keep full control over this self-image and to protect it from all forms of analysis and appropriation. She wants to be loved exclusively for her pure immediacy, her absolute presence, her essential self – a person, as it were, that exists independent from her past, her background, and her position in society. But this strategy is just as harmful to love as is the limiting, objectifying gaze from which it is supposed to protect her. Denying her lovers the openness required for love, and thus denying them the possibility to love an image that is, although it may be limited, firmly rooted in her true self, she eventually forces them to impose a far more constructed image upon her. Surely, Dan and Larry are anything but 'ideal husbands' and it does not follow necessarily from their failure to love Alice in the way she wants to be loved that the play agrees with my evaluation of her concept as an outright impossibility. Maybe, one might read between the lines, a better lover could love her as an absolute alterity. In any case, however, and to give Alice the last word on the matter, "It's a big want" (14).

5 "if you're not with me I feel less like a person": Sex, Drugs, and the Myth of Self-Sufficiency in Mark Ravenhill's *Shopping and Fucking*

Like *Closer*, Mark Ravenhill's *Shopping and Fucking* (1996) is a child of its time, a play that is firmly rooted in the social context of the decade and, together with Marber's play and Sarah Kane's *Blasted*, "ranks as one of the most significant plays of the 1990s" (Saunders, "Mark Ravenhill" 168). And like *Closer*, it nevertheless remains topical because, on the one hand, its atmosphere of unbounded capitalism, consumerism, and egoism, in which love is engaging in an almost hopeless struggle, is not a specific feature of the 1990s but of modern late-capitalist society in general and, on the other hand, because within this atmosphere it evokes the timeless discourses of love's compensatory function and precariousness.

In many ways, and in many details, the characters of *Shopping and Fucking* belong to a specific generation or 'cohort' often called 'Generation X' or 'Thatcher's Children' and are representatives of the 1990s adolescent and young adult generation that inspired and produced the phenomenon of in-yer-face theatre. But, in a play that relies in no small measure on metaphorical and metonymical imagery, they also stand in for an entire society characterised by a sense of desolation and disorientation in a way that is not restricted to the last decade of the twentieth century. As far as love is concerned, the troubles of the characters are not those of a specific generation but those of a postmodern late-capitalist Western society. The 'Generation X,' depicted exaggeratedly but nonetheless faithfully in the play, is not only a sociological reality of the 1990s but a metaphorical image of a civilisation lacking security, closeness, and orientation – of a society, to use Amelia Howe Kritzer's description of the world created by Ravenhill, where "[e]motional emptiness points towards a deeper void in isolated lives that draw nothing from social, moral, or historical sources of meaning" (39). One argument underlying the interpretations in this chapter is that, despite all its experientiality typical of in-yer-face theatre, *Shopping and Fucking* is also a play of ideas exploring the possibilities, promises, and chances of love in such a society. After a brief synopsis, I will thus first examine the play's image of a disoriented and solipsistic consumer society before analysing its treatment of love. In the latter part, I am going to argue that in spite of its apparent postmodern playfulness regarding ideas and concepts from the discursive field of love, the play eventually adopts a surprisingly clear posture concerning sense and nonsense of some of these concepts and the question of ethical responsibility in love. "The dichotomy in critical response that the play has attracted has been one of its principal features" (170), Saunders writes, and it is thus not surprising that my final interpretation stands in direct opposition to that of Leslie A. Wade, who argues that Ravenhill advocates an ideal of self-sufficiency. In my view, quite to the contrary, the play dramatises the futility of any attempt at self-sufficiency, emphasising the inevitable neediness and interdependence of human beings, while stressing at the same time the precariousness inherent in this situation.

5.1 Synopsis

The play starts with an unsuccessful act of care and affection when Lulu and Robbie try to feed heroin addicted Mark, who has exhausted himself with an attempt of self-detoxification and vomits the takeaway food onto the stage. Mark tells his flatmates that he is going to leave them for a while to undergo a professional detoxification programme, which hurts Robbie, who is in love with Mark and hoped that the help he and Lulu could provide would be enough. The next scene shows Lulu in a job interview for commercial TV during which Brian, the interviewer, first lets her take off her jacket and blouse and then gives her a bag with three hundred ecstasy pills in order to test her selling skills. Meanwhile, Robbie has been attacked with a plastic fork by an angry customer and has lost his job in a burger chain store, which puts him and Lulu under some financial pressure and urges them to pull off the drug deal. Mark enters, who has been dismissed from the rehabilitation clinic because he violated one of the rules forbidding sexual relationships during the programme. After a first kiss with Robbie, Mark reveals that, his dismissal notwithstanding, he has decided to follow the idea of the programme and to avoid dependencies of all sorts, including not only drugs but also "emotional dependencies" (17). He thus rejects Robbie, with whom there is too much "baggage" and with whom kissing "would mean something" (19) and declares that he is going to restrain himself from emotional intimacy. In the next scene, Mark tries to put this plan into practice with teenage rent boy Gary. When Gary asks for Mark's sexual preferences, he explains that "[t]he important thing for me right now, for my needs, is that this doesn't actually mean anything, you know? Which is why I wanted something that was a transaction. Because I thought if I pay then it won't mean anything" (24–25). Soon, however, Mark forms an attachment to Gary, who tells him about his desire for a caring father, the brutal sexual abuse by his stepfather, and his fantastic hope of a mysterious "big" and "rich bloke" (26) who is "[l]ooking out for me.

He'll come and collect me. Take me to his big house" (42). Gary invites Mark to stay at his flat and they go shopping for designer wear together. When they kiss in the changing room, Mark realises that he has fallen in love with Gary who, however, does not feel the same way. In the meantime, Robbie has tried to sell the ecstasy alone because Lulu was too exhausted after witnessing a shop assistant being stabbed by another enraged customer in a convenience store. However, breaking the first rule of drug dealing – "He who sells shall not use" (31) – Robbie has given away all the ecstasy for free in a drug induced state of euphoria, benevolence, and anti-capitalism. As he tells the thunderstruck Lulu,

> I was looking down on this planet, Spaceman over this earth. And I see this kid in Rwanda, crying, but he doesn't know why. And this granny in Kiev, selling everything she's ever owned. And this president in Bogota or ... South America, And I see the suffering. And the wars. And the grab, grab, grab. And I think: Fuck Money. Fuck it. This selling. This buying. This system. Fuck the bitching world and let's be ... beautiful. Beautiful. And happy. You see? (39)

Brian grants them seven days to pay back the three thousand pounds Robbie has thus lost, not without showing them a video tape of "someone who failed his test" (50) being threatened (and probably killed) with a power drill. To raise the money, Lulu and Robbie set up an astonishingly thriving phone sex business (Lulu: "Why are there so many sad people in this world?" [52]) which, however, becomes emotionally too stressful for Lulu when she realises that one customer is masturbating to a security camera's videotape of the stabbing she had witnessed earlier. Mark and Gary enter the flat and when Gary appeases the jealous Robbie by assuring him that he is not in love with Mark but waiting for his mysterious stranger, Robbie sees a chance to secure the rest of the money. He offers Gary the opportunity to pay for a kind of sexual role play during which he can live out his fantasy of being blindfolded and brought into a room as sex slave to the mysterious bloke. During the 'game,' both Robbie and Mark rape Gary and Mark even starts to hit him, but it turns out that Gary's fantasy goes even further: "It doesn't end like this. He's always got something. He gets me in the room, blindfolds me. But he doesn't fuck me. Well not him, not his dick. It's the knife. He fucks me – yeah – but with a knife" (84). While this is too much for Lulu and Robbie, Mark eventually agrees to fulfil Gary's wish. In the last scene, Brian surprisingly allows Robbie and Lulu to keep the three thousand pounds they have collected because they have learnt the lesson that "Civilization is money. Money is civilization" (87) and the play ends in a final tableau reminiscent of the opening scene, but with a more optimistic touch, when Mark, Robbie, and Lulu *"take it in turns to feed each other as the lights fade to black"* (91).

5.2 "Why are there so many sad people in this world?" The Play's Society

In his 2004 essay "A Tear in the Fabric" Mark Ravenhill describes how, in hindsight, it was the abduction and brutal murder of two-year-old James Bulger by two ten-year-old boys in 1993 that turned him into a playwright and set the tone for his early plays. Not that he was writing directly about the case, but what he wanted to approach was "the feeling inside me – and the people around me – that the murder engendered" (309) because it served as a focal point onto which "a sort of public grief [was] projected [...] – grief and guilt for the decade that had passed," a decade that was above all characterised by "greed and neediness" and the preference of "self-interest" over "the public good" (309). In the video image of a security camera showing James being led away from Bootle Shopping Centre by his later killers, Ravenhill speculates, culminated the grief and guilt of a nation sensing that "enough is enough, something has to change" (309). In his own plays, at least, and particularly in *Shopping and Fucking*, he retrospectively discovers the haunting images of the Bulger murder: "Shops, videos, children killed by children" and the painful absence of parental guidance (312). Consciously influenced by North American contemporary novelists such as Douglas Coupland, who wrote "about middle-class kids whose life had no meaning, with an overwhelming death wish" (311), and subconsciously urged by the Bulger case as a deathly image for a society of disoriented or misguided children lost in a shopping centre, he created plays inhabited by such characters. "Nobody in these plays is fully adult. They are all needy, greedy, wounded, only fleetingly able to connect with the world around them. Consumerism, late capitalism – whatever we call it – has created an environment of the infant 'me,' where it is difficult to grow into the adult 'us'" (311–12). As he told Nils Tabert in an interview, in *Shopping and Fucking* Ravenhill intended to portray a generation that is ignorant of other values than that of the market (cf. Ravenhill, "Interview" 68–69), a point also emphasised by Howe Kritzer's observation of the characters' "sense of loss in the lack of identity, relatedness, and values not rooted in business or the profit motive" (39).

With market values as the only principles ordering their world, the characters in *Shopping and Fucking* have not much to root their existence in. "The people in the play are just trying to make sense of a world without religion or ideology," Sierz quotes Ravenhill. "They're kids without parental guidance – they're out there on their own having to discover a morality and a way to live as they go along" (qtd. in *In-Yer-Face* 130). Indeed, the absence of parents from the play is even more conspicuous than in *Closer*. While in Marber's play parents or other relatives are merely absent from the stage in order to stress the characters' urban isolation, in *Shopping and Fucking* they seem not even to exist. Only Lulu mentions her parents once when she tells Brian that they "spend Christmas together" (8), apparently the only family contact any of the four protagonists has apart from Gary's abuse through his stepfather. The characters' loneliness and abandonment thus seem to be of a more essential nature. Lulu, Robbie, Mark, and Gary are not only younger than their counterparts in *Closer*, they are also even less rooted in any form of community, family, or regular working life. As Sierz writes, "*Shopping and Fucking* offers a snapshot of Generation X. [...] Young people have been abandoned. However funky and uninhibited, they are dazed, confused and boiling over. With all adults corrupt, there is little to relieve the pain and the tedium except shopping and fucking" (*In-Yer-Face* 132). The corrupt adult world is represented by Brian, who is moved to tears by his son playing the cello and by Disney's *The Lion King* and who reminisces about the experience of beauty as a glimpse of lost paradise, but who candidly asserts that the one thing "behind beauty, behind God, behind paradise" is money (48). Brian represents the generation – or the part of society – that still holds on to master narratives: "We need something. A guide. A talisman. A set of rules. A compass to steer us through this everlasting night. [...] / Something that gives us meaning" (86) – and for him it is money which is deified and endowed with this guiding function. Lulu, Robbie, Mark, and Gary, on the other hand, belong to a generation/society that can only nostalgically look back to the days when master narratives offered guidance for entire nations or people. As Robbie says in an often-quoted passage:

> And I think a long time ago there were big stories. Stories so big you could live your whole life in them. The Powerful Hands of the Gods and Fate. The Journey to Enlightenment. The March of Socialism. But they all died or the world grew up or grew senile or forgot them, so now we're all making up our own stories. Little stories. It comes out in different ways. But we've each got one. (66)

The characters, then, are presented as unguided and disoriented in a double sense. For one thing, as Dan Rebellato writes in his introduction to the play, "[t]he steady dismantling of those social arrangements which might once have fostered our desire and ability to live together have left these characters without the common bonds to help them do so" (xi). Without families and without the capability to successfully connect, they are threatened by extreme isolation and loneliness and even if, as Wade observes, Mark, Lulu, and Robbie "have bonded to form a surrogate family" (110) their situation is a far cry from the security offered by intact family structures. Secondly, as Wade rightly points out, the characters experience the absence of orientation that is inherent in the postmodern condition: "No master narrative serves to enliven or organize the world

5.2 "Why are there so many sad people in this world?" The Play's Society — 129

of these characters, so it is incumbent on each to self-fashion, to reject dependencies of various forms, and to assume a validity that is self-proclaimed" (112). While I will later disagree with Wade that the situation thus described represents the "understanding of human connection and communal responsibility […] advocated by the playwright himself" (112), I think it is an apposite description of the situation the characters find themselves in at the beginning of the play: Robbie's speech sketches a society in which the incredulity towards master narratives gives rise to individual micro narratives, narratives that are not only free from collective ideological dogmas but are also thoroughly individualised and self-centred. And if the play, for this reason, "well dramatizes the confusions, impasses, and emotional vertigo of the postmodern condition" (Wade 109), it is necessary to observe that this is not only because the characters' living conditions are a microcosmic image of the postmodern condition, but also because the characters themselves are, rather than realistic individuals, metaphorical figures representing a society thoroughly suffused with this condition. Sierz hints at this figure conception when he first criticises Ravenhill's "unrealistic idea of character" but then claims that "[w]hat audiences did feel was that Gary was central to the play, a symbol of neglect, abuse, urban drift" (*In-Yer-Face* 131), or that Brian functions as a representative of the "grown-ups" who combine "a vestige of old values" with "the most excessive spirit of capitalism" (132). Viewing the characters as mainly metaphorical or representative is more satisfactory than arguing about the obvious lack of realism or plausibility in the play. Brian's rapid mood swings and role changes between loving father, sentimental uneducated art lover, ruthless capitalist, gangster boss, and generous donor, Robbie's rather unexpected lecture on master narratives, the inconsistency of Mark's wish "to know if there are any feelings left" (Ravenhill, *SF* 34) with his decision to avoid emotional commitment, the incredible success of Lulu and Robbie's phone sex business, a social worker who hands fourteen-year-old Gary a leaflet when he reports the sexual abuse through his stepfather – all these things are better seen as contributing to a figurative image of an entire society than as realistic individual actions. This also ties in with the author's own explanation of his aesthetic approach in the interview with Tabert, where he makes clear that the play was designed as an assemblage of extreme situations and characters that could work as universal metaphors (cf. 69). The simplicity, lack of context and subtext, and the tendency towards amplification in the presentation of the characters and their actions, he adds, were inspired by the aesthetics of fairy tales (cf. 74–75). Hence, Sean Carney is right when he holds that Mark, Robbie, and Lulu are not so much a group of three individuals sharing a flat but an image of society or humanity in general: "It isn't clear why their lives have fallen apart: this isn't realism, and we will find out little about these characters' histories. Heroin is mentioned, but on a less specific level what seems broken is the general ability of human beings to make meaningful connections with one another" (240).

Bearing this metaphorical nature of the play in mind, the treatment of love in the play can be analysed as what it is: not as realistic depiction of the words and actions of concrete individuals but as pointed dramatisation and visualisation of selected elements of the discursive field of love, spread out for scrutiny. The importance of love in *Shopping and Fucking* has been noted before. Sierz, for instance, writes that the optimistic ending insinuates "that love, mutual caring and the search for new values are possible" (*In-Yer-Face* 134) and Caridad Svich thinks that "Ravenhill at heart seems to be making a plea for a world in which love can transcend the violence and hatred of a society that has been run into the ground by the consumerist values of a wayward class" (405). What both are mainly focussing on, however, is not erotic love but the relation between Lulu, Robbie, and Mark, which indeed appears to have developed in the course of the play into a community of loving solidarity where they share, as Ravenhill himself points out, the individual-sized ready meals, the play's most frequent symbol of isolation, which Lulu had previously considered impossible to share (cf. Sierz 134). In the following pages, however, I will focus on another love story in the play – Mark's erotic love for Gary.

5.3 Addicted to Love

Starting with the iconic title, the constant alignment or equation of love, sex, and consumerism is maybe the most distinguishing trademark of the play. Through repeated coupling of images, it achieves a "thematic unity" (Sierz, *In-Yer-Face* 129) and creates "a world where sex is a commercial transaction and consumption sexually arousing" (128). Right at the beginning, Mark offers one of the most prominent examples for this coupling of images, when he tells Lulu and Robbie "the shopping story" (4), a fictitious account of how he purchased them both from a "fat man" in a supermarket and took them home to "live out our days fat and content and happy" (5), promising, as Lulu remembers, "I love you both and I want to look after you forever" (4). The motivation to buy two 'dependants' in order to care for them already indicates the desire to be needed, which will reoccur later as a crucial factor in Mark's emotional life. In any case, however, love and money are brought in close connection right from the start, and people are presented as commodities in this story, which is repeated with variations another two times in the play. The second and most central combination of love/sex and consumerism are Mark's 'transactions,' first with the offstage character Wayne during his stay at the rehabilitation clinic and then with Gary, manifesting his attempt to purchase erotic human contact without becoming emotionally dependent. Besides these two examples, there is a series of less directly significant moments indicating the seemingly inextricable connection of love, sex, and money: the phone sex episode, in which Robbie and Lulu sell sexual fantasies but also comfort and company for the multitude of "sad people in this world" (52); Brian's love for his son, which materialises as money for "boarding fees and the uniforms, the gear, the music, skiing" (49); the *"distant sound of coins clattering"* (25) that provides the background to Mark and Gary's meetings in the latter's flat above an amusement arcade; and the very fact that Mark confesses his love for Gary in a luxury fashion store.[37] While the sum of all these images engenders the specific atmosphere of the play in which even the most intimate feelings are in danger of commodification and "the degradation of human experience is portrayed as a seemingly ubiquitous, dystopian condition" (Carney 239), it is Mark's futile attempt to turn his amorous activities into mere transactions – and the motivation behind this experiment – that leads into the play's surprisingly profound exploration of erotic love.

When Mark returns from the clinic, he has thoroughly ingested the central tenet of the therapy that "emotional dependencies [...] are just as addictive" as drugs (17) and should hence be avoided. This curious conviction is, again, not so much the idea of a single individual but a pointed satire on a set of ideas circulating in pop psychology and therapy for which Mark acts here as a mouthpiece. As Sierz astutely observes, the play "makes frequent use of discourses," and just as Robbie copies the postmodern discourse about *grand récits*, Mark "parodies therapy-speak" (*In-Yer-Face* 132) when he explains his situation to Gary:

> I have this personality, you see? Part of me that gets addicted. I have a tendency to define myself purely in terms of my relationship to others. I have no definition of myself you see. So I attach myself to others as a means of avoidance, of avoiding knowing the self. Which is actually potentially very destructive. For me – destructive for me. (32–33)

In the typical language of therapy and self-help books,[38] the idea of a shared identity appears as a destructive threat to the self, which has to be entirely self-fashioned to be healthy and stable. It is a crime against the self, in this view, to defer identity development too much into the phase of a relationship. Instead, through introspection and self-examination, the self must be shaped and defined

---

[37] Rebellato provides a similar list of images visualising the connection of 'shopping' and 'fucking' in his article "Love and Information."
[38] As David Alderson puts it, "Mark [...] can only conceive of independence in the terms indebted to counselling-speak" (865).

before entering the amorous adventure. In this "regime of authenticity," as Illouz calls the pattern of rules following from this central conviction, "commitment does not precede but rather follow emotions that are felt by the subject and that become the alternate motivation for the commitment" (*Why Love Hurts* 31). Prior to any emotional commitment, the subject has to scrutinise and classify his or her emotions, needs, and preferences and arrive at an authentic, autochthonous, 'essentialised' self which can then be the starting point for the selection of a compatible partner. Mark is in line with these ideas. He does not rule out intimate relationships forever but only for as long as he does not know himself well enough. Only once he is in possession of an entirely self-fashioned identity will he be strong enough to avoid emotional dependency within a relationship. Until that moment, he considers love just as destructive as drugs because it is a potential source of the most horrid threat the self can be exposed to: dependency.

The *tertium comparationis* in this metaphorical equation of love and drug addiction is the loss of control and self-sufficiency. In both the subject feels an urgent need for something the supply of which is beyond control; without it, there can be no happiness or well-being and anything will be done to secure it. To avoid this addiction, Mark longs for a relationship that is "sexual but not personal, or at least not needy" (25), a relationship that shuts out the precariousness of love. When he then falls in love with Gary despite himself, the play seems to showcase the nature of love as passion, as overwhelming, uncontrollable force overpowering human will. Even more, however, the play stresses the futility of any attempt to elude love's precariousness. Right from the start it is obvious that Mark, like the lovers in Plato's Aristophanes myth, is looking for more than mere sexual gratification. He deludes himself when he thinks that sexual transactions can provide what he desires or that he actually wants something "that doesn't really mean anything" (24). The least he wants, as he admits early on, is "to find out [...] if there are any feelings left" (34) after the numbing effects of capitalism, consumerism, and drug abuse. Soon, however, he realises his wants go further. In the shopping scene with Gary, he delivers a little speech that is, once more, a collage of ready-made discourses, this time harking back to the romantic notion of compensation originating in the Aristophanes myth, followed by an immediate recognition of the egoism inherent in this concept.

Gary: Do you love me? Is that what it is? Love?
Mark: I don't know. How would you define that word? There's a physical thing, yes. A sort of wanting, which isn't love is it? No, That's well, desire. But then, yes, there's an attachment I suppose. There's also that. Which means I want to be with you, Now, here, when you're with me I feel like a person and if you're not with me I feel less like a person.

Gary: So is it love, then?
Say what you mean.
Mark: Yes.
I love you.
Gary: See.
Mark: But what I'd like to do – now that I've said that which was probably very foolish – what I'd like to do is move forward from this point and try to develop a relationship that is mutual, in which there's a respect, a recognition of the other's needs. (55–56)

Like his parody of 'therapy-speak' and Robbie's Lyotardian lecture before, this speech is another example of how characters in Ravenhill's plays sometimes tend to become spokespersons – not necessarily for the author but for a kind of imagined Olympian observer who analyses their own and their society's condition with recourse to a certain discourse. Clare Wallace refers to this as a "display of knowingness" (*Suspect Cultures* 93), drawing upon Dominic Dromgoole's argument that Ravenhill's "characters look into themselves, and at their peak, find a way of describing themselves. This is compelling, but not alive. It's perilously close to soap, where everyone knows and describes what they are feeling" (236). What Wallace and Dromgoole describe here is what Manfred Pfister has called 'transpsychological' figure conception, where a character's "level of self-awareness transcends the level of what is psychologically plausible" so that "the figure is able to discuss itself and its situation with a degree of explicitness and self-awareness that it could not possibly have acquired from 'own' experiences alone" (182). According to Pfister, such figure conception tends to produce characters that are less individuals and more representatives, and I would agree that Mark speaks here not as a concrete individual but represents philosophical discourses of love that elucidate his current situation but would not have been available with such clarity and precision to a realistically conceived character.

With the first part of the speech, Mark renounces all claims to self-sufficiency he might have had before. Love is now no longer a threat to self-sufficiency, but the natural human lack of self-sufficiency is one of love's motivations. Love does not threaten or diminish the self, it adds to it. Mark has to admit his fundamental neediness, the insufficiency of his solitary being which makes him "feel less like a person." In the second part of the speech, he then draws on the long tradition of critique that denounces the immanent selfishness of all concepts in which love's primary function is compensatory and the beloved only a means towards an ultimately egoistic end. Therefore, he wants to elevate his love to a level where he fulfils Gary's needs as much as vice versa. Of course, this wish to fulfil Gary's needs is not entirely altruistic either. It corresponds to Lacan's desire for the other's desire, the fundamental need to be needed, which manifests in the wish to care for someone that, as Ravenhill points out, characterises most of the figures throughout the play (Ravenhill, "Interview" 71). Mark has not been consciously aware of this need to give and to satisfy because he was too much concerned avoiding dependence on what others might give *him*, but both the unwillingness to be needy and the need to give were maybe subconsciously at work when he earlier paid Gary not for satisfying him but for allowing him to pleasure Gary orally. Now, at any rate, he definitely recognises his desire for Gary's desire, his wish to be there and care for Gary, which has been triggered by Gary's vulnerability, desolation, and repeated calls for a father. This insight is accompanied, however, by a simultaneous evidence of love's precariousness. Gary, although unmistakably needy and driven by a very specific desire, does not allow Mark to become the object of this desire. "I'm not after love," he tells him. "I want to be owned. I want someone to look after me. And I want him to fuck me. Really fuck me. Not like that, not like him [his stepfather]. And, yeah, it'll hurt. But a good hurt" (56). This is not what Mark can or wants to give him. It is not what he wants to be desired for. If this is what Gary really wants then Mark, it seems, cannot become his object of desire.

What Mark wants is not sex or power but love, and what he has to learn is that there is no risk-free love, no love without precariousness, no way to shut out all forms of neediness and dependence. But, to return to the love-drug-metaphor, could he not have tried to be a little more consistent in his effort to shun all forms of dependency or addiction? Could he not have relinquished love altogether instead of seeking love without risk? I think the reason why the answer has to be 'no' lies in the way the play depicts all characters as searching for ways to escape the ugliness of the world and to come to terms with its absurdity. *Shopping and Fucking* offers its characters three possible choices to reach this end – money, drugs, and love – and Mark has tried them all. First, he was a businessman: "I traded. I made money. Tic Tac. And when I made money I was happy, when I lost money I was unhappy" (33). Then came the heroin. Now it is love. All three options are shown to be utterly precarious – the availability of the desired object is never fully controllable – but to relinquish them all is not an option either. As Freud argues in *Civilization and its Discontents*, mankind has developed various 'palliative measures' to make life endurable, of which the most effective are, among others, drugs, love, and religion (see 2.1.3). Some such measure is inevitable, says Freud, and considering Brian's apotheosis of capitalism, the measures offered by Freud and Ravenhill are quite in accordance: drugs, love, and (the consumerist) religion. Brian, moreover, provides another connection to Freud's text when he is moved to tears by his son's playing the cello: "You feel it like – like something you knew. Something so beautiful that you've lost but you'd forgotten that you've lost it. Then you hear this. / [...] Hear this and know what you've ... l-l-looost" (45). This is more than just slightly reminiscent of Freud's description of an 'oceanic feeling' as a reminder of the blissful state of oneness before or immediately after birth, a memory of a beatific paradise for the loss of which we seek compensation. Both Mark and Brian thus experience the urgent need to compensate for some unspecified but deep-felt loss or insufficiency. The same structure of desire can be found in Gary.

5.4 Gary's Desire

If Gary represents a position in the discursive field of love it is that of the desire for an unattainable object. From the speeches of Aristophanes and Alcibiades to Courtly and Romantic Love and to psychoanalysis, this has always been a pivotal source of love's precariousness, either because the desired person was out of reach for the lover or due to the insight that the desired person was no more than a surrogate for the real object of desire which is forever unattainable in this world. Gary's desire is of the second kind. What he desires, as he repeatedly declares, although not in precisely these terms, is a feeling of absolute security and belonging. "I want a dad. I want to be watched. All the time, someone watching me" (33), he sobs to Mark, and he pins his hopes on his desperate fantasy of a rich, big 'bloke' who will take possession of him. This longing for someone who is always there, always watching, bears clear religious undertones, evoking the image of an omnipresent God. Together with his wish to submit himself entirely to this towering, powerful father figure, Gary's desire takes the form of the kind of religious devotion described by Simon May in which the self-sacrifice of surrendering completely to an almighty God is rewarded with feelings of security and ontological rootedness. It resembles the love for the God of the Old Testament, who is all powerful but also unpredictable and uncontrollable, sometimes unreliable, and often fearsome. The people of Israel, May argues, "are nonetheless to love him with all their heart and soul and might, though they can never count on his presence when they need him" and they do so because "[l]love is evoked not by beauty or moral goodness (in the sense of kindness) but by the mysterious promise of the loved one to anchor and sustain one's life, such that one can feel at home in the world" (36). In extreme cases, this power to provide a sense of ontological rootedness "is the power over life or death itself" (36) that characterises, for example, gods, absolute rulers, or hostage-takers, who all can serve as objects of devotional love. Love as the response to the promise of ontological rootedness "involves the willingness to pay a tremendously high price for this supreme good – a price that might seem perverse to an outsider but is of little consequence to one who experiences the promise of ontological rootedness" (37). For Gary, the mysterious, fearsome, cruel father figure, who offers a 'good hurt' and total surveillance in return for submission, embodies this promise of ontological security, and in Gary's fantasy of being raped with a knife the figure is indeed endowed with a god-like power over life and death.

Gary is by no means singular in uttering a desire of this kind. From the very beginning, when Mark tells the first version of the 'shopping story,' the motif of people as commodities is accompanied by the undisguised wish to be obtained and possessed on the part of some of the characters. As Wallace rightly articulates, "Lulu and Robbie are more than quiescent in this imagined transaction, as is illustrated by their demand at the beginning of the play that Mark reiterate the story," and she is also right that "it is easily recognisable as a somewhat warped love-at-first-sight narrative that exaggerates a vocabulary of possession commonly and popularly applied to love" (*Suspect Cultures* 100). When Mark describes Lulu and Robbie's reaction in the narrative as "You see me and you know sort of straight away that I'm going to have you. You know you don't have a choice. No control" (5), he indeed evokes, for example, Barthes' comparison of "the myth of 'love at first sight'" (190) and of "love-as-passion" with the rape of the Sabine women as a kind of emotional "ravishment" (188). A desire for such ravishment may also be discerned in Gary's fantasy of the big bloke ("And he'll fetch me. Take me away" [66]) and in the way he responds positively when Lulu and Robbie make him the object of transaction in the second version of the shopping story (cf. 78 – 80). But more important than the overwhelming power of passionate love that shines through the story is the feeling of security that is implied in its image of self-abandonment and possession. For Lulu, Robbie, and especially for Gary, an unreserved entry into a relationship of absolute dependency, the complete surrender of freedom in exchange for a sense of security, belonging, and at-homeness, seems more promising than any attempt at defining and defending their own identity along other lines than either possessing or being possessed. As Wallace argues, in the world of *Shopping and Fucking* "identity is considered in terms of ownership of oneself or others" (100), and for those who are excluded from ownership, being possessed is the only available source of self-reassurance. Thus, for Carney, "Gary's rhetoric of slavery [...] sounds [...] like a comforting Nietzschean phantasy of a clear, unambiguous relationship of cruelty, of owner and owned, a fundamental apprehension of difference as a ground to a stable selfhood" (243). Gary prefers the certainty of the simple relationship between master and slave, in which he is allotted a clear position and in which the renunciation of freedom and 'self-possession' is rewarded with a sense of belonging and security, to Mark's offer of love as mutual recognition. Only, as the play makes increasingly obvious, this absolute security is not to be had in this world and Gary's desire is thus directed towards an unattainable object. "The legitimate tragedy of *Shopping and Fucking*," Carney writes, "is that this compensating phantasy of affirming cruelty is not available to Gary" (243), a recognition that manifests itself on a superficial level in Gary's reluctant acceptance that "he's not out there" (85) – that the shadowy figure of his cruel but kind saviour is only a wishful imagination. On a deeper level, however, this acceptance is the acceptance of the fact that the 'bloke,' whether he is 'out there' or not, has always only been a surrogate for the real object of desire and that this real object is not available to the living. Once Gary realises this, it becomes clear both for him and the audience that his real desire amounts to a death wish.

In my discussion of Hegel, Freud, Bataille, and Lacan I have delineated in parts a tradition of thinking that understands the process of individuation, the moment of birth, or even the act of procreation as an incident of great loss. What is lost is 'immortality' or 'pure life' (Lacan), or the continuity of being, which is replaced by the discontinuity of individual existence (Bataille), and a fundamental human desire consists in the wish to return to the state preceding this loss. Bataille's ideas are particularly fruitful as far as the connection between the desire for this lost state of wholeness – which ultimately is a desire for death – and sexuality is concerned. For Gary, death and sex seem to be as closely related as for Bataille; the violence of the sexual act seems to have the same transgressive function of approaching and crossing the border between discontinuity and continuity. Realising eventually that his sexual desire is not erotic but lethal, that is, that it is not balancing on the dividing line between discontinuity and continuity but rather seeks only the latter, his fantasy of rough sex becomes a death wish. Traumatised and incurably wounded – "I've got this unhappiness. This big sadness swelling like it's gonna burst. I'm sick and I'm never going to be well" (85) – he has lost the will to preserve his discontinuous existence that usually balances the wish to return to the continuity of death. When his (and also Lulu and Robbie's) earlier wish to be possessed already indicated an inclination to dissolve the individuated self, his wish to be raped with a knife ("Robbie: It'll kill you. / Gary: It's what I want" [84]) is an unambiguous demand for the violent destruction of the isolated discontinuous self in order to reach the continuity of death. The instrument to reach this end is Mark, who first refuses to take part in the role play staged by Lulu and Robbie but then is the only one who is prepared to satisfy Gary's ultimate desire. That he does so even though only moments before he still tried to convince Gary of his concept of love as mutual care and recognition and that he "must like that. Just to be loved" (81), is certainly disappointing, but not wholly surprising. It is, after all, Mark's only opportunity to obtain what he desires: Gary's desire. As a consequence of Gary's persistent refusal to be loved according to Mark's new ideal, Mark, in his desire for Gary's desire, gives in to his demand and kills him.[39]

This climax and the ending, in which Mark relates a third version of the shopping story before he, Robbie, and Lulu feed each other from their ready meals, have provoked controversial interpretations. The careful optimism inherent in the final tableau (cf. Sierz, *In-Yer-Face* 134) is clouded by Gary's death, which easily appears as the price for the reconciliatory ending, not least because it was probably his money that enabled Lulu and Robbie to content Brian. For Carney, "[t]he violence perpetrated against Gary seems disturbingly regenerative to the trio: it is difficult to shake the sensation that what they are nourishing each other with, in the final moments of the play, is Gary himself" (240). Mark's final story (89–90) is equally ambiguous. Set in a post-apocalyptic future, Mark buys a "tanned and blonde" mutant with "pecs" and a "three foot long dick" – an alluring sex slave strongly reminiscent of Rocky in Richard O'Brian's *Rocky Horror Show* – just to set him free immediately. The mutant begs to be kept in possession ("I can't feed myself. I've been a slave all my life. I've never had a thought of my own. I'll be dead in a week") but Mark insists on his gift of freedom ("That's a risk I'm prepared to take"). For Wade, the story epitomises the play's ethical position of radical freedom and self-fashioning. When Mark declares that he is prepared to take the risk that the slave will not survive his freedom, Wade argues, he "voices his preference for self-sufficiency over co-dependence, even if death is the consequence. It is here that Ravenhill sets forth the basic ethic of the piece, one that valorizes freedom, rejects external social or moral claims, and esteems self-actualization above all else" (113–14). I think that this interpretation rests on two mistakes. The first is the undue equation of the playwright with one of his figures. As Rebellato warns, it is crucial not to overlook the sceptical attitude Ravenhill adopts towards his characters and their self-consciously uttered maxims and wisdoms (cf. xiv). In contrast to Wade, Rebellato is unsympathetic to the ideals of self-sufficiency and self-fashioning to which Mark pays lip service. For him, they only mask a profound "failure to make contact" and he ascribes the "mist of pop psychology and vacuous new agery" that "tries to validate this failure" (xi) to the characters who are infused with these discourses, but not to the author. Ravenhill may be "the playwright of the E-generation," he writes. "But he speaks to this generation, not

---

[39] Even though the playscript leaves it open whether Mark indeed performs the act, most commentators assume that he does and that it is lethal for Gary, based on the facts that Gary is missing in the final scene, that Robbie and Lulu have the money to pay Brian, and that Mark is stained with "a bit of blood" (90) when he re-enters the stage. Moreover, in "A Tear in the Fabric" Ravenhill himself writes that "Gary is murdered by anal penetration with a knife" (311).

for them" (xx). Therefore, despite – or rather in contrast to – Mark's initial ideal of self-sufficiency, "everywhere in the work Ravenhill affirms our fundamentally social character, that we are only ourselves when we are with others, forming human social bonds that are not driven by economic exchange" (xviii). For Rebellato, whose view I share, *Shopping and Fucking*, like all of Ravenhill's earlier plays, is full of "moments where characters are forced to admit their need for one another," offering a deep-felt expression of "our profound yearnings for each other" (xix). In fact, Mark is the character who learns this lesson most urgently when he realises that he needs Gary to feel whole and complete. That he has apparently forgotten all this and eventually tells a story that indeed seems to prefer self-sufficiency over co-dependence, points toward the second factor that Wade fails to see in his interpretation: Mark's notorious inconsistency. The character who wanted to know if there are any feelings left but simultaneously tried to shun emotional relationships, who objected to Lulu and Robbie's role play but then took on the central role of the murderous rapist, here once more says one thing and does another. In his tale, he resists the temptation of a three-foot penis, refuses the slave's wish to be owned, and forces him to live as a free individual. Only moments before, however, the real Mark was unable to resist the temptation of embodying Gary's deepest desire, gave in to his wish to be possessed, and allowed him to exit the realm of discontinuous individuals. For Elizabeth Kuti, who reads the play as a tragedy in an almost "classically Aristotelian sense" (460), the rape of Gary marks "the absolute failure of Mark's hero-quest to extricate himself from the world of consumption, dependency, purchase, exchange and commodification" (462–63). After his "moment of *anagnorisis*" (461) – the recognition of his love for Gary, of his own fundamental neediness, and of the necessity to replace the patterns of ownership, egoism, and one-sided dependence with a vision of co-dependence and mutual love – "Mark is finally and horribly dragged back down into the economy of transaction" (462). Instead of living up to his ideal of love that seeks the good for the other – even against the other's will, if necessary – Mark resorts to the safety of the transaction, the *quid pro quo*, where every service must be reciprocated. More than money, the currency in this final transaction is Gary's love and desire that Mark hopes to obtain in return for his service. "Do it. Do it and I'll say 'I love you'" (85) are Gary's last words, for which Mark is willing to sell his ideal. "Tragedy," Kuti writes, "examines an individual at a particular moment of crisis making a terrible decision, but it also shows how close he or she came to making the opposite decision, and how forces – global, national, historical, political – operated on that person in the process of making a choice" (469). In this sense, *Shopping and Fucking* is indeed tragic: pressed by the logic of consumerism and his egoistic desire to appropriate Gary's love, Mark makes the decision of which he knows, as he demonstrates with his final version of the shopping story, that it was wrong. The shopping story does neither represent Mark's ideal of mutual love nor does it express a "preference of self-sufficiency over co-dependence," as Wade proposes. What the story suggests is that it is better to end relations marked by one-sided dependence and unequal power, that it is better to lose the other than to possess him, that the other's freedom is a risky but necessary precondition for the ideal of mutual love – but it does not berate human neediness and co-dependence as such, which the play treats as ineluctable properties of the human condition rather than denouncing them as pathologies. There is a difference between Gary's or the mutant's wish to be owned like a slave and the acceptance of the dependence on and the desire for intimacy and communality with others. When Marissia Fragoku argues that the play "clearly inscribes love and intimacy as cruel attachments which fail to transcend consumerist values" (31), I think she neglects to make this distinction. In her interpretation, which aligns the play with Lauren Berlant's concept of 'cruel optimism,' Gary's desire, the mutant's wish to remain a slave, and the final tableau with Lulu, Robbie, and Mark feeding each other, are all put on the same level and understood as exemplifications of how the desire for intimacy and relationality leads to situations of possession and dependence which reinforce the market system that has made the desire so pressing in the first place. "Instead of offering the idea of moving forward," Fragoku writes, "the characters remain static in an infantile dependency which carries on 'feeding' the same capitalist system" (32). Intriguing as this approach is with regard to Gary's and the mutant's wish to be owned, it is not a convincing interpretation of the final scene and the entire play, unless absolute autonomy and self-sufficiency are declared reachable and desirable goals while all forms of neediness and dependence are castigated indiscriminately as the bedrock of capitalism. To rephrase Martha Nussbaum in this context, it is not dependency that is infantile but the desire to overcome all dependency, as if it was possible to return to the imaginary status of infantile omnipotence (see 2.2.2). Not at all a static character, Mark has realised the futility of his dream of emotional self-sufficiency and has exchanged it for the search of mutual love. Telling the final version of the shopping story, Mark is aware of his own insufficiency but realises that he cannot compensate for it with a 'slave' but only with a free individual with whom genuine mutuality is possible. In the story, he fantasises about resisting the temptation to exploit another's neediness and appropriate his desire and instead enforces freedom upon someone who does not even share the ideal of mutual love. To put this theory into actual practice, however, was beyond his power.

5.5 Conclusion

Shopping and Fucking, with its plot improbabilities, its sometimes inconsistent, sometimes overly insightful and articulate characters, its quotation and juxtaposition of various discourses, and its "aesthetics of directness" typical of in-yer-face theatre (Stöckl 221), is obviously no piece of social realism. The play is not mimetic in the sense of imitating the actual words and actions of real-life people. Nevertheless, however, it is mimetic in the sense that it turns its characters into mouthpieces for a series of discourses that do shape and influence everyday lives in our empirical reality, even though they are not usually simultaneously performed, recognised, and articulated by those who act under their sway. The 'realism' of *Shopping and Fucking* lies in its faithful depiction of discourses, not in the depiction of their articulation in words and action.

Apart from that, its overt discussion of postmodernity's incredulity towards metanarratives implies a crucial question, Kuti argues: "What kinds of collective narratives, or archetypal dramatic plots, could make any sense to us any more in our fractured, introspective, isolated lives" (460)? A narrative, it might be replied, that tackles the very ideological foundations of precisely this conception of the isolated modern individual, a counter-narrative to the ideal of the self-sufficient, solipsistic monad that is so inextricably entwined with the capitalist logic. The play, I think, proposes this kind of grand narrative, the narrative of humans as social creatures, always needy, always insufficient, and therefore always dependent on others. For all its conscious citationality and its playful use of discursive snippets, the play is quite unambiguous in its call for solidarity that follows from this conception of mankind, and love is the play's thematic core that is used to bring home this message – erotic love, as the emotionally most violent symptom of need and insufficiency, and love as friendship, solidarity, and mutual care as the indispensable backbone of any society. Both kinds of love are suggested as desirable alternatives to the isolation and alienation that permeates the characters' fictional world. But the play also foregrounds the precariousness of love, especially in an environment where the logic of market economy spreads through all areas of life, where insufficiency and co-dependency are castigated as individual weakness, and where the safety of transactions and the security of ownership relations are preferred to the riskiness of mutuality.

6 Autopsies of Love: Sarah Kane's Erotic Plays

6.1 The Love Experience

Of all the plays Aleks Sierz discusses in his *In-Yer-Face Theatre*, the works of Sarah Kane probably correspond best to his definition of this "new sensibility" within British theatre of the 1990s as a form of drama that "takes the audience by the scruff of the neck and shakes it until it gets the message" and that is "experiential, not speculative" (4).[40] The two plays I have selected, *Phaedra's Love* (1996) and *Cleansed* (1998), present both the sensational and visceral impact evoked by graphic stage imagery and the abstention from subtlety in plot and character conception implied in Sierz' formulation. Moreover, they contain all of Sierz' ingredients of genuine in-yer-face theatre – filthy language, unmentionable subjects, nudity, sex, violence, humiliation (cf. 5) – to their full degree, and even if the employment of this "aesthetics of extremism" (5) probably still falls short of giving the audience the "startling feeling of having lived through the experience being represented" (7) due to the improbability of experiencing literally the *same* emotions represented on stage, it is an attempt at sending the receiver into a state of emotional turmoil that is comparable to that of the dramatically represented emotion. With regard to my research, this means that the audience probably is not made to feel love or lovesickness. Rather, the plays' creation of suspense, shock, and fear within the spectator may be said to be not unlike feelings and experiences engendered by and surrounding love, such as hope and aspiration, desperation and desolation. The representational aspect of theatre – the artistic representation "*of* something or someone, *by* something or someone, *to* someone" (Mitchell 12) by means of iconic (i.e. mimetic), symbolic, or indexical reference (cf. 14) – is thus accompanied, if not replaced, by a presentational function that is not concerned with the representation of reality but with the immersion of the spectator in the experience of the performance and with the direct evocation of intense emotions. As Piet Defraeye argues, provocative theatre such as Kane's can lead to a "moment in which the representational aspect is dissolved and, in Artaudian fashion, becomes a presentation, a communal ritual or collective act" (82) bringing about "the overwhelming effect [...] of integration and absorption" (83). The plays are experiential in their attempt to reinforce their discussion and representation of love with a presentational experience of

---

40 Sierz borrows the expression from Kane herself who, regarding the hostile reactions to *Blasted*, surmised that "the press outrage was due to the play being experiential rather than speculative" (qtd. in Sierz 98).

the emotional upheavals caused by love. And they are largely non-speculative because they do not ask the audience to weigh arguments against each other or to compare, reconstruct, or deconstruct the discourses of love they utilise. They forgo subtlety in figure characterisation, character motivation, or the translation of abstract discourses of love into plausible speech acts. Rather, prevailing discourses are expressed in blunt statements and graphic images (cf. Stöckl 217–19), and this not in a postmodern ironic or citational tone but in a very straight-forward manner as a method of drawing a comprehensive outline of the discursive field of love that is eventually subsumed under the overarching "message" (Sierz 4) brought home unmistakably to the audience: that love is utterly precarious, painful, and dangerous, but that it is our only hope and, ultimately, invincible.[41]

While Kane explores the "landscape of love" in all her five full-length plays (Greig ix), *Phaedra's Love* and *Cleansed* most persistently combine love as thematic core with a dramatic form. Unlike in *Blasted* and *4.48 Psychosis*, love is not only part of the plays but constitutes their undisputed centre, and unlike *Crave* and *4.48 Psychosis*, the two selected plays still belong within the realms of dramatic theatre, featuring scenic presentation, recognisable characters, and a delineable narrative, even though the latter two elements are reduced to a minimum in *Cleansed*. In fact, *Cleansed* marks a borderline in my selection of plays for this study which, after all, seeks to investigate how contemporary drama translates prevailing discourses of love into 'contemporary' plots, stage imagery, and character speeches. With its total rejection of realism, its reduction of plot to a series of narratively loosely connected, highly metaphorical images, and its minimalistic language, the methods of 'translation' cannot be compared to those applied in the other plays discussed in this book, where the discourses are woven into and buried under comparatively realistic plots and speech utterances and have to be reclaimed in acts of interpretation that search for the abstract discourses underlying the concrete surfaces of the plays. By contrast, in the two plays by Kane, and in *Cleansed* in particular, what is required instead of a back-translation from realistic speech and plot into abstract discourses is

---

[41] Some scholars refute this optimistic reading of *Phaedra's Love* and *Cleansed*, arguing that the plays instead depict love as a destructive force that brings about a loss of self. The optimistic reading, on the other hand, is backed up by Kane's comments on *Cleansed* in an interview with Nils Tabert, where she explained that she wanted to write about how love and hope can be kept alive in the midst of violence, how love works as the last hope of salvation in states of total desolation. It was only with her next play *Crave*, she reports, that she abandoned her belief in the redemptive power of love (cf. 19–20). A more detailed discussion of this point follows in the subchapter "Heaven or Hell."

an understanding of Kane's imagery as the attempt at a direct visualisation of these discourses, with the characters' speeches serving as "captions" (Stöckl 219) for these images rather than asking for interpretation themselves. Kane's language, in other words, provides not 'translations' but 'declarations' of discourses of love, which are illustrated with the most vivid and intense stage imagery. Kane's theatre is one that, above all, produces images, and it does so without realism's self-restriction of producing realistic images following logically from a plausible plot or of producing dialogue that imitates real-life communication. Instead, her plays create gripping metaphorical images that are less motivated by plot than by the playwright's desire to visualise what cannot be turned into convincing realistic dialogue, and when her characters speak they (often) do not imitate real people but provide the linguistic skeleton that is fleshed out by the images.

This understanding of Kane's imagery as direct visualisation of discourses allows for application of my analytic approach to her plays despite the fact that the significance of experientiality and the visceral impression they make on a live audience might seem at odds with the methods of discourse analysis. At the same time, the presence of graphic, physical stage imagery in *Phaedra's Love* and *Cleansed* provides the target for a discourse analysis that is focused on *dramatic* texts – texts, that is, in which the translation of discourses into art/literature is still, if only loosely, bound to a narrative and the scenic presentation of interacting characters, most of which is absent in her later plays. While *Crave* is already "more of a poem than a play" (Sierz, *In-Yer-Face* 119) where "the borderlines of character evaporate entirely and her imagery moves from physical to textual realisation" (Greig xiv), *4.48 Psychosis* goes even further in denying any obvious traces of plot and character. While it is a special aesthetic feature of *Cleansed* as a dramatic text that its language acquires a poetic minimalism and a declarative directness that is unusual in drama and does not correspond to traditional psychologically realistic character conception, the language in Kane's later plays is entirely free of any conventions and restrictions that come along with the dramatic form. Put differently, *Cleansed* is a special case of drama, one that borders on different art forms and demonstrates both the possibilities (the impact of stage images) and limitations (speech bound to characters) of drama; *Crave* and *4.48 Psychosis* have, in my view, crossed this border of the dramatic and have relinquished the creation of physical stage images for an absolute freedom of poetic language. In the following, I will first reflect in more detail upon the aesthetics of the two plays under consideration before I will analyse them separately but with an emphasis on the main element connecting them: the intertwining of love with forms of self-loss and the unshakeable optimism that underlies the plays – whether despite or because of this intertwining.

6.2 The Aesthetics of Kane's Dramatic Plays

Laurens De Vos is certainly right in claiming that "it is not hard to recognize love as one of the most prominent concerns in Kane's plays" (22), but David Greig points toward an important fact when he writes in his introduction to the *Complete Plays* that *Phaedra's Love* was "the first of Kane's plays to deal explicitly with what was to become her main theme: love" (xi). Even if love and desire had already played a role in her first play *Blasted*, it was with her idiosyncratic adaptation of Seneca's version of the Phaedra myth[42] that it developed into Kane's central topic, rendering *Phaedra's Love* and *Cleansed* love plays in the strict sense that they investigate the very nature of this emotion. In *Crave*, love is still the central theme (cf. Lublin 122–23) and in *4.48 Psychosis* it is presented, along with medicine and suicide, as one of three solutions to "relieve the tension of living" (124), but as I do not regard them as actual plays,[43] I will now focus on the aesthetics of those two works which can be seen as Kane's proper love plays.

While I argued above that the plays of Sarah Kane like few others live up to Sierz' definition of in-yer-face theatre, associating Kane with the group of playwrights commonly subsumed under this term is nevertheless problematic. Ken Urban, for instance, sees Kane "as the most far-reaching experimentalist" (40) of the new playwrights of the 1990s. Where other newcomers such as Ravenhill, Penhall, Butterworth, or McDonagh still largely preserved the quintessential elements of critical realism, "her plays represent the most devastating overturning of that form" (40), especially in their reliance on images instead of dialogue. Similarly, Saunders sees her "rejection, or at least manipulation, of the conventions of realism" as the "key distinguishing feature of [her] dramatic strategy" (*Love Me* 9), which sets her apart from most of her contemporaries, whom Saunders, following David Edgar, "in terms of *dramatic form*" regards as "conserva-

---

42 For details about Kane's reasons to select and adapt this ancient myth see Sierz (*In-Yer-Face* 109) and Kane's interview with Tabert (10–12). For differences between Kane's and Seneca's versions, see Barry and Giannopoulou.
43 Neither did Kane, as the following quotation from the *Guardian* article "Drama with Balls" (20 August 1998) demonstrates: "Increasingly, I'm finding performance much more interesting than acting; theatre more compelling than plays. Unusually for me, I'm encouraging my friends to see my play *Crave* before reading it, because I think of it more as text for performance than as a play" (qtd. in Saunders, *About Kane* 95). Kane thus seems to concur with fellow playwright Phyllis Nagy, who said about Kane in a conversation with Graham Saunders (17 July 2000): "When you abandon character you abandon drama, so for me she has effectively abandoned drama. [...T]here is a diminishment of dramatically viable image structure in both of the last two plays, which renders them, for me, viable works of experimental literature rather than viable works of drama" (qtd. in Saunders, *Love Me* 159–60).

tives" (7). Catherine Rees, too, points out that, unlike most other examples of in-yer-face theatre, Kane's plays are not set in a recognisable world of social realism (cf. 113). She also reports a "growing tendency to view her work in the postdramatic sense, that is, as theatre based on experiential emotions rather than traditional text" (127) – a view that rests on the "fragmented narrative, lack of distinct characters and rejection of social realist structures" in her writing (135). Yet, as Rees reminds us, "Kane's plays are not pieces of performance art; they do still hang on to structures or representation and mimesis (albeit without regard for traditional conventions of naturalism), and as such are not truly postdramatic" (135). At least with regard to the two plays under discussion in this chapter, I agree with Rees that Kane's plays are still properly dramatic, the strong presentational elements in *Cleansed* notwithstanding. Therefore, as long as naturalist aesthetics, realistic characters, or a linear structure are not declared defining features of in-yer-face theatre, I think that Kane's dramatic plays can be subsumed under this label, even if they certainly deviate aesthetically from the majority of in-yer-face theatre and new writing from the nineties in general. In my view, the most significant differences to other plays of the period – but also of the following decades – can be pinned down to figure conception[44] and the reduction of language in favour of imagery.

To be sure, the figure conception in *Phaedra's Love* differs immensely from that in *Cleansed*. In the former play, the characters do have recognisable and, to some extent, explicable personalities that serve as understandable motivations for their actions, while comprehensible motivation and character psychology are largely absent in *Cleansed*. But despite this difference, both plays showcase Kane's increasing unconcern about the creation of realistic characters and the provision of plausible motivation or discernible subtext underlying the figures' actions and utterances. Where these elements are still to be found, they are marked by a conspicuous lack of subtlety that can be attributed to Kane's increasing unwillingness to create consistent and psychologically convincing figures by means of skilful characterisation, observable in the gradual disappear-

---

[44] I have argued that the characters in the plays of Marber and Ravenhill sometimes appear rather type-like and/or psychologically unrealistic. As the following discussion of Kane's aesthetics will hopefully demonstrate, however, these playwrights' deviations from realistic figure conception do not compare with Kane's outright rejection of mimetic and psychologically plausible characterisation which, moreover, is motivated by a peculiar objective that is distinctive of her plays.

ance of dramatic characters from her plays.⁴⁵ Ehren Fordyce writes with regard to this matter:

> In a move towards performance, Kane's plays gradually eradicate dramatic subtext. *Phaedra's Love* offers Kane's last example where characters proceed more by recognisable Stanislavkian subtext than straightforward declaration. The character Phaedra, notably, is a mess of subtext, and the first scenes in the play are among Kane's clunkier in terms of technique. The doctor and Strophe dole out exposition, telling the audience (in the guise of obviously leading questions to Phaedra) about the queen's subtextual passion for Hippolytus. Dramaturgically, playing the game of subtextual hide and seek requires authorial subterfuge. In order to reveal character, Kane must play at concealing it (and throughout her work, it is hard to ignore an urge to say, 'Enough of covering your arse!'). Therefore, from *Cleansed* onwards, Kane abandons subtextual characterisation. (110–11)

Fordyce's observation of a development from rather clumsy figure characterisation and motivation to an aesthetics of outright declaration where dramatic figures – or, in her later plays, mere voices – simply and straightforwardly express the words and perform the actions that Kane wants to be staged, draws attention to her priorities. Her focus is on the creation of powerful theatrical images, both visual and verbal, and not on a realistic or plausible narrative leading up to these images.⁴⁶ While *Phaedra's Love* is already written in a tone that ostenta-

---

45 Not surprisingly, critics prepared to treat Kane's plays as innovative or at least untraditional forms of theatrical expression were hardly bothered by the absence of subtle characterisation, while many of the hostile reviews reveal the disappointment of critics resulting from Kane's nonchalant frustration of naturalist theatre expectations. Thus, while Sierz in his favourable review for the *Tribune* greeted *Cleansed* with "Bye, bye naturalism; hi there, live art" and stressed its essential "symbolism" (568), Charles Spencer complains that the play "entirely fails to touch the heart" because the "one-dimensional characters seem little more than shadows in an unhealthy imagination" (565), a criticism echoed by Sam Marlowe who, too, thinks that the undeveloped characters render the play "unmoving" because audiences are not allowed to "gain any insight to their complexity as human beings" (567). Sheridan Morley excoriates the playwright as "a naughty schoolgirl desperately trying to shock an increasingly bored and languid audience [by] pil[ing] horror upon horror without ever bothering to give us a character or a situation to care about" and the play as one "in which any real skill of characterisation or plotting is simply replaced by yet another bloody amputation" (568). Even David Benedict in his rather benevolent review that lauds the power of Kane's imagery and appreciates how "her handling of image and metaphor sets her apart from almost every other playwright of her generation" and "makes so much contemporary dialogue-driven young writing look limply unambitious" (564) then goes on to lament as a "clear weakness in the writing" that most of the characters are "fatally underwritten" (565).
46 In an interview with Saunders, James Macdonald, who directed *Blasted* and *Cleansed* (and *4.48 Psychosis*) at the Royal Court Theatre, recalls that "[i]n both plays she was very concerned to tell a story through images, but *Cleansed* confirmed for me that the images are there to tell the

tiously prioritises what is said and done on stage over the seemingly uninspired explanations behind it, *Cleansed* marks the step in her work where she openly abandoned traditional figure characterisation that equips figures with consistent background stories and personalities to motivate their utterances and actions. For observers from a more or less traditionally realistic angle, this style of writing may indeed appear as disappointing and inferior, as it does for Klaus Peter Müller, who writes about *Phaedra's Love* that its "extremely one-sided characters" (99) prevent the development of a truly dramatic conflict so that the deaths in the play "do not raise any questions and remain simply banal, as the characters and their actions reveal no motivation, intention, complexity, or serious conflicts" (101).

As Saunders writes, this peculiar conception of character, "in which character became more an expression of emotion than the outward manifestations of psychology and social interaction" (*Love Me* 88) is closely linked to Kane's use of language, which he describes as a "project to pare language down to a stark minimum" (88). Many scholars have taken note of the uncommon language in Kane's plays. Christopher Innes, for instance, thinks that "it is possible to see the language of Kane's plays as a type of free verse" (534), a viewpoint that is supported by Kane's own statement about *Phaedra's Love:* "I wanted to keep the classical concerns of Greek theatre – love, hate, death, revenge, suicide – but use a completely contemporary urban poetry. I see the writing as poetic. Just not verse" (qtd. in Saunders, *About Kane* 68).[47] Apart from its poetic quality, the language of Kane's characters is marked by a sharp minimalism. As Sierz observes, Kane thoroughly avoids long speeches because she argues that "[i]f each character can only say nine or ten words at a time, they become incredibly articulate and precise" (qtd. in Sierz, *In-Yer-Face* 101). This mixture of poetry, minimalism, and precision is, on the one hand, the linguistic aspect of her overall rejection of mimetic realism – and it befits the aesthetics of declaration already observed in her figure characterisation. To the same degree that she is not interested in motivating and explaining her figures' actions and utterances through an imitation of psychologically plausible characters, she is not interested in imitating the way people speak in reality. Relieved of the vagueness and awkwardness of everyday speech, Kane's figures speak from their hearts with infallible accuracy, displaying degrees of self-knowledge and verbal accuracy that are far removed from realistic figure conception. On the other hand, her linguistic

---

story more powerfully and immediately than the text. [...] Her work always seeks to engage an audience by the most direct route possible" (Saunders, *Love Me* 121–22).

47 Benedict, David. "What Sarah Did Next." *Independent* 23 January 1996.

minimalism can be seen as a symptom of what I consider one of Kane's most important objectives. One reason for the reduced language in *Phaedra's Love* and *Cleansed* can be found in the attempt underlying these two plays to provide access to the non-verbal realm of feelings behind discourses of love. Dissatisfied with imitating how these discourses are verbalised and communicated between people in reality, Kane tries to approach what lies beneath verbal articulation. Her aesthetic style, which prioritises theatrical images and supports them with precise, minimalistic, caption-like figure speech, is supposed to facilitate a non-verbal, experiential approach to the content matter of discourses of love. Again, this is not to say that an audience is made to directly experience the precariousness or compensatory function of love. It means that *Phaedra's Love* and *Cleansed* are attempts to transpose these discourses into gripping, shocking, and absorbing stage images so that they are not only, as in other plays, made recognisable and understandable as dominant discourses of our reality, but supplied with an affective potential and made experienceable on a non-discursive level (cf. Stöckl 222–23). While the recognisability of the discourses of love is guaranteed by the figures' verbal declarations, the images, which without the text could only seldom be assorted to the motif of love, enable the experiential effect and hence the access, as it were, to larger or 'purer' truths than could be attained by the mere reproduction of language. Due to this replacement of social realism with an aesthetics that attempts to approach the grand, even universal questions of mankind, an approach that tackles metaphysics through an immensely physical theatre, Saunders likens her plays to classical and Jacobean drama:

> Certainly, in comparison with many of the domestic settings and adherence to realism in the work of her contemporaries, Kane's drama is in possession of an overreaching feel in its grand attempt to make sense of the world rather than a specific event [...] or social agenda. Moreover, like the Jacobean drama of William Shakespeare, Thomas Middleton, and John Webster, Kane manages to condense great themes such as war and human salvation down to a series of stark memorable theatrical images. (*Love Me* 20)

Her plays, then, are not narratives about particularities but metaphors for larger issues. They do not tell stories about individuals but try to present universal human concerns. When they are about love, they are not about a particular person's love for another person, but about love as such; when they repeat recurrent discourses of love, they are not about a character's specific use of language, but about the meaning that lies buried deep beneath those words and can never be fully expressed verbally. Kane's love plays are attempts to express or articulate something non- or pre-verbal and it is here, I think, that the often-made comparison between her work and the theories of Antonin Artaud can be most reasonably applied.

In an interview with her German translator Nils Tabert in February of 1998, Kane declared that until very recently she had never read Artaud (nor any comparable theory of this kind) and that she was surprised, now that she had started reading his essays, to see how much they tied in with her own work (cf. 19). Despite this clear statement, Laurens De Vos is convinced that "she was definitely influenced by Artaud" (18) and that all of her five plays, "although they are all extremely different in style, [...] are permeated with Artaud's ideas" (23). To prove his point, De Vos endeavours a minute comparison of Artaud's theories for the theatre with three of Kane's plays (*Phaedra's Love*, *Cleansed*, *4.48 Psychosis*), pointing out analogies both in terms of aesthetics and content matter. The similarities on the thematic level, in particular Kane's linking of love to experiences of a loss or disintegration of the individual self, will be addressed further below. Regarding the congruence in Kane's and Artaud's aesthetics that points towards the shared objective of expressing non- or preverbal content, what characterises both is a painful awareness of the inadequacy of language to communicate what they perceive as the only genuine topic of theatre: human emotions. As Martin Esslin writes in his short monograph about the French theatre theorist and practitioner, his previous attempts as a poet had increasingly convinced Artaud "that it is a profound mistake to equate all human consciousness with that part of it capable of verbal expression" (68). Far removed from conceptions of all human thought and consciousness as verbal, he would hold that large parts of human consciousness are never translated into words, above all many sensations of the body, and that it is here where art and literature should seek their genuine field of activity.

> To a poet, Artaud would argue, it is precisely that non-verbal element of consciousness which is of supreme importance. For it is closely bound up with the very stuff and matter of poetry itself: human emotion. [...] And if we analyse the nature of emotion, we shall find that it is, however sublimated, however intensified, part and parcel of those very body sensations; that it differs only in degree from such lowly feelings as an over-full stomach or, indeed, Artaud's nagging headaches. (69)

To this preference for the non- or preverbal elements of the human consciousness as 'the very stuff and matter of poetry' comes his profound scepticism towards language as an adequate means of expression for such extralinguistic truths. As De Vos writes about Artaud's struggle "to divulge the only real, metaphysical truths" (63), he had to acknowledge that "the truth is not communicable via normal language. [...] Artaud is desperate to say everything, and hates to be impeded by the limits of language" (64). In his search for means of expression outside language, Artaud turns to non-verbal signs – symbolic and metaphoric images – which he, moreover, desires to be unconventionalised, signs, that is, that are not already part of a system of signs which would only be another kind of language.

> Artaud wants a sign to be characterized by singularity. No sign should be endowed with meaning as a result of its repeated use. Rather than attempting to convey reality as faithfully as possible, the theater must turn to signs and symbols that stand completely on their own, without having recourse to conventional codes that are applicable to the world outside. (45)

The rejection of ordinary language and conventional signs results in Artaud's 'Theatre of Cruelty' driven, as Esslin writes, by "the demand for an impact on the audience which amounts to an imposition of suffering on them, i. e. cruelty; the demand that the theatre should be able to communicate deep, subconscious and therefore not yet verbalised emotion directly to the spectator by means that amount to nothing less than magic" (79). This is not the place to list the technical features of Artaud's theatre of cruelty and compare them to Kane's plays (which would uncover more differences than similarities). What I would like to emphasise, instead, is the similarity of the goal their theatres are supposed to reach: in both, the function of language is minimised in favour of unconventional, surprising, often shocking images in order to create a direct, visceral, experiential impact on the audience as the only way of communicating the non-verbal truths of human emotion. And in both the body – of the actor and the spectator – is endowed with a significance it largely lacks in traditional theatre: it is the primal medium of communication, the most important ingredient of the theatrical images with which emotions are expressed, and it is the spectator's receptor inasmuch as the theatre experience has a truly bodily – as opposed to purely cognitive or intellectual – impact. In Esslin's words, "[t]o re-establish contact with the true meta-physical basis of human existence it is the body which must be re-awakened and re-activated: in other words to reach the metaphysical sphere we have to become more physical" (81). Kane may not actually share Artaud's metaphysical beliefs that align him with figures such as Bergson, Shaw, Nietzsche, or Freud and their variations of a concept of 'life force' (cf. Esslin 80), but if, despite all biology and neuroscience, we consider the still mysterious realm of human emotions, and love in particular, as the metaphysics of our time, *Phaedra's Love* and *Cleansed* can well be described as plays that address metaphysical questions with a very physical, Artaudian approach.[48]

---

48 Concerning the links between Kane and Artaud, see also Quay 301–19, who sees a clear (if probably unwitting) parallel in Kane's "Rückgriff auf eine archetypisch geprägte Bildlichkeit" (308).

The aesthetics of reduced characters and language and graphic imagery is visible in different degrees in the two plays. *Phaedra's Love* is certainly the more 'traditional' of the two, deviating less ostensibly from the realistic norm (cf. Saunders, *About Kane* 24; Urban 42). Nevertheless, the first reviews already demonstrate the critics' sense that Kane's style clearly differs from conventional drama. Kate Bassett notes that "[s]peech is terse, truncated" (651) and Sierz writes that "the dialogue veers from exchanges that are genuinely disturbing to [...] bonehead declarations" and blames her for a "complete lack of discrimination between what works on stage and what's maddeningly banal" (651). Sarah Hemming's conclusion that the text "is not a reading that anyone could accuse of subtlety" (653) probably summarises the reservations most critics shared against a play that was unanimously certified to have a 'visceral impact' but that was perceived to lack skilful plotting and characterisation. Very likely, one reason for the largely unfavourable reviews was the way in which the play, with its linear narrative, naturalistic production, recognisable characters, and provision of expository information, raised certain expectations towards dramatic realism which were then gradually disappointed in the course of the play. It may actually be regarded a weakness of the play that it does not foreground its abandonment of realism as visibly as *Cleansed*, so that the lack of subtlety in dialogue and figure characterisation sometimes appears like a lack in artistry. The same is true for the plot, which is a disturbing hybrid form of a linear, causally connected narrative in which the causal connections are often weak and unconvincing and where actions seem to lack motivation. However, it makes more sense to follow Greig's assessment that Kane's second play "saw her continue the process of fragmenting naturalism" (x), to see it as a play, that is, which deliberately neglects the rules of naturalistic plotting and characterisation. In this view, the ungainly exposition, the characters' declarative statements, and the unmotivated or underexplained plot developments are not examples of poor writing but part of a style that prioritises and emphasises the creation of memorable stage images. In *Phaedra's Love*, this pertains above all to the moment of Hippolytus' death and 'dissolution,' aptly described by Stefani Brusberg-Kiermeier as having "a great artificiality as he continues talking after all these things have been done to his body, like an opera singer who goes on and on singing after his character has been mortally wounded" ("Rewriting" 169), thus stressing the play's renunciation of realism quite plainly in the final tableau.

Although generally received more favourably than her first two plays, *Cleansed* still was met with similar accusations by some critics. Robert Gore-Langton, for example, judges that "it's not a great text" and that "[a]s a wordsmith, Miss Kane is a non-starter" (563). Nicholas de Jongh complains that

"Kane's trite visions of love [...] keep being overwhelmed by the motiveless intrusion of violence" (563), Charles Spencer faults the writing for its "dreary, linguistically impoverished flatness" (565), and Michael Billington criticises that the play "lacks circumstantial detail" and hence remains ineffective as a political argument (566). However, much more than *Phaedra's Love*, Kane's third play leaves no doubt that it is futile to analyse language, figure characterisation, and motivation according to the standards of dramatic realism (cf. also Saunders, *Love Me* 86–88). In this play, Greig notes, "Kane stripped away the mechanics of explanatory narrative and presented the audience with a series of poetic images and pared dialogue" (xi–xii). As Heiner Zimmermann observes, while some critics "deplored the play's comic strip technique, the absence of conventional psychological characterisation and the incoherence of this collage of episodes," the majority acknowledged "the powerful physical impact of Kane's theatrical imagery which was as important as the dialogue" and "gave the production the character of an installation" (178). Zimmermann himself then stresses the "metaphorical character of the play's atrocities" and its conception as "physical representation of psychic experience" that tries "to give bodily expression to desire and mental suffering" (179). Similarly, for Margarete Rubik the play is "emphatically post-modern and non-humanist, rejecting unity of character and consistency of motivation, a logical story line and mimetic imitation of reality" (131), and asks "for a metaphorical rather than realistic interpretation" (133). Even though the often "unperformable" stage directions sometimes contain narrative information that can only inadequately be transferred onto the stage, giving "an odd quality of a closet drama to an otherwise extremely theatrical and visual play" (133), she argues that "*Cleansed* achieves its effect primarily through haunting visual images of love, suffering and dying" (134), images that are "symbolic and surreal rather than realistic" (134). In the same vein, Christopher Innes notes that the play possesses "none of the elements – a plot line to shape audience response, credible characterization, an identifiable setting – that conventionally give drama social relevance. Instead the action and the figures only have validity as symbolic expressions" (530). Sean Carney writes that "*Cleansed* takes place in an entirely expressionistic environment where the literal and the figurative, the material world and the inner landscape of the self are indiscernible from one another" (275) and Sierz calls the play a "parable of love in a time of madness" that is "full of metaphors of addiction, need, loss, and suffering" (*In-Yer-Face* 114) – the very same metaphors, one could add, that are used to talk about love in Ravenhill's *Shopping and Fucking*.

There is no doubt, then, that *Cleansed* is a highly metaphorical play,[49] with its focus not on plot, character, and dialogue, but on the non-verbal language of its imagery, and conceived not as an imitation of outward reality but as "a journey through the subconscious" (De Vos 124). It is this understanding of Kane's images as metaphors for an inward rather than an outward reality that is absent from reviews such as Susannah Clapp's, who on the one hand applauds the play as "full of visual ideas" and "a series of vividly lit, cunningly designed tableaux," but then laments "that the piece wears its heart on its many limbless sleeves; that its moral position is too evident, too simple" as it amounts to little more than declaring that "being warm towards each other" is preferable to "chopping each other to pieces," "that the weak and the unconventional are tortured," and "that love has a redemptive power" (566). If that were indeed all that *Cleansed* tried to convey, much of the criticism accusing it of banality and lack of subtlety would be justified. However, as I will try to demonstrate in the following, the images of both *Cleansed* and *Phaedra's Love* lend themselves to a much deeper analysis and even though the final message of love's invincibility and redemptive power remains central to both plays, it is preceded by a remorseless vivisection of love that delves deep into the recesses of the human heart.

6.3 "A spear in my side, burning": Precarious Desire in *Phaedra's Love*

The plot of *Phaedra's Love* can be summarised in a few sentences. Phaedra, the queen of an apparently contemporary British kingdom,[50] who has not seen her husband Theseus since their wedding night, is hopelessly in love with her depressed and apathetic stepson Hippolytus, who listlessly kills time with junk food, television, electronic toys, masturbation, and casual sex, unable to derive pleasure from any of these pastimes. Despite the warnings of her daughter Strophe, Phaedra confesses her love to Hippolytus and performs oral sex on him in a futile attempt to stir him emotionally. But instead of breaking his apathy, she only learns that Hippolytus is adamant in remaining emotionally seclud-

---

[49] In the interview with Tabert, Kane herself explains that in *Cleansed* she used violence even more metaphorically than in her previous plays (cf. 20). Regarding the necessity to read *Cleansed* metaphorically instead of naturalistically, see also Deubner (e.g. 123, 127–28).

[50] Although place and time are not specified throughout the play, objects like Hippolytus' electronic toys point to the present time and the fact that one of the men in the crowd of people who kill Hippolytus in the final scene declares to be from Newcastle (cf. 98) point to England as the play's setting.

ed, that both he and Theseus have slept with Strophe, and that Phaedra might have contracted gonorrhoea. In the next scene, Strophe informs Hippolytus that Phaedra has hanged herself and left a note accusing Hippolytus of rape. Although Strophe and a priest implore him to deny the imputation for the sake of the (royal) family and national stability, he enthusiastically embraces the accusation as a present and sign of her affection. In the final scene, Hippolytus is led by policemen past an infuriated mob who have gathered outside the court, the disguised Theseus among them and spurring their bloodlust. Hippolytus "*hurls himself into the crowd*" (100) and is strangled and beaten. When Strophe, who is also present in disguise, tries to protect him, Theseus, not recognising her, rapes and kills her before he partakes in the slaughter of his son. Hippolytus is castrated, disembowelled, "*kicked and stoned and spat on*" (101), and his genitals and intestines are thrown onto a barbecue. Theseus recognises the dead Strophe and cuts his own throat while vultures are circling above Hippolytus' remains, who smilingly sighs his last words, "If there could have been more moments like this" (103).

The entire play unravels to lead up to this final tableau, which carries its most crucial image. On its way, however, the play addresses – in an often declarative and overt, sometimes in a covert manner – a series of topics from the discursive field of love. In my reading, three major issues are brought to discussion. First, the notion of love as irresistible, delusive, illusory passion, which is contrasted with Hippolytus' disillusionment, apathy, and depression. Second, Phaedra's romantic belief in the redemptive, transformative, and compensatory function of love, which is both crushed by Hippolytus' brutal rejection and, at least partly, corroborated by the transformation he undergoes after her suicide. And third, the intricate connection between love and forms of self-loss, which is at the heart of both of Kane's love plays.

6.3.1 "No one burns me": Hippolytus' Refusal of Love

In general, Robert I. Lublin states, "the characters in Sarah Kane's plays suffer from overpowering, irresistible desire" (116), a recurrent motif that manifests in *Phaedra's Love* in "the eponymous heroine's incestuous, uncontrollable need for her stepson Hippolytus" (120). As Paula Deubner observes, the urgency of this passion is repeatedly evoked by verbal images of 'burning' which, Deubner argues, is Kane's prevalent metaphor for love in the play (cf. 110). With vivid words from the well-known semantic field of a romantic love that is both sensual and spiritual, Phaedra describes her feelings for Hippolytus to Strophe – "A spear in my side, burning" (69); "Can feel him through the walls. Sense him.

Feel his heartbeat from a mile" (70); "There's a thing between us, an awesome fucking thing, can you feel it? It burns. Meant to be. We were. Meant to be" (71); "Can't switch this off. Can't crush it. Can't. Wake up with it, burning me. Think I'll crack open I want him so much" (71) – and to Hippolytus – "You burn me" (84) –, unmistakably bringing home to the audience the overwhelming and uncontrollable nature of her passion. The obsessive nature of Phaedra's love that is bordering on "total self-abnegation" (Greig xi) is further emphasised by the fact that Hippolytus is portrayed as a very unattractive and hardly loveable young man to whom she is nevertheless devoted – unconditionally, as it were (cf. Quay 256). In one of many typically declarative passages, Kane conveys this irrationality of Phaedra's passion:

Phaedra: I love you.
Silence.
Hippolytus: Why?
Phaedra: You're difficult. Moody, cynical, bitter, fat, decadent, spoilt. You stay in bed all day then watch TV all night, you crash around this house with sleep in your eyes and not a thought for anyone. You're in pain. I adore you.
Hippolytus: Not very logical.
Phaedra: Love isn't. (78–79)

For no obvious reason, then, Phaedra has stumbled into complete emotional dependency on her repulsive stepson, and her obsessive, passionate, but unrequited love eventually culminates in her act of suicide.

Phaedra's burning desire contrasts with Hippolytus' apathy. The languid prince has shut himself off from all desire, striving for nothing and living a life devoid of any commitment. For Brusberg-Kiermeier, this depiction of Hippolytus is one of the most successful strategies Kane uses "to transfer the myth into a post-modern British context" ("Re-writing" 168). By changing the character "from a spoilt young man who shuns love to a spoilt young man who can get nothing out of television, food or sex any more because he has it all in abundance" (170), Kane adapts the reason for Hippolytus' rejection of love to a modern audience. Instead of the self-imposed chastity in Seneca's version it is the numbing effect of "living in affluence" (170) that prevents the promiscuous prince in Kane's play from developing any desire for his infatuated stepmother.

To this notion of the death of desire through affluence, Kane adds the notion of a wilful spurning of desire as an attempt to attain self-sufficiency. In the straight-forward manner of Kane's dialogues, Hippolytus counters Phaedra's declaration of her burning love with "No one burns me" (82), and when Phaedra asks whether this was also true concerning a woman called Lena he all of a sudden becomes violently aggressive, grabbing Phaedra by the throat, forbidding her to mention Lena again, and repeating his resolution: "No one burns me, no one fucking touches me. So don't try" (83). The short scene implies that Hippolytus' detachment from the world is a deliberate attempt to avoid dependency on other human beings, to avoid the precariousness of emotional commitment that he has obviously experienced in the past. Like Mark in *Shopping and Fucking*, he has made the decision to eschew love and the dependency and precariousness it entails. As Greig puts it, he "is driven to preserve his self inviolate. Emotions, love in particular, and need of any type are an unbearable threat to him" (xi). Unlike Mark, however, Hippolytus succeeds in keeping up his emotional detachment throughout the many meaningless sexual encounters he has. He does not fall in love, does not become emotionally dependent on another person. But the price he pays for his successful avoidance of precariousness is a life full of "tat" and "bric-a-brac" (90) and utter boredom – a life, actually, that is so void of striving, tension, or aspiration that Hippolytus himself does not consider it a life at all but rather as "Filling up time" (79). His apparent self-sufficiency gets back at him, turning his existence into a listless state of stagnation that threatens his mental health and even his human nature. For Lublin, "Hippolytus' love for Lena perverts into his attempt to suppress all desire and avoid further pain. But desire constitutes human subjectivity. Attempts to quash it are akin to killing oneself while still breathing" (120). Hippolytus' negation of desire is thus anything but a healthy form of resistance against affluence and the abundance of choice. Rather, the ability to have almost everything he wants and his decision not to desire what he cannot have (Lena) have robbed the prince of an essential ingredient of human life. As Elizabeth Barry notes in her comparison of Kane's play with Seneca's version of the myth, Hippolytus' rejection of desire and commitment, which is an embodiment of Stoic philosophy in Seneca's play, is pathologised in Kane's revival: "Hippolytus's detachment from the world itself is brought under the category of an illness of the soul. For Kane, in line with the modern sensibility her play reflects, the Stoic view, given voice in Hippolytus, that nothing external is worthy of serious concern, now seems a pathological one" (125).

That Hippolytus' desireless existence is neither healthy nor pleasurable is illustrated not only by his repellent appearance, the sorry description of his daily routine, and the fact that he "hate[s] people" (77), especially those who are happily in love (cf. 80), but is also directly announced with the very first words spoken in the play: "He's depressed" (65). The royal doctor's diagnosis opens up another level of meaning of Hippolytus' condition. Phaedra's uncontrollable passion is not only contrasted with Hippolytus' apathy but also with his state of depression, which amounts to a contrast between the delusion of love and a presumed clear-sightedness of depression (cf. Fisher 172). For Christine

Quay, Hippolytus' depression is the expression of a radically nihilistic outlook on life in which any activity is merely a form of passing time given the meaninglessness of existence (cf. 257–58). This stance makes it impossible for Hippolytus to accept love as a source of meaning in life, which would not only mean to fall victim to the deluded perception of others that is observable in Phaedra, who adores her unlikeable stepson, but also to delude himself about the meaninglessness of the human condition in general. Kane herself stresses this gap between love and depression in her conversation with Nils Tabert, where she recalls that she initially wanted to write a play about depression because she found herself in such a state at the time of writing, but also wanted to include the second half of her split personality, which found expression in Phaedra's blind and unconditional love – two psychic and emotional states that she considers oppositional extreme conditions as she associates depression with a perfectly realistic perception of reality (cf. 12). Following this logic, which juxtaposes the allegedly realistic view of depressives with the illusions and self-delusions of lovers, Hippolytus is unable to love in the way he is loved by Phaedra since, in line with his commitment to an utterly realistic view of life and other people, he is unwilling or unable to effect the self-delusion that is obviously facilitating Phaedra's love for him.

In two senses, then, Hippolytus and Phaedra mark "the two poles that are the extremes of the human response to love" (Greig xi). Phaedra falls victim to the delusive power of love that draws her towards her disagreeable stepson and she unreservedly gives in to her burning desire. Hippolytus, by contrast, is immunised against love's blindness by his depression and, moreover, has banned all desire from his life in an attempt to reach self-sufficiency and invulnerability. However, the wretchedness of his daily life is a clear indication of the futility of this endeavour. Repression of desire is not satisfaction and attempts to kill it off do not lead to happiness.

6.3.2 "Don't imagine you can cure him": The Need for Compensation

Phaedra's love, although largely irrational, is at least partly inspired by one comprehensible conviction: her romantic belief in the power of love to change and rescue Hippolytus – her belief, that is, in the compensatory function of love. That some form of compensation is needed is repeatedly indicated in the play. Christopher Innes, for instance, suggests that Hippolytus' "[p]hysical hunger" can be understood "as an analogue for unsatisfied and unfulfillable desire." In this reading, he "stuffs himself continually with junk food as a substitute for the personal involvement he is unable to feel" (531). Bereft of any friends, denying himself the emotional commitment of love, and, as a spoilt prince, exempt from the task of earning a living and securing the satisfaction of his material needs, there is nothing in his life that requires involvement as a person. The Doctor further speculates that Hippolytus "is missing his father" or "his real mother," questioning Phaedra's "abilities as a substitute" (68) and thus encouraging, without following this thread any further, psychoanalytic interpretations of Hippolytus' condition as one of hopeless desire for a lost object for which all worldly objects can only stand in as deficient surrogates. Barry, too, argues that a "framing metaphor" of Kane's (and also Seneca's) play is "that of the need for a medicine of the soul" and that at the play's beginning all characters "seek a 'cure' for ills that do not originate in the body (although they may manifest their effects there)," most importantly for "the existential pain, with no fixed cause or object, that Hippolytus feels" (124). The 'cure' for this 'existential pain,' Phaedra hopes, will be her ardent love.

The play thus opens up the possibility for the presentation of love as a compensating, transformative, redemptive force – the force that reawakens Hippolytus to life and endows his existence with involvement and meaning in an otherwise meaningless world. According to Deubner, for instance, the mainspring of Kane's drama is the deeply romantic conception of love as compensation for the lack of meaning and identity resulting from the deterioration of social ties and structures (cf. 68), while for Quay, love's promise of compensation relates to a more metaphysical level, where it stands in for faith and mythology as sources of meaning (cf. 270–71). In any case, however, it is important to bear in mind that the plot development of *Phaedra's Love* disallows a too idealistic interpretation of love's redemptive power. Strophe's warning towards Phaedra – "Don't imagine you can cure him" (71) – is justified by Hippolytus' brutal rejection and humiliation of Phaedra after her confession of love. Neither does he fall in love with her, nor does he regard her love as anything but irrational nonsense. There is no direct effect of love on Hippolytus, no 'contagion' with love that transforms the cynical nihilist into a romantic believer. If Hippolytus is indeed transformed from an apathetic phlegmatic into someone who enthusiastically embraces his destiny, it is because Phaedra's suicide and the ensuing prosecution as a rapist stir up his uneventful life, tear him out of his lethargy, and allow him to live by his ideal of absolute honesty up to its final and fatal consequence. Even if one understands Phaedra's suicide and accusation, as do both Hippolytus and several critics, not as an act of desperation and revenge but as a "sign of love" that affords Hippolytus the "possibility to exercise his free will" (Brusberg-Kiermeier, "Re-writing" 171), the conclusion that "Kane constructs a phantasm of female love that serves to deliver the loved man from a meaningless life" (171) is only acceptable with the reservation that this 'service' is indirect. In other words, Hippolytus is not rescued by love but by Phaedra's self-sacrifice, for which her unrequited but undying love was the motivation. Without turning him into a lover, Phaedra's love enables Hippolytus to fill his existence with life and experience personal involvement. What remains to be answered, however, is the question why Hippolytus' prosecution and execution, caused by Phaedra's love-induced suicide, work as sources of meaning and pleasure and, ultimately, as his true objects of desire.

6.3.3 "If there could have been more moments like this": The Joy of Death

When Hippolytus learns about Phaedra's accusation, his reaction is strangely joyful. Delighted with the expectation of excitement and not thinking a moment about running away or defending himself against the legally untenable charge, he turns himself in. "Not many people get a chance like this" (90), he rejoices, but it does not become immediately clear why he is so enthusiastic about the impending prosecution. Instead of a simple answer, I think there is a threefold explanation why Hippolytus so cheerfully seeks his own destruction.

One factor is certainly that Phaedra's accusation puts an end to the boredom of his daily existence. "Life at last" (90), he sighs, welcoming the diversion the coming events will provide. As the sole explanation, however, this would be unsatisfactory. There is no reason why Hippolytus had to wait for an event like this to change his daily routine. Either suicide or the actual commitment of a crime, to name just two possibilities, could have put an end to his languid existence at any time. These two routes of escape, however, are blocked by the second reason for Hippolytus' eager acceptance of his imminent doom, which is elaborated in the dialogue between Hippolytus and the priest who has come to his prison cell. The priest exhorts Hippolytus to confess his sins before God but deny the rape in court and in public for the sake of the moral and political stability of the nation. Hippolytus, however, has no intention of saving his life and the monarchy in this way. Rather, he sees Phaedra's accusation as a chance to live up to his ideal of absolute honesty and integrity in a situation that is neither 'tat' nor 'bric-a-brac' but serious and where his ruthless honesty not only hurts others but has consequences that involve him personally. Neither will he confess to a God he does not believe in, nor will he deny a rape that he has committed, even if not in the literal sense of the word. From the priest's standpoint, what Hippolytus did to Phaedra is a matter for religious confession, not a legal case. For Hippolytus, who does not believe in God, his 'moral' rape of Phaedra deserves punishment just as much as an actual rape would, and he refuses to be spared worldly punishment and ask for forgiveness from a God he does not recognise. "I lived by honesty," he tells the priest, "let me die by it" (95). And for him, admitting a crime he has not committed in order to be punished for a crime that would remain unpunished otherwise is closer to honesty than saving his life by sticking to the letter of the law.

Besides Hippolytus' rejection of God, the conversation with the priest also reveals that he still, despite Phaedra's sacrifice for him, does not believe in love. Rather, he maintains the belief in his own self-sufficiency.

Priest: So where do you find your joy?
Hippolytus: Within.
Priest: I find that hard to believe.
Hippolytus: Course you do. You think life has no meaning unless we have another person in it to torture us.
Priest: I have no one to torture me.
Hippolytus: You have the worst lover of all. Not only does he think he's perfect, he is. I'm satisfied to be alone.
Priest: Self-satisfaction is a contradiction in terms.
Hippolytus: I can rely on me. I never let me down.
Priest: True satisfaction comes from love.
Hippolytus: What when love dies? Alarm clock rings it's time to wake up, what then? (93)

Instead of God or love Hippolytus follows his own metanarrative of absolute honesty, which is how he invests his life with meaning. It is a metanarrative that befits his self-chosen isolation and self-reliance as he depends on no one but himself in his project of always being honest. Until Phaedra's suicide, however, his honesty has manifested in nothing but emotional brutality to others (cf. Barry 125) and contempt for the hypocrisy of other people that turned him into a proper misanthropist. Only after the accusation of rape does he become personally involved in his ideal of honesty and is determined to pursue it until the bitter end.

The third explanation why Hippolytus seeks his own destruction so relentlessly connects Kane's play to the writings of Bataille and, again, Artaud, however this time on a thematic more than on an aesthetic level. The way Hippolytus desires his violent end recalls the desire for the violent destruction and dissolution of the individual as providing the pleasurable experience of pure life and the momentary reunion of mind and body that is described by Bataille and, slightly differently, by Artaud. As Quay argues, drawing on Kane's own statements in her conversation with Tabert (cf. 10), the moment of Hippolytus' death is the moment in which he surmounts the Cartesian mind-body dualism and experiences life in its full authenticity (cf. Quay 260). Similarly, Müller writes that, in the "most important scene" of the play, "Hippolytus dies and experiences a moment of satisfaction and freedom, a unique unity between body and mind that he does not even think of when he is alive" (98). This dissolution of the border between mind and body, which is effectuated by the violent opening of the body, is part and parcel, for Bataille, of the always desired but never entirely accomplished transgression of the discontinuous individual to the state of continuity – the ravishing rupture of the *principium individuationis*, in other words, that had already fascinated Schopenhauer and Nietzsche (cf. Wiechens 71)[51] and that reappears in the writings of Bataille, Artaud and Lacan.

In death, this usually temporary transgression is carried out permanently. The moment of death is the first time Hippolytus is truly alive, which gives full meaning to his "Life at last" (90) and his "If there could have been more moments like this" (103). For the detached, isolated prince, true life is only possible in the face of death. While in Bataille's account people usually can make their discontinuous existence bearable through repeated moments of transgression and excess – moments when rules and taboos are broken in ritualised or temporarily sanctioned transgressive acts of sex and violence –, Hippolytus is bereft of this possibility since for him, who can have everything and everyone in a world without sexual taboos, transgression and excess are hardly possible. The opening scene that shows him surrounded by junk food and expensive toys, masturbating, and watching a "*particularly violent*" (65) film in total apathy indicates the degree to which saturation and affluence have numbed his capacity to experience excess or the transgression of taboos. His inability to derive pleasure from sex (cf. 83) and the borderless promiscuity of the entire royal family demonstrate that sex has lost all its transgressive nature and thus its eroticism. Only his violent death can reconnect Hippolytus to life.

The step from Bataille to Artaud is only a small one. Edward Scheer calls Bataille "one of the very few who can be said to be in any way a fellow traveller with Artaud" and regards the works of the "surrealist ethnographer," for example his *Eroticism*, "an important discursive frame for considering some of Artaud's considerably less systematic formulations" since his "notion of transgression can be seen as a key to explicating some of Artaud's [theatrical] strategies" (16). Scheer goes as far as to suggest that "[p]erhaps the theatre of cruelty is exactly what Bataille had in mind with the notion of transgression" (16), which might not be too far-fetched considering Artaud's intention of re-connecting audiences to their lost natural lives through the violation of taboos in the theatre. De Vos shows in his study of Artaud's influence on Beckett and Kane that the

---

[51] In *The Birth of Tragedy*, Nietzsche speaks of this ambivalently joyful and destructive experience as "die wonnevolle Verzückung [...] die bei demselben Zerbrechen des principii individuationis aus dem innersten Grunde des Menschen, ja der Natur emporsteigt" (28).

desire for a reunion of body and mind lay at the heart of Artaud's thoughts and "marked the beginning of his theatrical career throughout as a quest for wholeness and unity, a quest for the real" (28). De Vos draws on the close parallels between Artaud and Lacan, in whose theoretical system 'the real' denotes (among other things) the "undifferentiated state of perfection" which the child experiences in the "harmonious unity with its mother" (29) before it is torn out of this blissfulness through entering the symbolic order (cf. 33–34). As Lacan indicates elsewhere (see 2.1.3), this loss is already preceded by another loss which happens at the moment of birth, and it is here, De Vos argues, that for Artaud the story of suffering begins. "Artaud himself has expressed several times his wish to return to the prenatal," De Vos writes, a realm where "life resembles in nothing reality as we know it; it is rather a bundle of forces, a surge of energy [...] freely flowing, merging, burning, spinning, gyrating, exploding" (38) – a realm, in other words, where the *principium individuationis* and feelings of demarcation, separation, isolation have no place. De Vos summarises the two experiences of loss that characterise Lacan's and Artaud's conception of human development as follows: "From an undifferentiated world in the womb, man has been expelled to a reality made up of differences and appearances. Moreover, he has been stolen from himself. Once introduced into the linguistic order, he has been cut off from what cannot be rendered in language" (46). This double experience of loss has installed in the human subject an essential and insatiable desire to return to the prelinguistic/prenatal state of the 'real.'

This desire is usually diverted to surrogate objects – a life-saving diversion, as De Vos points out, since "the approach toward the real brings about the risk of the subject's elimination. After all, the lack belongs fundamentally to the subject and its erasure will inevitably put an end to his existence too" (53). In Kane's Hippolytus, however, we have a figure who twice behaves in opposition to this description of 'normal' existence: while he is virtually dehumanised by his lack of desire in the first half of the play, his burning desire in the second half is not directed towards a surrogate object but presents an 'approach toward the real' which indeed ends with 'the subject's elimination.' As De Vos aptly formulates, Hippolytus "is gradually killing himself as a 'normal' subject. In fact, the prince is on the verge of collapsing into the abyss of a desireless vegetative existence and it is Phaedra who makes several attempts to prevent him from pulling out completely" (93). Put differently, Phaedra is offering herself as a surrogate object for a desire Hippolytus does not have in the first place. The Doctor's speculations about her "abilities as a surrogate" (*PL* 68) acquire a deeper sense in this context. Her inability to raise Hippolytus' desire does indeed not disqualify her as a desirable surrogate object. Rather, the prince has numbed himself out of any desire. His existence is marked only by the void which the fundamental experience of loss has created, but not by the desire that usually arises from it. Only after her suicide and accusation, which indeed work as her gifts of love for Hippolytus, does he develop – or rediscover – the longing and desire he had once felt and directed to Lena and which he now does not divert to a surrogate object but follows directly, with all the consequences this entails. Suddenly clear-sighted about the true goal of human desire, he seeks the dissolution of his discontinuous, separated individuality which he finds in his violent death. "In this short flash," De Vos writes, "he embraces the abolition of his fragmentation as a human being and welcomes the unity he has been longing for since Phaedra's sacrifice" (96). The forceful destruction of his body facilitates, for a brief moment, the unity of body and mind and replaces his demarcated, isolated existence with a feeling of unity with the whole of existence from which he was torn by birth and the entry into the symbolic order. This is what he has been longing for, and it is a fundamentally precarious desire. While Phaedra's love was precarious in the sense that her object of desire was incontrollable and unattainable, Hippolytus' desire is precarious because its 'object' resides in the liminal sphere between continuity and discontinuity where lack and desire mean life, and consummation means death.

6.4 *Cleansed:* Fragments from the Laboratory of Love

Apart from the fact that love is their central topic, *Phaedra's Love* and *Cleansed* also share a close affinity to the ideas of Artaud. But while in the largely naturalistic *Phaedra's Love* this affinity is observable above all on a thematic level – Hippolytus' desire for the reunion of body and mind in the dissolution of the demarcated, individual subject echoes the yearning underlying Artaud's writing that connects him to Lacan and Bataille –, Kane's next play comes close to Artaud's concepts in terms of aesthetics, too (cf. Saunders, *Love Me* 91; Quay 308). Scholars, critics, and colleagues have pointed out the play's proximity to surrealism (cf. Rubik 132; Innes 533) and expressionism (Greig xiii; Macaulay 567) and its dreamlike (Carney 275; Saunders, *Love Me* 94) or nightmarish (Peter 564) atmosphere. When Artaud demands from theatre that it "provides the spectator with the truthful precipitates of dreams in which his taste for crime, his erotic obsessions, his savagery, his fantasies, his utopian sense of life and of things, even his cannibalism, pour out" (190), *Cleansed* seems to meet his expectations astonishingly well. This is how critic John Peter described his experience of *Cleansed:*

6.4 *Cleansed*: Fragments from the Laboratory of Love

> I came out of the theatre after Sarah Kane's new play, feeling bruised to the bone, tight in the stomach and hopeless. [...] *Cleansed* [...] is a nightmare of a play: like a nightmare, it unreels somewhere between the back of your eyes and the centre of your brain with an unpredictable but remorseless logic. As with a nightmare, you cannot shut it out because nightmares are experienced with your whole body. As with a nightmare, you feel that somebody else is dreaming it for you, spinning the images out of some need that you don't want to think of as your own. (564)

With the 'remorseless logic' of a dream, the play forces the spectator to confront secret and repressed desires and come to grips with the 'truths' transported by the symbolical and metaphorical images it presents. We do not so much follow a plot as try to digest the series of compelling images that do not tell a coherent story but are thematically connected as they evolve around the vicissitudes of love. Accordingly, it does not make much sense to try to summarise the story of this episodic play in a linear narrative. What can be done instead is to describe each of the interwoven plotlines on its own.

The figure who does most to connect the different episodes and plotlines is Tinker, who is in charge of the institution that serves as the setting and of which all the other characters are inmates. Tinker variously tortures, mutilates, or kills other figures in situations that seem designed to punish and prevent love. However, he is himself searching for love and is torn between the unattainable Grace and a female stripper on whom he projects his love for Grace. Grace has come to the institution searching for her dead brother Graham, who was killed by Tinker with an injection of heroin into his eyeball. Graham's ghost, however, remains present on stage, visible only to the audience and Grace, whom he accompanies throughout the rest of the play as a kind of guardian angel. Driven by incestuous desire, Grace not only sleeps with Graham but gradually resembles him more and more in an attempt to virtually 'become' Graham. Unable to deter her from this project through torture and rape, Tinker eventually grants her wish and performs an amateurish but nevertheless successful sex change. The necessary body part is taken from Carl, whose vow of eternal love to his lover Rod at the beginning of the play was subsequently tested by Tinker with torture and a series of amputations. The last figure is Robin, a childlike nineteen-year-old boy, who develops an unrequited love for Grace. Grace feels motherly affection for the boy but reserves her erotic love for Graham. Robin hangs himself after, in an attempt to impress Grace, he has learned to calculate with an abacus and realises how many days of imprisonment are still ahead of him.

What connects the twenty scenes more than a coherent storyline is the repetition and mirroring of certain themes, symbolic gestures and actions: Tinker puts a pill into Grace's mouth in Scene Three, and he puts the ring symbolising Carl and Rod's love into Carl's mouth in Scene Four, each time with the single command "Swallow" (114, 118); Carl's promise of eternal love to Rod in Scene Two, interrupted by Rod's incredulous laughter and gruesomely tested by Tinker in Scene Four, is repeated almost verbatim in Scene Six, where Tinker promises unshaking fidelity ("I won't let you down"; "I won't turn away from you") to the unnamed stripper, who responds with laughter (122); this rejection of Tinker by Woman is mirrored in the following scene, where Grace does not reciprocate Robin's infatuation; in the same scene, Grace teaches Robin to read and write and his attempt to express his feelings for her by writing her name in a way that resembles his perception of her (he represents her as a flower because "She smells like a flower" [129]) is burnt by Tinker, who in the next scene cuts off Carl's hands with which he has just written a message for Rod in the mud; the same attempt at destroying all means to communicate feelings (and love in particular) is observable in Scene Thirteen, where Tinker cuts off Carl's legs after he has performed a dance of love for Rod, and in Scene Fifteen, where Tinker forces Robin to eat an entire box of chocolates that were intended as a present for Grace; another dance of love has already been performed by Graham for Grace (Scene Five), whose love for each other is moreover symbolised in Scenes Six and Ten by flowers that suddenly start to grow on the stage; in Scene Nine, the unnamed stripper, on whom Tinker starts to project his love for Grace, asks Tinker to "save" her (130), and in Scene Ten he approaches Grace, who has just been beaten, raped, and shot "*by an unseen group of men whose* **Voices** *we hear*" (130) with the words, "I'm here to save you" (133); the 'Saviour' is also evoked when Carl betrays Rod in Scene Four, which is reminiscent of Peter's betrayal of Jesus the night before his death (the analogy is enforced when Rod, threatened to be tortured and killed with a pole pushed up his anus, cries out "Jesus" in response to Tinker's question, "What's your boyfriend's name?" [117]), and in Scene Ten, where Grace, heavily beaten by the unseen men, cries "Graham Jesus save me Christ" (131) and Graham, touching her body, "*begins to bleed in the same places*" (132) in an image of reversed stigmatisation.[52] The most important recurrent motif, however, that gives a form of coherence to the entire play and, moreover, also connects it to *Phaedra's Love*, is that of the obstacle preventing the consummation of love.

---

[52] Christine Quay, too, identifies these and further parallels between Jesus and Graham, who 'rose again' from the dead, whose 'garments' are 'parted' after his execution (Tinker gives Graham's clothes to Robin before Grace demands them for herself), and who seems to have conquered death for Grace, too, who "*opens her eyes and looks at him*" (*Cleansed* 133) after minutes of automatic gunfire at her body, while daffodils, the flower connected to Easter as a symbol of rebirth, burst out of the ground and cover the entire stage in yellow (cf. Quay 274–75).

6.4.1 A Panopticon of Love

In a (rather superficial) sense, both *Cleansed* and *Phaedra's Love* can be read as variations of the classical romance pattern of illicit love faced with (almost) insurmountable obstacles that either bring about the lovers' tragic ending or serve to test and prove the invincibility of their love. Such a reading is only partly satisfactory. In *Phaedra's Love*, only Phaedra is ardently in love and only her death has a tragic quality, whereas Hippolytus' enthusiasm for his violent end turns the final catastrophe almost into a happy ending. And in *Cleansed*, where love indeed surmounts the most horrible obstacles, it also proves fatal for Rod, Robin, and, in a sense, Grace. Nevertheless, it is worthwhile to explore the obstacles in *Cleansed* in some detail, which amounts to an analysis of the dramatic situation of the play. What kind of 'institution' is Tinker running and is there an interpretation that can explain the series of obstacles that are put in love's way as more than a merely thematically connected series of stage images or installations circling around the precariousness of love?

The most widespread explanation of the dramatic situation is that which Sierz quotes from the blurb of the playtext: "In an institution designed to rid society of its undesirables, a group of inmates try to save themselves through love" (*In-Yer-Face* 112). Many critics have adopted this reading (Edwardes 563; Benedict 564; Spencer 565; Billington 566; Gross 567), which ascribes a strong political message to the play. Zimmermann summarises this interpretation when he writes that

> *Cleansed* focuses on the fascist exclusion from society of everyone who does not conform to its idea of 'normality': the drug addict, the homosexual, the psychotic, the mentally disabled, the alien. The means are examination, lobotomy, electroshock, chemical therapy by drugs, etc. which are seen as another form of brainwashing, torture and auto dafé, murder and eugenics used by the fascist state to 'purify' its society of dissenters and deviants. In a fundamentalist way the play criticizes and exposes as fascist the order of reason and morality which with the norm creates the deviation from the norm. (180)

David Ian Rabey and Christina Wald offer similar interpretations when they describe the play's setting as "an institution of social and probably ethnic cleansing, where inmates are spiritually and physically reduced on each occasion when they profess defiant action" (Rabey 206) or as "an ostensibly 'therapeutic' setting" where the "inmates are, as the play's ambivalent title indicates, either re-educated by drastic measures and thus 'purified,' or, if the re-education fails, killed and thus 'purged'" (Wald 198). But even if the play allows for a political interpretation of this kind, the text provides little support for it. Tinker nowhere specifies the deviations he allegedly punishes or the norm he seeks to preserve. Homosexual and incestuous love are penalised just as much as Robin's heterosexual desire for the person Tinker himself desires. Graham's death through an injection of heroin is demanded and appreciated by Graham. Tinker performs the sex change Grace asks for and the final scene shows Grace, who "*now looks and sounds exactly like* **Graham**" (149), and Carl, emasculated and wearing (Grace's) women's clothes – two rather obvious deviations from 'normality' that are not punished but produced by Tinker and his institution. For a political reading, I think, it is indeed inauspicious, as Billington suggests, that "you never learn who or what lies behind Kane's hermetic chamber of horrors. If it is meant as a political metaphor, it remains an extremely shadowy one" (566). Due to the lack of concreteness and logic coherence regarding the institutional violence, attempts at a political reading will probably yield no more than Peter's reasonable but very general and almost existentialist interpretation that "[t]he world is a prison, disguised as an educational institution, which trains you with the utmost brutality for nothing much else than dying" (564).

It might thus be more promising to read *Cleansed* not as a political play but as a play about the experience of powerful, shattering emotions – and Kane herself provides the link between this topic and the atmosphere of a fascist institution created by the play. In the conversation with Tabert, she reveals not only that the Holocaust, although having influenced her writing, is not the content of the play (cf. 15), but also that Roland Barthes' comparison of an unhappy lover with a prisoner at Dachau concentration camp in *A Lover's Discourse* contains an important clue to her play, which ultimately is about situations of total self-loss (cf. 16). It is worth quoting at length the passages in question from Barthes' collection. The comparison is part of the fragment entitled *catastrophe*, which he initially defines as "[v]iolent crisis during which the subject, experiencing the amorous situation as a definitive impasse, a trap from which he can never escape, sees himself doomed to total destruction" (48). Of course, the destruction of the subject that is at stake here is not physical, not fatal, but merely psychic or emotional, and Barthes himself questions the legitimacy of a comparison between lovesickness and the physical extermination of millions of people. "Yet," he argues, "these two situations have this in common: they are, literally, panic situations: situations without remainder, without return: I have projected myself into the other with such power that when I am without the other I cannot recover myself, regain myself: I am lost forever" (49). It is the threat of a complete loss of self in love, the precariousness that results from making an uncontrollable other an essential, indispensable part of one's self-worth and subjectivity, that Barthes broaches here and that plays a central role in Kane's play. Both use drastic, shocking, and concrete images to talk figuratively (in similes and metaphors) about the unpresentable inward side of human emotions. Both

6.4 *Cleansed*: Fragments from the Laboratory of Love — 169

evoke or create images of extreme pain, fear, desperation, and humiliation to exemplify their notion of love's precariousness, which resembles in quality, if not in degree, the situations of horror they choose.

In a reading that focuses not on possible political implications of the play but on its examination of love, it makes sense to interpret the play's setting in "*a university*" (107) in a way that supports this approach.[53] Instead of viewing this location with Michelene Wandor as implying clear criticism of our society's educational system and structures of socialisation as "sadistic, sexually punitive [...], and exclud[ing] any sense of female-gendered identity" (233), I find it much more convincing to follow Sean Carney's lucid interpretation of the setting as an allegory for the attempt to approach love in a scientific manner. He writes:

> The instruments of the university are still in Tinker's hands, namely the application of cold reason to the question of human feelings, as if the situation is a laboratory whose primary purpose is the reified, fascistic dissection of human emotions under controlled experimental situations. English reason performs a vivisection of the aspects of the human that said reason represses and puts behind bars: feelings. Love is the object of study for Tinker, attacked not so much because he wants to destroy it but because he wants to understand how it works, particularly in himself. (276)

Carney here not only recalls the mind-body dualism (here: reason-feeling dualism) that is so intricately connected with Kane's Artaudian approach, he also offers a very satisfactory explanation of Tinker's otherwise hardly reasonable actions. He is not punishing love or deviant behaviour, but he is examining and testing the conundrum of love that he wants to understand because he is himself affected by it. The result is a play that examines different varieties of love, focusing, however, not on differences but on the precariousness – particularly in the form of self-loss – that they all have in common.

This interpretation is supported, for example, by Robert Lublin, who writes that *Cleansed* sets out "to explore the nature of love in what appears to be a series of experiments" (121), and by Christine Quay, who is reminded of an experimental setup designed to analyse love, pain, and identity (cf. 245). For Quay, the play produces a "typology of love" (271) that encompasses sexual, affectionate, and spiritual love alike but that always draws attention to the danger of total self-renunciation if such love is absolutised (cf. 272). Sierz offers a similar account that sees the play as an analysis of different variants of love:

---

[53] Interestingly, the original production of *Cleansed* apparently did little to convey this stage direction to the audience. Several critics wrote that it was only after consulting the script that they found, to their surprise, that the play is set in a university (cf. Peter 564; Nightingale 565; Gross 567; Macaulay 567; Morley 568).

> In a play about love, each of the four main relationships is different, each symbolic. Grace and Graham represent the fantasy of incestuous identity-sharing twins; Carl and Rod are the classic couple, one member of which is idealistic, the other realistic; Tinker and the dancer represent domination and alienated love; Grace and Robin experience a teacher and pupil, mother and child rapport. (*In-Yer-Face* 114)

From this perspective, the play reproduces the very discourses of love that have preoccupied Western societies for so long: the idea of love as merging or returning to a form of lost wholeness, here represented by the undying desire of Grace and Graham to fuse into a single entity; the clash of idealism and realism, represented by Rod's incredulity towards Carl's promise of eternal love; the problem of unequal power, symbolised by Tinker's relationship with the female stripper who is a 'property' of Tinker's institution, who is exposed to his appropriating gaze, and on whom he projects his love for another, unattainable object of desire (Grace); and the nexus of motherly/affectionate and erotic love that dominated much psychoanalytical theory. All these forms of love are put on display and under test conditions, turning the play with its university setting into a panopticon of love. Some of the various obstacles to love are part of Tinker's experimental setup, others arise from within the characters. But what they mostly help to unfold is what is at the same time the play's primal topic: that absolute love is accompanied by a loss of self.

6.4.2 Heaven or Hell: *Cleansed*'s Ambivalent Eschatology of Love

Not least based on Kane's own statements (cf. "Interview" 20; see also the interviews quoted by Saunders in *About Kane* at 74 and 78), *Cleansed* has often been understood as a principally optimistic play. "While *Cleansed* is undoubtedly a dark play," Saunders for instance writes, "Kane also believed it is essentially hopeful in its central theme: on how love for all the characters can survive even the most extreme and savage of situations" (*Love Me* 91). This interpretation has not remained uncontested, however, mostly because of the instances of self-loss that accompany love in the play and, for many, mark it as a source of pain and loss rather than pleasure and hope. In this final part of the chapter, I will first collect the most important reasons for an optimistic reading of the play. I will then consider the nexus of love and self-loss, which is indeed the focal point of Kane's treatment of love, before finally asking the question whether this precarious connection of love and (self-)destruction does necessarily call for the rejection of optimistic interpretations of the play.

Love's Promises: Durability, Unity, Redemption

For Sierz, "*Cleansed*'s idealism lies in its conviction that love is the one basis of hope in an evil world" (*In-Yer-Face* 114), manifest in the refusal of love to vanish even in the face of extreme cruelty and terror. For Quay, the play contains the strongest expression of Kane's belief in the metaphysical power of love to grant meaning and redemption in an otherwise cruel and senseless universe (271) – even though this conviction, which Quay observes in all of Kane's plays before *Crave*, is already supplemented by traces of doubt in *Cleansed* (cf. 300). And to many critics, love's redemptive power and its capacity to survive exposure to horrific cruelty in a 'fascist institution' appeared as the play's main message (cf. Peter 564; Spencer 565; Nightingale 565; Billington 566; Clapp 566; Marlowe 567). The two plotlines that give rise to this view of love as an invincible source of hope and redemption are that about Carl and Rod and that about Grace and Graham.

The story of Carl and Rod starts out right away as a debate between an idealist and a realist. Knowing each other for no longer than three months, Carl wants to exchange rings with Rod as a sign of commitment, protesting against the latter's laughter and incredulity that he will always love him, never betray him, never lie to him, never turn away from him, and that he would even die for him (cf. 109–11). Rod answers with what Rebellato calls "the most genuinely romantic speech in contemporary British playwriting" ("Appreciation" 281):

> Listen. I'm saying this once.
> (*He puts the ring on* **Carl**'s *finger.*)
> I love you *now*.
> I'm with you *now*.
> I'll do my best, moment to moment, not to betray you.
> Now.
> That's it. No more. Don't make me lie to you. (111)

In an imitation of a traditional wedding ceremony, Rod refuses to make the promise of eternal love and fidelity that usually goes along with the exchange of rings and instead assures Carl only of his momentary affection. The speech is certainly another example of Kane's esteem for sincerity and absolute honesty that was already observable in her depiction of Hippolytus, but, with regard to Rebellato's assessment, it is arguably much more contemporary than romantic. With due respect to his honesty, Rod is also a representative of the commitment phobia described, for instance, by Illouz and Bauman that characterises contemporary society and relationships. His declaration of love separates two components that are traditionally united in the concept of romantic love. Whereas the romantic concept combines the overwhelming and uncontrollable nature of love with its durability, the former excludes the latter in Rod's concept. While according to the romantic argument the experience of irresistible attraction and passion serves as an assurance of the 'rightness' and thus durability of love, in Rod's version the uncontrollable force of passion prevents sincere commitment since he cannot guarantee not to fall out of love with Carl and in love with someone else: "Carl. Anyone you can think of, someone somewhere got bored with fucking them" (111). For Rod, undying love is ridiculous as an idea and untenable as a promise and a declaration of eternal love amounts to an outright lie.

Kane's play first seems to support Rod's anti-idealist view. Under torture and threatened with a horrible death through rectal impalement, Carl betrays his lover: "Not me   please not me   don't kill me   Rod not me   don't kill me ROD NOT ME   ROD NOT ME" (117). Tinker further emphasises the betrayal by cutting off Carl's tongue, the 'medium' of his betrayal, and forces him to swallow the ring he has put on Rod's finger as a sign of his invincible love. But the story does not end here, and neither does the love between Carl and Rod. In subsequent scenes, Carl tries to ask Rod for forgiveness, first by writing his plea in the mud with his hands and then by performing a dance of love. Both attempts are followed by Tinker's destruction of his means of communication, who first cuts off Carl's hands and then his feet. However, the 'Passion' of Carl is not in vain. In view of Carl's unfaltering love, Rod not only forgives the betrayal but even turns his declaration of momentary love into a promise of eternal love. Realising that "[t]here's only now. [...] That's all there's ever been" (142) and that it is his own decision, moment for moment, to love Carl, he repeats Carl's previous vow of love:

> I will always love you.
> I will never lie to you.
> I will never betray you.
> On my life. (142)

The scene, which develops into an intense image of the lovers' attempt at merging – they make love, Rod makes Carl swallow the second ring so that both rings are (re)united inside him, and they finally "*hug tightly, then go to sleep wrapped around each other*" (142) –, ends with Tinker's last test of love. When Rod, asked to make a decision, is prepared to die for Carl, Tinker cuts his throat.

Rod has developed from a cynical realist into an idealist who dies for love. Carl's initial betrayal is outweighed by the proof of unyielding love that is produced through Tinker's series of tests. As Brusberg-Kiermeier puts it, "Carl's continuous suffering becomes a proof of his love for Rod, which in turn enables Rod to admit the special quality of their love and to sacrifice his life for Carl" ("Cruelty" 86). Even if Carl is not prepared to die for love, his love never dies as long as he is alive – and this is enough to convince Rod of love's durability. Thus, as Zimmermann writes, "Tinker's torture of Carl reveals the weakness as well as the invincible strength of Carl and Rod's love" (180). Love may not bring every lover to sacrifice his life for the beloved, and keeping the promises made in declarations of love is, moment to moment, a decision that depends on more than on (still) being in love. But this does not mean that love does not at least have the capability to survive both betrayal and the extremities of life.

In *Cleansed*, love does not even end with the death of one of the lovers. The story of Grace and Graham is the story of (incestuous) love that is both spiritual and deeply sensual and stays alive beyond death. Or rather, it seems, their love can only reach its desired end after one of them is dead since their desire to unite their mind(s) within one body presupposes the physical death of one body. Thus, instead of psycho-pathologising Grace's state of mind and seeing Graham's presence as a projection of her troubled psyche, I understand the story of the woman who makes love to her dead brother and gradually resembles him more and more until she eventually 'becomes' him as Kane's dramatic visualisation of an erotic love that is both spiritual and sexual and that is driven, above all, by the desire to overcome the separation that is imposed on the lovers by their individual bodies. In other words, the Grace/Graham plot can be seen as Kane's modified version of the Aristophanes myth and her comment on the ancient idea of merging.

Wearing Graham's clothes, imitating his movements and way of speaking, and finally turned into a man through Tinker's operation, Grace apparently transforms into her brother in the course of the play. It would be misleading, however, to understand Grace's development merely as a change of body and personality. Together with Quay and De Vos, I understand her desire to change her body so that "it looked like it feels. Graham outside like Graham inside" (126) not as a desire to abandon her old self in favour of a new one but as a desire for complete and perfect unity. Driven by the "discomfort that comes from the separation between body and mind," De Vos argues, "she aspires to eventually merge both components into a flawless whole" (137). Similarly, Quay writes that the sex change enables the unity of body and mind that Grace is longing for and that allows her to transcend her existence (cf. 276). What Grace seeks to attain is the union of body with a mind that hitherto seemed to stretch over two bodies. Feeling one with Graham, Grace experiences her body as an annoying limitation,[54] as the most tenacious obstacle to an experience of unity and whole-

---

[54] As Deubner observes, Graham, too, feels impaired by the limitation of his body. His "I know my limits. Please" (107) when he asks Tinker for the overdose of heroin can be read as the pain-

ness. Grace 'becoming' Graham in the play is then not an image for one becoming the other but for two becoming one. As Lublin holds, "Grace and Graham are the object of and resolution to each other's desire. Consequently, each serves in large part to constitute the other's subjectivity, and it becomes difficult to determine where one ends and the other begins" (122). Even if at the end Grace "*looks and sounds exactly like* **Graham**" (149), I do not think that she has simply turned into her brother. Rather, the annihilation of Graham's body and the opening and modification of Grace's body allow for the union of mind within this new, modified body. Maybe the strongest image for Grace and Graham's desire to become one (again) is their love-making scene, or rather its description in the stage directions:

> *They slowly embrace.*
> *They begin to make love, slowly at first, then hard, fast, urgent, finding each other's rhythm is the same as their own.*
> *They come together.*
> *They hold each other, him inside her, not moving.* (120)

Images of merging abound in this description and culminate in the momentary state of tranquillity immediately following the sexual act when, for a brief moment, all desire has ceased in a temporary feeling of complete wholeness. But, like in the Aristophanes myth, sex is only a temporary relief from the burning desire that haunts all human beings as long as their distinct and separate bodies prevent a permanent union of body and mind. It is only through Tinker, who plays at Hephaestus in Kane's version of the myth, that the two separate halves are eventually welded together in a new body.

A third possible reason for optimistic interpretations of the play is Tinker's emotional transformation as a further consequence of love's redemptive power. It has been variously noted that Tinker, who tries the love of the inmates so mercilessly, is himself in "desperate need for affection" (Urban 43) and is "yearning for friendship and love" (Zimmermann 179). But De Vos shows that Tinker's conscious search for love is not present from the start but is the result of a considerable change of character during the play. Initially, Tinker is a stranger to human emotions. Whether, as De Vos suggests, "he attempts to amputate Carl and Rod's love" (129), or whether, as I have argued above, he is executing a scientific vivisection of a mystery he is as yet unable to understand, in any case De Vos is right that far into the play Tinker is struck with blindness regarding the

---

ful recognition of the limitation of his bodily existence that he seeks to transcend through the drug and, ultimately, through death (cf. Deubner 136).

realm of emotions. "He is literally incapable of looking at the signs of human vulnerability, affection, friendliness, or sexuality" (130), De Vos observes and supports his reading with scenes from the play demonstrating that Tinker repeatedly avoids the sight or experience of human vulnerability and affection in situations where he is not in control. He looks away or at the floor when Graham thankfully smiles at him for granting his death wish (cf. 108), when Grace undresses to put on Graham's clothes (113), and when he visits the peep-show booth for the first time (121). To perform signs of affection like stroking Grace's and Carl's hair (cf. 113; 117), holding Grace's hand (133), or kissing Carl's face (116) is only possible for him in situations where he is absolutely in control, the targets of his affection being tied up or (almost) tortured into unconsciousness. For De Vos, "Tinker's psychological problem consists in the fact that these instances that attest to a more human contact might threaten his control and power. In order to be constantly in full control, he can never give in to his feelings" (130).

Not unlike Hippolytus, then, Tinker initially bans the precarious dependence that comes along with love from his life in order to maintain his position of total power and control. Knowing that love necessarily entails a loss of power, he is afraid of its effect on his sovereignty. At the same time, however, his 'scientific' curiosity towards love and his longing for human affection and contact, which becomes most obvious in his visits to the stripper, undermine any attempt at self-sufficiency. Quite in line with his wish to retain a position of power, his careful approaches to love at first take place in situations he can control: both the torture chamber and the peep-show booth are places where he is in charge. Quite in line with the incontrollable nature of love, however, he is falling in love, almost ironically, with an unattainable 'Lady,' Grace, whose indestructible ties with Graham he has to accept eventually. His position of superior power is of no use. While he can humiliate Robin, in whom he sees a potential rival, he cannot destroy the love between Grace and Graham, and his performance of the requested sex change is his ultimate recognition that he cannot have her.

The experience of unrequited love does not lead to Tinker's further embitterment and brutalisation. To the contrary, the play provides an utterly hopeful twist in that Tinker seems to be reformed into someone who is now able to love. Throughout the play, Tinker has been projecting his love for Grace onto the unnamed woman: he calls her Grace, he comes to "save" Grace (133) after the woman has asked him to save her (130), and he furiously urges her to admit her femaleness after Grace has voiced her wish for a sex change (137). Giving up his desire for Grace enables Tinker to see the unnamed woman no longer as an inadequate substitute for his real object of desire but as a distinct personality that can answer the desperate need for love he has come to accept as a result of his 'scientific analysis' and personal experience of the irresistible power of love. As De Vos aptly notes (cf. 132–33), it is only after Grace's 'disappearance' that the unnamed woman properly 'appears' when she "*opens the partition and comes through to **Tinker***'s *side*" (*Cleansed* 147) for the first time in the penultimate scene. That the following love-making scene both visually and verbally resembles the previous, highly idealised love-making between Grace and Graham – again, for example, the final stage direction reads, "*They hold each other, him inside her, not moving*" (149) – further testifies to "the sincerity of this relationship" (De Vos 132). "At the end of the day," De Vos concludes, "Tinker seems the one to have achieved real catharsis" (132). Having witnessed the strength of love in his test series, having admitted to himself that his need for love and affection is more urgent than his need for power and control, and having relinquished his desire for an unattainable object, the initially inhuman torturer is now capable of opening himself to love in an utterly precarious and thus fully human relationship.

Love's Precarious Downside: The Loss of Self
While the precariousness of love in the form of the loss or unattainability of the other is surely addressed in the play – Robin commits suicide not only because he realises how long his sentence is but also because Grace does not reciprocate his love; Carl has lost his lover Rod and is crying throughout the last scene –, the main focus lies on images of self-loss. For Quay, the largely optimistic view on love's redemptive power is somewhat disturbed by the illustration of the strong link between absolute love and the disintegration of the self that she observes in Grace and Carl (cf. 272). Many other scholars have commented on this in a similar fashion. For Greig, "the exploration of love's assault upon the wholeness of the self" (xiv) is the binding motif of the play. Clare Wallace notes that "[t]he theme of self-loss is played out from different perspectives" in *Cleansed*, where both in the Carl/Rod plot and in the Grace/Graham plot "loss of self is mapped out on the body" ("Sarah Kane" 94). Carl is dismembered and Rod is killed as a result of their love, and "Grace 'becomes one' with her brother/lover only by the erasure of her own identity" (94). In the same vein, Maria Elani Capitani understands Grace's "absolute mental/bodily fusion" with Graham as a traumatic "loss of self" (139). With its various images of mutilated and fragmented bodies, Capitani argues, the play "suggests that absolute love implies a dissolution of identity" and the "atomisation of the self" (140). De Vos, too, writes that "Grace loses her identity as a result of her search for an unspoiled and flawless self" (137), and according to Zimmermann, the combination of Tinker's tortures and Grace's desire to become Graham "leads her to a complete loss of herself" (179).

While it is more than convincing to read the images of physical dismemberment, dissolution, and transformation as metaphors for internal, psychic processes experienced in and through love, it is less clear how love's disintegration of the self is actually evaluated by the play. In the writing of most commentators, the concept of self-loss in love seems to be connoted negatively by default. This posture befits the post-romantic scepticism regarding the concept of merging. Throughout his work, Irving Singer repeatedly attacks the idea as both unfeasible and undesirable in any literal sense that would imply the disintegration of distinct selves (see e.g. *The Modern World* 406–17, *Pursuit* 23–25) and his trilogy describes a general development in the history of ideas during which the ideal of fusion with the beloved into a form of higher existence was largely replaced by the insistence on the preservation of individuality and the integrity of the self. From this vantage point, *Cleansed* can only appear ambivalent due to the apparent "duality within the play between tenderness and affirmation in love placed against annihilation and loss of self-hood," as Saunders observes (*Love Me* 93). The optimism of the play, manifest in the images of love's invincible strength and Tinker's transformation, is stained, in this interpretation, by the constant foregrounding of self-loss through love.

It is questionable, however, whether self-loss is the menacing factor in Kane's play that it seems to be. In particular, I doubt that Grace's transformation, which, as I have explained above, I read more as an act of merging or fusion than as a change from one into another person, has necessarily to be understood as a story of loss and thus, as Quay suggests, as a trace of Kane's doubt in the redemptive power of love and as a warning against the potential destructiveness of absolute love (cf. 274–76). Despite the truly ambivalent ending of the play – Grace/Graham's final monologue allows as much for optimistic and pessimistic readings as the final tableau, where it stops raining, the sun comes out, and Grace/Graham is smiling but where Carl is crying and the sunlight and the squeaking of the rats are increased to an unpleasant level – and despite the tendency in scholarly criticism to prefer ambivalent interpretations to closed readings, it is not unreasonable to understand *Cleansed* as a play that stresses the precariousness of love with stark images but that ultimately eulogises its invincibility and redemptive power. As Lublin writes about Grace and Graham: "With this relationship specifically and the play generally, Kane suggests that love offers the possibility of answering the demands of desire in a full and complete way" (122). Such an interpretation ties in with Kane's own self-analysis. In the interview with Tabert, she stated that "[p]robably all my characters in some way are completely Romantic" and that she considered herself "a complete and utter Romantic, in the tradition of Keats and Wilfred Owen" (qtd. in Saunders, *Love Me* 64). In view of this self-assessment, Keats' famous lines to

Fanny Brawne sound downright intertextual: "I cannot exist without you," Keats writes to his beloved. "You have absorb'd me. I have a sensation at the moment as though I was dissolving. [...] I have been astonished that Men could die Martyrs for religion – I have shudder'd at it – I shudder no more – I could be martyr'd for my Religion – Love is my religion – I could die for that. I could die for you" (122). All the elements of Keats' experience of love as expressed in this letter are combined in the imagery of Kane's play: insatiable desire for the other, dissolution of the self, endurance of martyrdom and death for love, and the association of love with religious sentiments. Moreover, Kane's self-description as a Romantic opens up a possible link between her play and the characteristic Romantic striving for oneness or wholeness. The desire for merging in Romantic love is to be understood as a manifestation of a compensatory desire that seeks to overcome the separation, isolation, and alienation of individual existence and longs for a return to a state of unity and wholeness. The dissolution of the demarcated self implied in this desire is articulated, for instance, in Hegel's demand for 'complete surrender' to the 'living whole' that lovers are to form, and is enthusiastically embraced as a literal dissolution of bodily existence in all notions of *Liebestod*, in which the possibility of total union is restricted to the next world. In this light, Grace and Graham's desire for fusion appears like a genuinely Romantic striving and Grace/Graham's state of contented tranquillity at the end marks, like Lublin argues, full satisfaction and hence the end of desire.

Moreover, an interpretation of the play as a dramatised completion of (and through) love is in keeping with Kane's proximity to Artaud and also Bataille (who in turn, as suggested above, is in many respects indebted to Romanticism). De Vos, who focuses on the parallels between Kane and Artaud, analyses the final tableau with the "unified character Grace/Graham" who is bathed in ever brighter light as an image of total fulfilment: "This is no longer an image of a woman craving for help, but of a hermaphrodite who has ascended to an aura-surrounded divinity as the embodiment of the reconciliation of all oppositions" (129). Although De Vos, too, speaks of "annihilation" (129) and "sacrifice" (137) with regard to Grace's loss of her individual identity, evoking the same negative connotations that are triggered in most interpretations of Grace's transformation, he explicitly points out that in an Artaudian reading of Kane's play, the loss of the individual self is nothing to be resented. For Grace, he writes

> identity and individuality are obliterated once she has reunited with herself. The same pattern of ideas runs through Artaud's work. Inimical to individuality and identity, he canceled these out in order to obtain the unity of body and mind. In spite of some superficial differ-

ences that seem to alienate Kane from Artaud, they intimately share the same ideas on life. (138)

If that is true – and there is reason to believe it – the disintegration of the individual in Kane's plays is neither a horror scenario nor a warning against the harmful consequences of love. Asked by Tabert about her relation to Artaud, Kane denied any direct influence, but affirmed the parallels between their writings: "So I only started reading him very recently. And the more I read it I thought, 'Now this is a definition of sanity; this man is completely and utterly sane and I understand everything he's saying.' And I was amazed on how it connects completely with my work" (qtd. in Saunders, *Love Me* 16). If Kane and Artaud were indeed 'kindred spirits' as this suggests, it can be assumed that Kane at least partly shared Artaud's aversion to identity and individuality and their separating function. This would put her into stark contrast to the 1990s' cult of individuality. Individuality and its demarcating, distinguishing effect would then not be the sanctuary of modern humans but a source of alienation and separation from a more unified existence.

Grace and Graham reach the goal of their desire, which is to overcome their separate existence and merge into a unified wholeness, and I tend to read this as a happy ending, albeit a "surrealist" one, as Deubner puts it (139). Hence, I disagree with Carney, who also understands the play as a story of invincible love reaching its desired goal, but then goes on to argue "that the indestructability of love in the play is precisely what makes this place hell on earth" (276). His conclusion rests on an interesting shift of focus in the analysis of the Grace/Graham plot from self-loss to the loss of the other. "[I]n becoming her brother," Carney argues, Grace "has in fact lost her brother and her love for him altogether. Graham is no longer present to her, she can no longer hear his voice, and most importantly she can no longer feel, inside, the love that was also a form of perfect identification with him" (277). Perfect union with the beloved, the incorporation of the other, entails the loss of the other as a distinct object of desire. The argument is reminiscent of Simon May's sceptical view on Hephaestus' offer to weld the two halves together forever in the Aristophanes myth. It is questionable, May argues, whether this is really what the lovers want or what might make them happy. Rather, he suggests, "it would be a nightmare of perfect contentment, without appetite, desire, or motion. Not to be whole is torture; but once we have tasted desire and its delightful, if fleeting, satisfactions and torments, would it not also be torture to be whole?" (44). Carney seems to have something similar in mind when he writes that "Grace cannot love the loved one and also be the loved one: love cannot be fully fulfilled, perfected, or completely realized. Love must always be tragic in some way, flawed by a sense of loss, distance, or dissonance" (278). Without incompletion or imperfection there is no erotic love, no striving for wholeness, no longing for the other. If love is to survive, it must never fully reach its goal. Love, in other words, desires the suspension of its own precondition. Once it succeeds it ends.

Convincing as this argumentation is, I do not think it follows that Grace's final condition is 'hell on earth' and that the permanent satisfaction of her desire is a nightmarish vision. This would mean, after all, to declare desire – the tormenting experience of lack, the painful need for compensation – an end in itself and to value it higher than its goal. It is the idea of love for the sake of love, desire for the sake of feeling its sweet pain, followed to its utmost degree. It means to value an ailment higher than its absence or cure, to prefer hunger to food, thirst to drink, pain to relief. A 'nightmare of perfect contentment' is, literally, a contradiction in terms, and Grace/Graham's state of perfect contentment is not to be confused with Hippolytus' deadening apathy. The spoilt prince tries to reach self-sufficiency by shutting out his unfulfilled desire, but the attempt is bound to fail and he never reaches contentment as long as he retains his solitary existence as a discontinuous being. Grace/Graham, on the other hand, is not fruitlessly trying to quench the irrepressible desire to overcome separation but has already reached a state of true contentment where this desire is not repressed but satisfied. In the end, however, these speculations are futile, anyway. The solution Grace and Graham find for their love is inimitable. It is not a solution open to living human beings in a world outside Kane's dramatic vision. No matter if we consider Grace/Graham happy or not in their state of perfect wholeness, no lovers in our world will ever enjoy or suffer from this condition. Love will never cease to be precarious.

6.5 Conclusion

It may be somewhat uncommon to read two plays that abound with images of violence and the disintegration of bodies and individuals as ultimately optimistic. In particular, the dissolution of the individual self, symbolised with Hippolytus' violent death and the merging of Grace/Graham, is rarely considered a desirable achievement in a culture and *zeitgeist* that consecrates individuality, declares the preservation of its integrity life's greatest good, and tends to stress its self-sufficiency rather than its dependence and neediness – a culture where, in the wake of psychology as a discipline, "love – based on self-sacrifice, fusion, and longing for absoluteness – came to be viewed as the symptom of an incomplete emotional development" (Illouz, *Why Love Hurts* 164). But Kane's plays are far removed from this conception of human life. Her characters are anything but self-sufficient and they resent rather than cherish their existence as separate individuals. As Kane says about the characters in *Cleansed:* "They are all emanating this great love and need and going after what they need, and the obstacles in their way are all extremely unpleasant but that's not what the play is about. What drives people is need, not the obstacle" (qtd. in Saunders, *Love Me* 91).[55] If what drives them is need, two things are clear: they are neither self-sufficient, nor do they prefer lack to fulfilment. It is not the obstacle they are secretly seeking because they fear the satisfaction of their desire, as might be inferred from Carney's notion that fulfilment means hell. Rather, in both plays the utterly needy characters are seeking a way out of an existence that leaves them incomplete and dissatisfied and, once they have found this exit, they walk through it unhesitatingly and without remorse. Hippolytus is trapped in his existence as a solitary, isolated discontinuous being, trying to make his life bearable by numbing his senses with all sorts of junk. Motivated by her overwhelming love, Phaedra awakens him to life with her sacrifice, but it is only in his violent destruction that Hippolytus feels reconnected to life as a whole. Only through the disintegration of his separated, discontinuous individuality does he reach the re-integration into the continuity of being that he was longing for all the time. That this reintegration, if permanent, means death is a price he is eager to pay. Similarly, Grace is driven by the desire to overcome her separate identity and merge into a new, complete form of being with her dead brother Graham. Experiencing their discontinuous existence as a separation of one mind into two bodies, she wilfully embraces the violation of the integrity of her body and individuality.

Two features, then, connect *Phaedra's Love* and *Cleansed* and make the two plays comparable on an aesthetic and a thematic level. In terms of aesthetics, both plays use the human body as the metaphor for their central topic. The distinct, demarcated body stands for the discontinuous individual, the tortures it undergoes for the turmoil of love, and its (violent) dissolution for the effect love has on its absolute integrity. To a considerable extent, Kane's theatrical images correspond to Nancy's ideas in "Shattered Love." Love does not leave the autonomous self intact, and in Kane's plays this is visualised as the destruction of bodies. As I have discussed above, for Nancy the experience of love is that of a cut or violation of the autonomous self: "it cuts, it breaks, and it exposes" ("Shattered Love" 97). Love opens the monadic self to and for the other; it breaks the solitary and secluded existence of the individual in an act of violence. And

---

[55] Sarah Kane. Interview with Dan Rebellato. Department of Drama and Theatre, Royal Holloway University of London 3 November 1998.

this is the main content of Kane's metaphorical images. There is, however, no notion of merging in Nancy's philosophy. For him, love opens the self but does not initiate its dissolution or fusion into a new form of existence. This is where Kane's plays differ from his concept. In both plays, the individual self is not only opened indirectly (*Phaedra's Love*) or directly (*Cleansed*) by love but is also dissolved as such and transformed into a state of oneness or union. For Hippolytus and Grace alike, fulfilment of their desire for oneness means 'death' as distinct individuals. It is indeed justified to regard this as a pessimistic message if applied to the real world. Grace/Graham's solution is an impossibility and Hippolytus' solution is death. The play offers no feasible solution for the desire of real people so that the indestructibility of love and desire might really appear like a cause for despair instead of joy. But then, the pieces are no 'well-made plays' offering plausible solutions to everyday problems. They are powerful metaphors for the romantic and erotic idea that what drives human beings is need: a need to compensate the feeling of incompleteness and separation that is the result of individuated existence. The endings of both plays may be unfeasible in reality. Within the logic of the plays, however, they are happy endings.

7 "Not saying I don't want things though": Emotional and Material Desires in Dennis Kelly's *Love and Money*

Dennis Kelly's thrilling *Love and Money* (2006) displays some striking parallels with the plays from the 1990s discussed in the preceding chapters. Like *Closer*, it depicts characters afflicted by loneliness and lack of integration seeking contact and connection with another human being. Like *Shopping and Fucking*, it portrays these characters and their love as infused with and inhibited by consumerism and economic logic. Like *Cleansed*, it scrutinises the strength and durability of love in extreme situations. Like all these plays, it examines love both as a source of vulnerability and precariousness and as a source of meaning and security. It is thus comprehensible why Vicky Angelaki, for instance, brings up the idea of "*Love and Money* as the *Closer* (1997) of the '00s" (84) or why Dan Rebellato finds that "[i]n its satirical elements it resembles some of Mark Ravenhill's work" and points out that "the title resembles a loose translation of *Shopping and Fucking*" ("New Theatre Writing" 606). These observations are certainly justified. The parallels to the plays of the preceding decade are undeniable, bearing testimony to the unbroken topicality of the same or comparable questions some ten years later. The specifics of Kelly's play, however, are best approached by focussing on where it deviates from its forerunners. In this chapter, I will examine the key themes *Love and Money* shares with the plays by Kane, Ravenhill, and Marber – the longing for connection, the resilience of love, and the desire for meaning –, but I will also emphasise the thematic and aesthetic differences that characterise Kelly's take on these issues. After all, even though *Love and Money* in some moments "combines naturalism and expressionism" (Angelaki 82), it has nothing of the directness, experientiality, and reliance on stage images characterising Kane's plays or in-yer-face theatre in general. Instead of creating metaphorical images for the expression of inner states, it relies almost exclusively on realistic language (cf. Stöckl 225).

Furthermore, the reason for the failure of love is not, as in *Closer*, that type-like characters incapable of compromise and commitment take up opposing extreme positions. Nor is the situation identical to that in *Shopping and Fucking*, where the protagonists are so enmeshed in the logic of the market that they first have to learn that there are higher values than money and that solidarity and love can redeem their alienated existence. Rather, Kelly's play, though formally experimental, is thoroughly realistic in its use of language and characterisation of figures, who represent society not as exaggerated types but as middle

https://doi.org/10.1515/9783110714708-008

characters steering a middle course at first until hurled into a vortex that gyrates them into extreme situations. The tragedy of their story consists not so much in the (expectable) frustration of naïve expectations and idealised concepts of love but in the fact that the consumerist obsession that leads to the final catastrophe is already immanent in at least one of the figures (Jess) from the very beginning and, at the same time, seems so utterly avoidable since both protagonists start out with an aversion to the cold world of capitalism and consumerism, with an awareness of the artificiality and constructedness of (their) consumerist desires, and with a clear sense of the superior value of non-material goals. The play thus demonstrates how the best contemporary British drama does indeed not repeat the simple story of the frustration of naïve and idealistic expectations, which has been told so emphatically by sociologists like Beck/Beck-Gernsheim and Hondrich that it works no longer as an enlightened counter-narrative to the cliché of romantic love but has turned into a cliché itself. Rather, *Love and Money* tackles the complex phenomenon of postmodern love as set forth by Catherine Belsey and Eva Illouz where a strong belief in the substitute religion of love is accompanied by distrust in its power and where the idea of love as the last bulwark against capitalism is undermined by love's suffusion with consumerist values. Both David and Jess, the protagonists of the play, are aware that their consumerist desires are running counter to their self-chosen and self-proclaimed ideal of love, but they are at the same time unwilling or afraid to rely exclusively on this ideal as the only source of meaning, security, and groundedness. The tragedy of their story, I will argue, is not that love did not do for them what love could possibly do but that the promises of love were never all they expected from life. Love, in other words, was never all they needed.

7.1 Synopsis

Part of the saddening effect of the play results from the fact that it presents its story in reversed chronological order, enhancing the last scene, in which the recently engaged Jess soliloquises about how love has filled her life with happiness and meaning, with a distressing load of dramatic irony. After all, the audience has just witnessed the episodic disintegration of her love and marriage, culminating in her death. The play's opening scene shows David in an email exchange with his new love interest Sandrine during which he eventually confesses that he has killed his wife Jess. Coming home one day after test-driving a car he could never afford due to the massive debts accumulated by Jess, David had found her unconscious after an unsuccessful suicide attempt with an overdose of anxiolytics. He reports how, driven by the prospect of getting rid of the debt and being able to spend his earnings on luxuries he had been denied until now, he had fed her a lethal amount of Vodka. Scene Two appears like an interview (but with no interviewer present) with a mother and a father first bemoaning the untimely death of their daughter and then justifying the demolition and desecration of the monumental tomb erected by a Greek husband for his late wife that was overshadowing the more modestly decorated grave of their daughter. Scene Three is set much earlier and shows David, obviously in desperate need for quick money, applying for a new job in the telecommunication business. The conversation is acutely embarrassing, not only because the former teacher and amateur writer David applies for a job he detests but also because he is forced to make himself dependent on his future boss Val, a former girlfriend still resenting their breakup, who does everything to humiliate him, including the offer of a ridiculously low starting salary and the suggestion to consider gay porn as an alternative. While this scene is presented as a conventional dramatic dialogue, the next is reminiscent of Martin Crimp's use of unidentified speakers in *Attempts on her Life* or of a "Kanean play of voices" (Rebellato, "New Theatre Writing" 607). Five voices, in a kind of partitioned monologue rendered in second person narrative, talk about how "You work for a credit card company" and "you want to do well for your family" (242) and "one day you have this idea" (243) of a shady credit scheme that will yield enormous profit for your company. As the scheme is especially designed for applicants with low credit standing and demands high interest repayments, it produces a considerable number of "defaulters" (247) who are plunged into insurmountable debt and from time to time, the voices say, "you might think about these people" (247) not as numbers and figures but as human individuals. At this point, the abstract narrative of the voices materialises as it is brought together with Jess' story, whose short monologues intersperse the voices' tale and uncover her shopping addiction. In the shared consciousness of the voices, the abstract business scheme turns into a concrete human tragedy as they imagine David, alone at home while his wife is recovering in hospital from a shopping-related nervous breakdown, discovering that Jess has fallen prey to their credit plan.

>**2:** and you imagine this man, you imagine him
>**5:** this teacher
>**4:** you imagine him ...
>**1:** He sits there in their living room.
>**3:** He spreads out everything on the floor.
>**2:** He has a pad and a pen and a calculator that he nicked from work, statements, bills, invoices even.
>**3:** that weekend, that very night, after having to leave his wife in that place, and he's

2: He's shocked.
1: again he's shocked because he understands that all the numbers and figures and pounds and red letters add up to a void
2: a void in her that he should be filling
5: shouldn't he?
1: And then he finds your great idea.
2: And he finds your beautiful idea.
4: and he cries.
[...]
5: He resolves to get through it.
[...]
5: He's going to get both of them through it.
[...]
5: He's gonna get a job that pays, get her back home, tell her how much he loves her and start putting things right
2: He loves her
5: and she loves him. (251–52)

The scene, which Angelaki calls the "converging point of the play backwards and forwards" (88), marks, or rather narrates, the turning point of this non-linear tragedy. It is the point where Jess and David still love each other, but where it is unveiled that love is not enough for Jess to fill the 'void' inside her and where David, out of love, resolves to take the first of many steps that will eventually alienate him from his wife.

Scene Five, where "the sleazy Duncan is persuading young Debby to work in porn" (Sierz, *Rewriting* 108), is, apart from one tangential point, only related to the main plot through a series of thematic interferences. Like the rest of the play, the conversation between Duncan and Debby is characterised by a longing for connection and mutual completion that finds a blunt expression in Duncan's wish to possess an item of Debbie's clothing and to have her chewing gum "because I think that by chewing on something that someone like you has chewed on something of you will become some part of me" (262). Another recurrent theme of the play is the ambiguous or ironic attitude towards, broadly speaking, metaphysical sources of meaning, which transpires in Duncan's declaration that the best thing about modern life is "belief. The absolute conviction that all this is right. [...] Only we don't really, do we Debs. That's the great secret. Not in our heart of hearts" (259–60). While Duncan here seems to express a sort of bad conscience regarding his means of making money, the problem is turned on its head in the last scene, where Jess' monologue betrays that in her 'heart of hearts' she is not fully convinced of the superiority and sufficiency of immaterial values. A third thematic parallel is, of course, the exploitative situation of the entire scene. Presumably driven by financial pressure, Debbie has to put herself into this potentially humiliating and coercive relationship. The tangential point that directly links the scene to the story of Jess and David is a photograph Duncan produces, showing "a man with a cock in his mouth" (256). His explanations – "He was a teacher. [...] Nice fella. Married. In debt" (262) – and the fact that Val had suggested this "way of making money" (239) during the job interview in Scene Three force the interpretation that the man is indeed David. Scene Six lays open how David's self-sacrificial attempts to save them both have corroded love and trust as the pillars of their marriage. Jess has just tried to administer first aid to a man who had been stabbed in the chest before her eyes and is now waiting in a hospital waiting room for David to pick her up. When she tells him what happened, David is most concerned with the fact that the assault took place in Oxford Street, which was not on her way back home from Poland Street where she had an interview for a second job. In a truly agonising dialogue, instead of giving her the emotional support she desperately needs, David starts questioning Jess about her "detour to Oxford Street" (274) and whether she did "buy something" (275). Unlike in the opening scene, however, where David's confession of murder for monetary reasons can only be appalling, his cold and suspicious behaviour in this scene is at least comprehensible due to the information about David's sacrifices the play has provided by now and which he recalls in their dialogue:

> Is that what you did, you went shopping?
> [...]
> After everything we've been through? After you being in hospital?
> **Jess:** Don't bring that up.
> **David:** In a mental hospital, Jess, in a fucking, after all the shit and fucking, after me getting a crappy, shitty job, after all the things, things I've done that you don't even, the financial fucking, things you don't even
> **Jess:** stop going on about money
> **David:** know about we have to talk about money
> [...]
> I want to talk about other things, I want to talk about the future and holidays and education, but instead we have to talk about money because (275)

In his angry reaction David mixes all the rage about the humiliating things he has brought down on himself with a panicking fear that all future plans of a normal family life will be forever rendered void by their never-ending money troubles. His reaction is understandable, but it is also grossly misplaced – not only because David fails to care for his wife in a situation where she needs solace and affection more than anything, but also because his suspicion that she indulged in another shopping frenzy on London's notorious shopping street is invalidated by her revelation that she "just bought some fucking CDs" (276). David's wordless reaction displays fury, helplessness, and a reluctant admission of guilt.

> *He goes to break the CDs. Doesn't.*
> *He goes to walk out. Doesn't.*
> *He goes to grab his hair. Doesn't.*
> *He goes to throw the CDs. Doesn't.*
> *He goes to say something. Doesn't.*
> *He goes to almost hit her. Doesn't.*
> *He goes to sit down. Doesn't.*
> *He goes to put the CDs down. Doesn't.*
> *He is in tears.* (277)

He seems to have realised in this moment that their marriage is lost without hope, that nothing he could *do* would make a difference now. He recognises his own suspiciousness as the symptom of a lack of trust which will always burden their relationship. His heartless behaviour, his unconcern regarding the psychical and emotional condition of his wife in this extreme situation, shows him that his love is no longer stronger than his financial anxiety. His inability to act or say something mirror his awareness that they have reached an impasse, trapped in a marriage from which mutual trust and emotional support are lacking. The scene reaches its sorrowful climax when a doctor enters and unintentionally confirms Jess' earlier intuition that her attempt to stop the stabbed man's bleeding had a worsening and in fact lethal effect. Without a word or gesture of comfort, David eventually walks out, leaving the wrecked and guilty Jess alone. The play ends with Scene Seven, a long monologue by Jess about the importance of love and connection, the superiority of immaterial values, the improbability of a universe so complex and perfectly tuned to have come into existence without some sort of "design" (283) or "purpose" (284), and the intensity of her love for David (who has just asked her to marry him) that she takes as a proof for spiritual connectedness and the cosmic force of a love that is 'meant to be.' What makes this last scene, and the entire play, truly tragic in an almost classical sense, however, is not the ironic contrast between Jess' great expectations and the bitter reality that is awaiting her, but the traces of an attitude that works as the 'tragic flaw' in this play and that, as I will argue below, retrospectively foreshadows the moribundity of their love (see also Stöckl 226–27). Before that, I want to analyse how the play repeats the 'same old songs' of love, the discourses of need, compensation, and precariousness, in its own special tune.

7.2 Everybody Needs Somebody

A decade after *Shopping and Fucking* and *Closer*, Kelly's play still treats alienated, isolated individuals longing for human contact and closeness in a cold, capitalist society, demonstrating, as Charles Spencer wrote in his 2006 review, the playwright's "profound understanding of the way we live now." Other critics agreed with this assessment of the play as a faithful snapshot of the status of human love and empathy in a late capitalist, materialist environment, and, reviewing the 2012 production in Chicago, Kris Vire called the play an "affluence-obsessed collage" that "feels only more current and vital six years later" (*Time Out*). Like Ravenhill and Marber, Kelly presents characters whose desire for connection leads them into a struggle not only with amorphous market forces but also with themselves, as they have internalised features of the capitalist logic. This struggle is central to the play and it, as has to be stressed again, results from the clash of the characters' longing for human connection with both external *and* internal adversary forces. It would thus be an undue reduction to see the play only as a depiction of a "world of easy credit, where the weak are tricked into debt and materialism muscles out things of real value" (Taylor, *Independent*) without adding, firstly, that the characters could not be 'tricked into debt' were it not for their own internalised materialism and, secondly, that their materialism does not prevent them from striving after 'things of real value.' One such goal whose value is unanimously accepted in the play is love – the intimate connection with another human being. As Angelaki writes with regard to a German review that discusses the characters' experience of *Sehnsucht* (which she translates as a "corporeal and emotional mix of desire and yearning"): "The feeling permeates the entire play, with characters experiencing a physical and emotional craving, a hunger, even, for connection" (86). Indeed, this craving for connection is a structural element of the play. Human closeness is repeatedly presented as a valuable goal, but also as inherently precarious.

Already the dramatic introduction with the email exchange between David and Sandrine displays the precarious desire for connection. What ends with David's confession of murder starts off as a flirtatious attempt to resume contact with a one-night stand he met at a business conference. But right from the beginning, David's advances are marked by fear of rejection and humiliation. Elżbieta Baraniecka offers a detailed and intriguing analysis of this scene ("Precariousness" 174–80), emphasising the "risk of rejection or ridicule" (174) and the moments of "self-exposure and vulnerability" (175) David accepts when he first contacts Sandrine in the hope of becoming closer. Pressured by his "longing for real connection" and Sandrine's demand of emotional openness, he "finally lets his guards down, opens himself, and tells Sandrine about his feelings" (177), but also about the death of his wife, which is more than Sandrine is prepared to hear. Consequently, "[t]he scene ends with an image of David's utter loneliness and dejection [...]: a desperate figure, obsessively checking his emails, and receiving none" (178–79). There is a bitter irony in this rejection of David. His attempt to establish a connection with another human being is thwarted by the revelation that he helped kill his wife for financial reasons, while at the same time it is motivated by the wish to escape, at least for moments, from the world of money which has made him the person capable of such a deed in the first place. As Baraniecka rightly observes, the fact that, after leaving his job as a teacher and entering sales business, David has "become part of the consumerist system that he used to disdain so passionately" (173) must not be omitted from an analysis of David's character. His wish to connect with Sandrine is an unsuccessful attempt to break out of this system again, to re-discover the non-materialist values he once cherished.

> I
> trust you, Sandrine. And I like you. I live this life here where everything is measured in pay grades and pension schemes and sales targets and people like Liam laugh when your orders are cancelled and you are scared of losing your job. And you
> Beat.
> I used to be a teacher, I ...
> Beat.
> You
> are outside of this. You are beautiful and outside of this and you inspire me and you make me believe in things
> that maybe it's not a good idea for me to believe in. (213)

David is quite aware of the contradiction between the consumerist system and the belief in higher values, which are represented here and throughout the play in the need for connection and closeness with intimate others. Solidarity, disinterestedness, trust, vulnerability – all these presuppositions for intimate relationships are at odds with the principles of the business world. His decision to try to connect with Sandrine amounts to an admission that there are indeed values higher than money, that consumerism does not compensate the feelings of loneliness and isolation, and that his need for an intimate relationship to another human being is more urgent than his fear of the precariousness of such relationships.

In the job interview between David and Val, this need for connection surfaces again. In Val, however, the longing for human contact is suppressed and replaced with a full dedication to materialism. She prefers loneliness and isolation to the vulnerability inherent in intimate relationships. During their conversation we learn that Val was David's girlfriend at college until he "dump[ed]" (235) her and thus left her not only without a lover but also without friends since, as Val notes, "[w]hen we split up they were all of a sudden your friends, not mine" (235). This experience of love's precariousness may well have contributed to her development into the cold, isolated businesswoman she now is. Like Ravenhill's Mark, she now prefers the safety of the inhuman market to the precariousness of human relations. Instead of making her happiness dependent on the volatility of someone else, she relies only on herself and her economic skills and has accordingly replaced her formerly strong Christian faith with the capitalist creed.

> Money. I believe in money.
> David.
> That's my thing now.
> David.
> And in the same way that a plant takes oxygen and nutrients and uses the process of photosynthesis to turn sunlight into energy, I take customers and employees and use the process of hard fucking work to produce cash.
> I am a photosynthesist of cash. (237)

What is contained in the image of the plant is the ideal of self-sufficiency after which Val is striving. Like a plant, and unlike humans and most animals, Val does not need her congeners for survival as she has cut off her need for company and affection. In her world, other human beings are only 'human resources,' the equivalents to oxygen and nutrients required for her photosynthesis of cash. As Baraniecka writes, "Val consciously rejects the whole experience and joys of love in favour of safety [...] and becomes an extremely confident but also cruel and lonely person in the process" ("Precariousness" 174). However, Baraniecka indicates that Val's rejection of human connection might only be skin deep. Like Angelaki, who uses the scene as an example for the "hunger [...] for connection" permeating the play (cf. 86–87), she quotes the end of the scene where Val wants to see the mole on David's forearm and then "*reaches out and touches it with her finger. Pushes it. Beat. Licks it*" (240). "This odd behaviour," Baraniecka holds, "may be seen as another way of asserting herself over David, but it may also be read as a lingering echo of Val's desire for human contact" (184–85). Deep down, the 'photosynthesist of cash' has probably not erased all traces of her human nature and her need for others.

Duncan's request for Debbie's chewing gum and knickers in Scene Five is another manifestation of the 'hunger for connection' characterising the play. Its clearest formulation, however, comes from Jess in her final monologue:

> I don't think we want to be alone, do we? Do we want that? Is that what we want? And sometimes you think that the only reason we do anything at all, anything, is to reach out and touch
> Just touch, just to
> feel
> something
> [...]
> Just to connect. (279)

For Jess, in fact, as I will analyse in more detail below, connection between human beings amounts to the meaning of life. She perceives of the love between her and David as unshakable proof for her conviction that the world is not absurd. As Baraniecka puts it, "[w]hat dominates the structure of Jess's character is her insatiable yearning for a real connection with another human being, a connection through feeling, which, by extension, she perceives as the thing that gives meaning and depth to human life" (180). Through its position at the end of the play, Jess' monologue about the importance of connection acquires almost the function of an epilogue, affirming the sense that, as Rebellato observes, "[t]hroughout there is a strange, perverse affirmation of human connection" ("New Theatre Writing" 607). Even though connection, intimacy, closeness, or love do not survive in the end, it seems beyond doubt that the play proclaims the necessity of striving for them despite their precariousness. The human need for others is the most significant antidote to the dehumanising effect of a culture of materialism. In the pointed imagery of the play, the moment David subordinates his need for another human (i.e. his wife) to his need for material goods is the moment that 'dehumanises' them both.

7.3 I Would Do Anything for Love ... But I Won't Do That

One central question *Love and Money* shares with Kane's *Cleansed* is that about the durability of love. While Kane uses the graphic imagery of Carl's torture to address the subject of love's limitations, Kelly uses the analytically revealed story of David's self-humiliating attempts at raising money. As Baraniecka argues, David's "idealistic attitude[] towards love [...] obliges him to renounce his own needs and sacrifice himself completely in order to save his wife and his marriage from financial crisis" ("Precariousness" 173). Like Carl, David is forced to endure insufferable things for his love which he eventually betrays. But in both cases, the protagonists show an admirable degree of commitment and willingness to fight for their love. Carl proves his commitment after the betrayal, in his persistent endeavour to ask Rod for forgiveness and to articulate his affection despite Tinker's effort to prevent their communication. David proves it before the betrayal, with the long period of self-sacrificial suffering he undergoes. I am thus sceptical about Angelaki's description of *Love and Money* as "a collection of people and things, evoking the lack of commitment to relationships that continuously gravitate around the central force of capital and profit" (85). Unlike *Closer*, Kelly's play is not only about egoism and commitment phobia. With Val, of course, there is a character who consciously shuns commitment to other people in the hope of protecting herself against the precariousness of intimate relationships. But with David, there is a character engaging in his relationship with a devotion that none of Marber's characters is capable of. Commitment is certainly not absent from the world of the play. Only, it is not absolute.

In fact, the play thematises two distinct limits, two things that different characters won't do for love. There is something for Val on the one hand and for David and Jess on the other that they are not prepared to give up and that prevents their full commitment. For Val, this sacred treasure is her emotional invulnerability and independence. As Baraniecka points out with reference to Jean-Luc Nancy (cf. 171–73), and as I have also concluded at the end of Chapter 2, risk-free love is hard to envisage. Love demands openness, vulnerability, and the admission of self-insufficiency and is thus inherently precarious. For Val, once deeply hurt when David ended their relationship, love is discredited as the road to happiness. Unwilling to take the risk again, she prefers to bank on money as the safer alternative, conceiving of herself as a self-sufficient 'photosynthesist of cash' who does not have to rely on others for the satisfaction of her needs. This posture links her to Hippolytus in *Phaedra's Love* and Mark in *Shopping and Fucking* and their similar attempts to avoid the precariousness of love by reducing intimate relationships to purely sexual encounters or transactions. They, too, consciously try to preclude emotional commitment in order to preserve their independent and inviolate selves, and while Mark immediately gives up this project as soon as he meets Gary, Hippolytus and Val sink deeper and deeper into their self-chosen isolation that makes them cruel, resentful, and insensitive to others.

While Val's refusal to give up emotional independence and invulnerability keeps her away from love and commitment entirely, David displays no such commitment phobia. He seems to enter the relationship with Jess without reservations. He does not expect the relationship to adapt to a completely self-determined, unchangeable individuality or 'essentialised self' but rather re-defines himself as no longer a singular, self-contained individual but a member of a co-dependent coalition of two. His resolution "to get both of them through it" (251) when he realises the full extent of Jess' debts signals how much he identifies with their marriage and how much he has made himself dependent and vulnerable by choosing to share a life with someone and accept responsibility for this shared life. In a sense, *Love and Money* is about emotional debts as much as about financial ones, and it asks the question how much lovers owe each other. As Angelaki comments on the job interview scene, due to their history as lovers "Val perceives of a debt between her and David, one that he must now repay at the cost of his dignity, as she offers him a post at low pay" (86). Similarly, David obviously feels indebted to his wife Jess and obliges himself to make painful sacrifices for their love. While there are only tenuous hints of the gay porn episode, David's application for a job in Val's company directly conveys the humiliation David is prepared to suffer. When he asks for the job, the passionate teacher has "just got this award for a short story" (230) and, although he does of course not admit it, Val's provocative questions indicate quite openly that David thinks that "sales is shit" (233) and that "telecommunications is beneath [him]" (234). Moreover, Val gives him to understand that he is not actually applying for a vacant post but that she "could make one" (232) in the stockroom and that, even if the starting salary is less than what he is currently earning, he has to consider this a "favour" (238) given that, with his "degree in English Literature" (239), he is not well qualified. Despite all this, David agrees, having the prospect of higher salaries "once you get into sales" (239) and being prepared to search for additional ways of making money until then.

But David's readiness to make sacrifices for his love does reach a limit. The episodic play does not present David's character development in a continuous, linear way, but at the end of the story (that is, at the beginning of the plot) he has become a member of the world of business and money whose materialist desires got the better of his former idealism. In his email confession to Sandrine, David presents himself as a man who is no longer willing to forgo all the conveniences and luxuries money can buy in order to save his marriage.

> I came home and she was lying on the bed and she'd taken –
> *Pause.*
> I'd been test driving this Audi. It was silver. It had ABS breaking, climate control, satellite navigation and roll bar as standard. It had these little push out trays for your drinks and things. It was a lovely car. I loved it. It really held the road
> *Beat.*
> well, and the thing was the bloke who's doing the test drive, you could just tell he was thinking 'you fucking waste of my time.' But the thing was I could afford it, but
> *Beat.*
> It felt so good that car. It felt like I'd earned that car, after the things I'd been through, the things I'd done
> it felt like the car was my right.
> But we had debts, big debts. My wife had debts.
> You see, he could see that I couldn't afford the car. Do you understand? It was visible on

> me. He could see it on me.
> *Pause.*
> So when I came home and saw her lying there I thought
> I'll be able to afford the car now. (215–16)

The love for his wife is replaced by desire for a material thing, a thing that promises more satisfaction and happiness than their ruined marriage. The car becomes for David what Sara Ahmed calls a "happy object," an object that is "attributed as the cause of happiness" (28) and hence becomes a "happiness means" (26). When he remembers that "it felt like the car was my right" he expresses the idea of a right to happiness – a right he has earned through his hard work and sacrifices but which he is denied because of the bond with his wife. Paul Taylor finds David's turn towards material goods as his happy objects not convincing. "I never really believed," he writes, "that a man who had been a teacher would buy into the values of this new world [...]" (*Independent*). But David's consumerist desire is only the logical consequence of his entry into the business world "where everything is measured in pay grades and pension schemes and sales targets" (Kelly 213) – an entry into which he was forced against his will but which nevertheless situated him in a new social environment whose values he had to adopt to prevent social isolation. As Ahmed states, "*we tend to like those who like the things we like*" (38; emphasis original). Thus, in order to be accepted in his new environment, David sooner or later had to make their happy objects his own. "To be affected in a good way by objects that are already evaluated as good is a way of belonging to an affective community," Ahmed writes. "We align ourselves with others by investing in the same objects as the cause of happiness" (38). The alternative, that is, to reject desire for the things that are accepted as desirable in a given community, bears the risk of social ostracism. "We become alienated – out of line with an affective community – when we do not experience pleasure from proximity to objects that are attributed as being good" (41). In other words, we run the risk of becoming "strangers, or affect aliens" (42) if we do not share the desires of our social environment. David's wish to possess a shiny new car is not an implausible inconsistency of character but a pointed depiction of how values may change with the entry into a new affective community.

David's character development also affects his concept of love. His former ideal of self-sacrificial, disinterested, and ultimately self-harming 'Gift-love' is replaced by a profit-oriented and risk-minimising concept that balances the advantages and disadvantages of a relationship. David switches from selfless commitment to the self-centred amorous business relation Zygmunt Bauman polemically describes as the relationship befitting a society thoroughly infused with market logics:

> A relationship [...] is an investment like all the others: you put in time, money, efforts that you could have turned to other aims but did not, hoping that you were doing the right thing and that what you've lost or refrained from otherwise enjoying would be in due course repaid – with profit. You buy stocks and hold them as long as they promise to grow in value, and promptly sell them when the profits begin to fall or when other stocks promise a higher income [...]. (13)

After all the effort David has invested in their relationship, and now that material goods which he has hitherto 'refrained from otherwise enjoying' have become his happy objects, he senses that the sacrifices he has to make to save their marriage will not be 'repaid.' In comparison to his disastrous marriage with Jess, other 'stocks' clearly appear more auspicious.

Lastly, another motivation for David's crime – one that shows how much he identifies with his new affective community of materialists and consumerists – is his demand for recognition of his professional career. When he once experienced his application for a job in sales business as a humiliating betrayal of his non-materialist ideals, he is now humiliated by the fact that his inability to afford a luxury car is 'visible' to the car dealer. While being loved is the most important recognition of the intimate or private self, David here craves the recognition of his 'social self,' of the role he plays and the position he occupies in a social hierarchy determined by income and financial liquidity. The car is not only supposed to be a convenience for him; it is supposed to function as a sign of his status, to attract admiring glances, and to elicit recognition for his hard and successful work. That the very act of confessing all this is part of an attempt to connect with another human being in an intimate relation that offers the chance for recognition of his intimate self implies that social recognition alone will not satisfy David in the long run. His radical openness in the conversation with Sandrine is a desperate cry for recognition not of the social role he plays but of his true self. His fateful deed, of course, has set the bar quite high for someone to love him now.

7.4 All You Need Is Love

David's newly developed and steadily increased consumerist desire is what, at one point, he is no longer willing to relinquish for love. Jess is confronted with a similar conflict. Within her, too, materialism and love are fighting a battle that the latter is doomed to lose. The difference between the two is that while

David was lured away by the temptations of the consumerist system from a position of extreme commitment, Jess was never prepared to commit so much to their relationship as to make her happiness entirely dependent on love. As I will argue, Jess considers love the most important source of meaning, groundedness, and happiness, and she views it as a value far superior to materialism. But as her final monologue reveals, this sincere conviction is from the start undermined by a trace of consumerist desire which she is neither willing nor able to suppress. Despite herself, as it were, Jess is incapable of relying only on love as her road to happiness. If, in other words, Jess indeed believes that 'love is all you need,' then it is also true that she wants more than she needs.

Rebellato observes about Kelly's plays that his "characters occasionally demonstrate an imaginative landscape of cosmological scale" ("New Theatre Writing" 605) and that, in these moments, "they all represent an inchoate reaching for a grandeur of scale that is conspicuously missing from their nose-to-nose encounters with everyday brutality" (606). What these characters experience is a clash between the world as it is and their desire for higher values and meaning, "the appalling discrepancy between a meaningless daily life and an unreachable beyond," "a deep disconnect between one's human needs and the values of the wider culture" (606). *Love and Money*, Rebellato argues, expresses this sensation with great force, especially in its final scene. Jess' monologue indeed unveils her deeply felt conviction that love is a form of spiritual, metaphysical connection that disproves the absurdity of the world and is far superior to materialism as a source of meaning and orientation. "I might be looking for meaning because I'm in love" (280), she admits towards the beginning of the monologue, but at the same time she takes the strength of her love as proof for the existence of meaning. Feeling this strong purposeful connection with David, and 'knowing' that he feels the same, she explains that "maybe this is why. Why I'm, you know, why I believe there is something more. / Because I have evidence of it. / I live in it. (286). This 'something more' she believes in is some sort of meaning or "design" (283) she deems indispensable in view of the complexity and perfectness of the universe. Just as her love feels 'meant to be,' the universe cannot be without purpose. What is more, Jess puts this idea of cosmic connectedness and purpose in direct contrast to what she perceives to be widespread but incomparably inferior alternative providers of meaning in our late capitalist age:

> I mean Jesus fucking Christ, you know, just what is wrong with purpose, what's wrong with, you know, fucking belonging or
> or
> or just, you know, having an idea that there is something, that there is a point and that maybe it's about more than just I have this pot of stuff here and that's got more in it than your pot of stuff over there, but I'm just talking about, maybe, I dunno, choosing a

> world that is more than numbers and quantities and saving and choosing a world that is
> flesh and bone and
> love or,
> more than just
> isn't it more than just
> money, mathematics, numbers, values, I don't know
> Isn't it?
> Don't you think it is?
> Isn't money dead? Or something?
> Isn't it? When you look around?
> Don't we know that in our heart of heart of hearts? (284)

Her speech simultaneously denies the meaninglessness of the world and discredits one of the most powerful 'surrogate religions' of post-religious societies. Evoking Nietzsche's 'God is dead,' her "Isn't money dead?" questions the suitability of materialism as a source of meaning or of capitalism as a guiding principle. Rather, in our 'heart of heart of hearts' we know that spiritual values are higher than material ones and that what gives meaning to life and, at the same time, proves life's meaningfulness, is love.

Due to the play's reversed chronology, however, her use of words inevitably recalls Duncan's sceptical statement in Scene Five that people, in their 'heart of hearts,' do not believe in their declared articles of faith (cf. 260). Just as Duncan does not really believe "that all this is right" (259) when he exploits Debbie's neediness, Jess, deep within herself, does not really believe that all you need is love. This is where I slightly disagree with Baraniecka's reading of the play, who argues that "[e]xpecting love to exclusively provide them with a sense of belonging and safety, they are sooner or later disillusioned" ("Precariousness" 173). In contrast to this view that the two protagonists "have extremely idealistic attitudes towards love and are equally disappointed in their high expectations" (176), I suggest that Jess' idealistic attitude towards love was never absolute and that, despite her denigration of materialism, she never pinned her hopes exclusively on love but always cultivated an ultimately corrosive element of materialism and consumerism as a sort of backup. If Jess, as Spencer writes, "feels something is missing, and so she shops to fill the void" (*Telegraph*), it is important to add that Jess is not disappointed that love did not fill this void. Lyn Gardner's analysis hits the nail on the head when she argues that "Dennis Kelly's play is a modern morality tale about debt and desire, the high cost of living and the things we buy to fill up the void. True happiness is not just love, but an MFI kitchen as well" (*Guardian*). This accurately describes Jess' position: love is the highest value and supplies life with meaning, but that is not enough for happiness; she denigrates materialism as a value system but does not exclude it from her system of values; she upholds love as the highest value but does not rely on this value alone but also on values she knows to be in opposition to it. Like Belsey's and Illouz' 'postmodern lover,' Jess is quite aware of her ambiguous attitude, of the fact, that is, that she does not fully trust in the idealist narrative of love, even though she has chosen it as her guiding principle. Moreover, she even knows that the consumerist desire, which she is unable to overcome due to her lack of confidence in the power of love to be entirely fulfilling, is medially constructed. All this becomes clear towards the very end of her monologue, just after she has so passionately argued for the superiority of love over materialism:

> Not saying I don't want things though.
> *She laughs.*
> I do. I do want things for us. I've begun to look at my life and say, well, it has been a bit scruffy
> and now I don't want that. I want it to be a bit neater, I want it to be a bit like it's supposed to be, I want it to be a bit like
> *She laughs.*
> a bit like it is on the telly.
> *She laughs.*
> I know, I know. What a cunt. (286)

Her embarrassed laughter and self-blame evince her awareness of the contradiction between her previously uttered idealism and this admission of materialism. But this contradiction, as Illouz shows, is typical of a society that cherishes love as the counter-principle to capitalism and at the same time is infused with "the romanticization of commodities and the commodification of romance" (*Consuming* 26). In a society where collective imaginations of love and relationships are influenced, if not determined, by the images disseminated by mass media and the entertainment and advertisement industries, Jess' wish to have 'things for us' and to lead a relationship that is 'a bit like it is on the telly' is symptomatic. She craves what Illouz calls "*emodities*," commodities that have been infused with emotional value through advertising and 'branding' (cf. "Einleitung" 23 and 30–35) and which appear necessary for a happy relationship and for 'life as it should be.' Apart from this media influence, there might be another personal motivation behind Jess' consumerism. As can be gathered from the conversation between Mother and Father in Scene Two (assuming that they are indeed Jess' parents, which is not clearly specified but very likely), the family was once rather wealthy, owning an "eight-bedroom house" (222) in the eighties. As Mother remembers, it "broke her [Jess'] heart moving from that house [...], wept from her soul, never the same again" (222). As a result of this traumatic childhood experience of material loss, Jess seems to be haunted by a constant fear of losing social status, and she seems to have an expectation that the marriage with David will put an end to a social descent that has damaged her self-worth. Like the car for David after his entry into the 'affective community' of materialists, the possession of material things seems to have a function of providing recognition and self-worth for Jess, too. Her story of an expensive bag she once bought because "it suddenly dawned on me that the bag was designed, not to hold things, but to hold me" (245) supports this idea. The bag, standing in for material possessions in general, is 'holding' her, not only in the existential sense of rooting her existence, but also in the sense of holding her in place and affirming her position in the social hierarchy that she is constantly afraid to lose. Her admission that "when [she] thought about it again that evening it just seemed ... stupid" (245) is again evidence of her awareness that binding her emotional health and happiness to material things of this kind is contrary to her rationality and her idealist convictions. But this awareness does evidently not prevent her from indulging her consumerist desire. While David is increasingly distracted from his total commitment by the temptations of materialism, Jess' tragic flaw is her general inability to rely fully on what she wants to believe in. At no point in their relationship does her commitment to love eclipse her desire for material things. Her idealism is flawed from the beginning by the unavoidable consumerism that is surrounding them. In the end, both attitudes turn out harmful. David, who believes that all you need is love, is "shocked" when he discovers the extent of Jess' shopping addiction because it points towards "a void in her that he should be filling" (251). Making himself responsible for the fact that his love is not enough to fill this void, he resolves to prepare at least the money that can apparently do what his love cannot and exposes himself to humiliation and suffering. Jess, who never sincerely believed that love is enough, makes it impossible for David to make her happy with his love alone and necessitates his entry into the world of business and consumerism. But the play is not interested in comparing and judging the two characters and their behaviour. In fact, both positions – David's unreserved, self-sacrificial, idealist commitment and Jess' realist refusal to make happiness entirely dependent on love – are equally legitimate and understandable. What the play shows is how an unhappy combination of the two positions breeds disastrous consequences. The temptations of the world of money, on the one hand, reverse David's loving commitment into its opposite and, on the other hand, hold back Jess from full commitment in the first place. It is 'either or' for David and 'neither nor' for Jess – two positions that do not make a good couple in the long run.

7.5 Conclusion

The two dominant discourses of love are as central in *Love and Money* as in the plays from the nineties with which I have compared it at the beginning of this chapter. Love is again expected to provide meaning and rootedness to otherwise disoriented individuals. Love and money appear as the two grand narratives of the present age. Diametrically opposed and yet inextricably entwined, they both are supposed to compensate for the absence of religious certainty and to satisfy the human desire for meaning. At the same time, they are both essentially precarious, threatening the individual with 'indebtedness,' dependence, and loss. Kelly's play, however, also differs from the earlier plays in a number of aspects. Unlike the typical in-yer-face plays by Kane and Ravenhill, it does not rely on the creation of powerful stage images and the use of metaphor. As regards the aesthetic representation of love's precariousness, in particular, the play derives its force from the strategic use of discrepant awareness rather than from the direct impact of visual effects. In the first scene, the audience is confronted by way of posterior narrative with the shocking information of David's largely incomprehensible and apparently ill-motivated crime and it takes the rest of the play to provide the story elements and steps of character development that lend a touch of the plausible and the tragic to his still repellent but less incomprehensible deed. Reversely, the audience's superior knowledge in the last scene endows Jess' monologue with an almost unbearable dramatic irony. Having witnessed the deterioration of their love and marriage, spectators inevitably contrast her idealistic illusion of love with its apparent fragility – with its dependence on unpredictable circumstances, personal decisions, and the impersonal forces swaying these decisions (cf. Stöckl 226–27).

In accordance with this abstention from unsettling imagery, language use and figure characterisation, too, are more subtle and less direct in Kelly's play. Although the characters repeatedly find themselves not in realistic dramatic situations but seem to talk to an unseen interviewer or simply to the audience, the way they talk is sharply realistic. This combination of authentic language and obscure speaking situation has the effect that, as Rebellato puts it, "his dialogue is both hyper-real and strangely alienated" (604). But no matter what the situation is, whether dramatic dialogue, narrative monologue, or something else, Kelly retains a "splinter-sharp alertness to the patterns of ordinary speech and the confused tumble of thoughts" and offers "tight linguistic close-ups on the minutiae of the everyday" (605). Even a speech like Jess' final monologue about love and the universe sounds realistic and plausible – like an unstructured heap of theories, facts, and intuitions gathered by an interested amateur from documentaries, magazine articles, and internet sources – and thus differs from the transpsychological insights and surprisingly eloquent declarations uttered at times by Ravenhill's and Kane's characters. Their more realistic use of language is a feature of the more realistic conception of characters. They are no type-like representatives of extreme positions but nuanced and subtly characterised plausible figures. What comes along with this figure conception is the persuasive degree of awareness concerning their inner conflicts demonstrated by Jess and David. They are both well aware of the contradictions between the two guiding principles in their lives: Jess is ashamed of her consumerist desires which are in opposition to her proclaimed idealism, and David hopes that Sandrine will rescue him from the cold materialism of the business world he has made his own, reluctantly at first and then with fatal determination. The two protagonists are neither naïve idealists frustrated that love is not all they need, nor are they unconsciously manipulated victims of the media and the consumerist system. They rather resemble the postmodern lovers described by Belsey and Illouz who distrust the idealism they harbour and know about the contents and discontents of both the romantic and the realist narratives of love. It is this awareness that makes them plausible and relatable characters and their story a contemporary tragedy.

8 "Love at first sight and the lost city of Atlantis": Penelope Skinner's *Eigengrau*, Or 'A Fairy Tale of Blind Love'

Eigengrau, Penelope Skinner's second play, premiered at the Bush Theatre in 2010. Reactions were favourable in general, but not enthusiastic, and some critics were too much revolted by the two shocking scenes in the play to appreciate the rest (cf. *Theatre Record* 30.6, 285–87). What is conspicuous is that even those reviews which were not dominated by expressions of repulsion in view of oral sex and self-mutilation, while unanimously praising the play's well written and darkly comical dialogues and its mercilessly acute observations of the battlefield of love, mostly shared some reservations concerning the plausibility of characters and plot and the play's ostentatious but mysterious symbolism. Explicit and implicit criticism of Skinner's play thus concentrated primarily on the climactic scene in which one of the characters blinds herself with the heel of her stiletto in a seemingly unmotivated act of obscure symbolism. However, as I will argue, both the construction of somewhat type-like characters and the use of symbolic imagery make sense as aesthetic strategies in an interpretation that reads the play not as a (thwarted) attempt at social realism in the strict sense of an imitation of the words and actions of real people but rather as a play that is 'realistic' insofar as it seeks to translate prevalent discourses of love into concrete dramatic actions and images. While the characters and their actions may indeed appear unconvincing as naturalistic emulations of reality, they are quite effective as visual concretisations of some of the abstract principles defining the discursive field of love. As in the plays by Marber, Ravenhill, and Kane, the characters are interesting not so much as unique individuals but as representatives of specific attitudes and dispositions, and their actions are understood not as psychologically plausible expressions of complex individuality but as deliberate dramatisations of theoretical ideas about love. The idea of love as compensation underlies the characterisation of all four figures, who are either searching for meaning or suffering from loss or loneliness. The focus of Skinner's play, however, is on love's precariousness. The unattainability and uncontrollability of the other as an object of love are stressed with graphic imagery, while it is also made obvious that, even more than the other him- or herself, it is the other's desire which is the most precarious object of love and the main source of pain and frustration. In Skinner's acerbic parody of the shallow but popular romcom genre, this precariousness works not merely as the unavoidable obstacle necessary for the story's arc of suspense that is as sure to occur as it is eventually overcome, but is painted in its full extremity.

8.1 Synopsis

Eigengrau tells the story of four Londoners in their mid-twenties falling in and out of love with one another in different constellations. As Robert Shore writes with some justification, the play sets in "as a fairly light comedy of types – the New Age kook, the confused feminist, the sentimental loser, the get-ahead materialist; four young Londoners trying to make a connection" (286). That the four characters are no paragons of psychological complexity is unambiguously indicated in the list of *dramatis personae* provided by Skinner, which openly displays their mono-dimensionality: we have Tim Muffin, "a fat bloke," Cassie, "a feminist activist," Rose, "a believer," and Mark, "the marketing guy" (Skinner 5). When the play starts, 'New Age kook' Rose is in a sexual relationship with Mark, who "make[s] eighty K a year" (23) and "spend[s] fifty quid a day on nothing" (21). Not surprisingly, their relationship is on unequal footing. For Rose it is true love – which she firmly believes in along with a litany of similarly miraculous phenomena, as she professes in her quasi-religious creed:

> I believe in
> Fairies gnomes elves cyclopses
> Leprechauns unicorns
> Pixies witches wizards
> Angels dwarves
> True love
> Love at first sight
> and the lost city of Atlantis. (45)

For Mark, it is a series of one-night stands which he is about to end rather sooner than later as he starts feeling both bored and slightly uneasy about the increasing obsessiveness of – as he calls her – "Crazy Rose" (25). The first longer scene shows Mark's first encounter with Cassie, Rose's flatmate, who is preparing a feminist speech while Rose is out for a job interview. Their conversation immediately becomes confrontational, Mark looking down on feminism and Cassie despising him for his success in marketing. However, their mutual attraction, which is obvious right from the beginning – Cassie "*finds a reflective surface and checks her appearance*" (12) as soon as Mark briefly leaves the room and Mark makes sure to stress that he is "not Rose's boyfriend" (16) – quickly wins over. Tellingly, when Cassie offers him tea, he changes his choice from Rosehip to Earl Grey – a decision the symbolic meaning of which is revealed shortly afterwards when Mark researches Cassie on the internet and finds that her second name is Grey (cf. 27).

The next two scenes introduce Mark's and Cassie's flatmates and produce a parallel figure constellation. Mark is living with Tim Muffin, his former fellow student, who descended rather than climbed the career ladder after they graduated from university together. Tim is currently working at a chicken fast food restaurant and half-heartedly follows his dream of becoming a professional carer, a vocation for which he considers himself qualified having cared for his grandmother before she died a few months ago. As becomes gradually more obvious, Tim has not yet overcome the loss of his grandmother, whose ashes he keeps in "*a large porcelain cat, with a removable head*" (9). Mark, who lives up to the stereotype of the heartless, calculating, profit-seeking 'marketing guy' in his treatment of women, seems to have a soft spot for Tim and tries to encourage him to follow his dream with more enthusiasm and commitment and to work on his appearance. Although the arrogance with which he treats Tim in their conversations is a genuine expression of his character, he is also credible when he says that he wants to help him:

Mark: I'm going to say something brutal now but I want you to know I'm doing it for your own good OK?
Tim: OK.
[...]
Mark: When you apply for a job
as a 'carer'
people look at you and think
that guy can't even look after himself.
Why would I want him to look after me? (22)

Tim is contrasted with Mark to produce two types: the attractive, energetic winner and the plain, sluggish loser. When Mark tells Tim that it "seems to me like nobody wants you" (22), it is to be taken as a rather reliable form of explicit characterisation that will be of major importance later in the play.

The figure constellation of contrastive flatmates is mirrored in the relation between Cassie and Rose. When Cassie, practising a speech about rape porn on the internet, is interrupted by Rose, who is wondering why she has not heard from Mark during the last few days, she warns her about the unreliability and potential violence of all men. The following dialogue brings to the fore Rose's tremendous naivety and Cassie's insecurity about her feminist principles.

Rose: Come on Cassie! You'll never find a man with that sort of attitude.
Cas: Oh my God. No? Well. Maybe I don't want to find a man.

Rose: Ah.
Cas: I mean. I don't
of course I'd like to find
someone.
Rose: A man?
Cas: Yes but not just any man. You know? Men
most men aren't very evolved. [...]
[...]
Rose: You mean maybe it's an emotional thing? Like
Cas: well
Rose: for example a fear of how quickly he's falling in love with me?
Cas: No! I mean because he's probably
along with ninety-nine per cent of men
for all their gun-toting
wife-beating
warmongering bullshit: a total emotional coward.
Delighted, Rose applauds.
What?
Rose: Was that from your speech?
Cas: No.
Rose: Thought you might be practising. (31–32)

That Rose cannot tell whether Cassie is really conversing with her or practising an activist speech not only ridicules Rose but also questions the authenticity of Cassie's feminism, which later in the play appears more like a porous façade of self-protection and identity formation than a deep, unshakable conviction. The rest of the scene discloses that Rose, who has just recently moved in with Cassie, has outstanding debts from her former accommodation and also already owes Cassie two months' rent for the shared flat which she hopes to be able to settle once she gets paid for her new job in a karaoke bar.

What the first five scenes with their schematic figure constellation have indicated almost as a necessary plot-development of the stereotypical romcom really seems to be confirmed in the following nine scenes. The initially antagonistic protagonists – the arrogant capitalist and the feminist activist – discover their love for each other, while their comical quirky sidekicks are recompensed for the loss of their initial partners when they find out that they, too, are made for each other. Apparently, therefore, the clearly incompatible Mark and Rose find seemingly more compatible partners in Cassie and Tim. In contrast to the romcom pattern, however, the process is exceedingly painful for Rose, who holds on to her hopeless infatuation with Mark too long, for Tim, who is in love with Rose while she still tries to win back Mark, and for Cassie, who has to find out that Mark really is as unscrupulous and insincere in matters of love as might be expected from 'the marketing guy.' Mark first succeeds in seducing Cassie by feigning interest and then an outright conversion to feminism. Although Cassie is suspicious and finds it "extremely unlikely for a a a / good-looking / white / *public schoolboy* to really give a fuck about any of this it's much more likely he's just trying to get in your pants" (40), she finally gives in. What wins her over is not only that he plays his role quite well, carrying feminist books with him, wearing feminist T-shirts, reproducing hackneyed feminist phrases, asking for permission to kiss her, and playing at gender trouble and male uncertainty: "I'm a man. And these days it's / It can be hard to know what that means. You know? I mean I think sometimes about men in my grandfather's time. Or men the men you see on the telly with the guns and the cause and the / The passion. I look at those men and I think yeah. You / You're men" (57). All this functions as recognition of her self-chosen personality as a feminist and flatters her just as much as his declaration of love at first sight, but the ultimate reason why Mark deceives her so effortlessly lies in the ambiguous attitude she entertains towards her own principles. Scene Eleven starts with a *"long and uncomfortable pause"* following an unsuccessful attempt at having sex thwarted by Mark's aversion to pubic hair. When Mark cleverly blames "the patriarchy" and "the media" (61) for manipulating his sexual preferences in spite of his sincere feminist convictions, Cassie admits to being in a similar situation:

> Because supposing you were [...]
> a feminist [...] and you believe in gender equality not just as an ideal but as a *necessity* and then at the same time you find yourself in a completely different sense or
> context
> wanting to be dominated in the bedroom. For example. You know? You'd start asking yourself wouldn't you? Why do I want that? [...]
> what if it's just inside me? This thing which I want which so completely
> betrays everything I really believe and makes me some kind of a
> deviant. Essentially. (62–63)

The question prompted here is the same Illouz arrives at when she tries to explain the spectacular success of the *Fifty Shades* trilogy, in which an autonomous, sovereign woman willingly submits herself to (above all sexual) male domination: Why is sexual submission so alluring that it is, as reliable studies estimate, part of the private fantasies of up to half of the female population (cf. *Hard-Core Romance* 57)? Why is there "an increasingly loud litany lamenting the fact that [gender] equality has brought the demise of sexual desire," or simply, "why are some of women's fantasies still caught in patriarchy" (58)? Illouz offers an answer that comprises "the *clarity* of the gender roles" (58) inherent in the unequal social system of patriarchy, the "'natural' mutual dependency and thick emotional glue" (58) ensuing from the manifestation of male power as protectiveness, and the spontaneity and immediacy of emotions facilitated by pre-established, stable gender roles. In contrast to these 'pleasures' provided by inequality, "equality is less pleasurable because it generates uncertainty and ambivalence" (58) and, in the absence of "well-scripted social roles" (59), constantly requires negotiation and communication. This is not to say, of course, that women secretly wish to reverse the achievements of feminism and to restore patriarchy to its former strength, but the clarity of roles associated with it and the 'naturalness' of eroticism issuing from such non-negotiable roles seem to possess some fascination. As Illouz puts it, "[t]he longing for sexual domination of men is not a longing for their social domination as such. Rather, it is a longing for a mode of sociality in which love and sexuality did not produce anxiety, negotiation, and uncertainty" (61). In this view, Cassie's wish to be dominated in the bedroom expresses above all a desire for clarity and certainty. However, since she associates this pleasurable clarity with the roles of strong masculinity and weak femininity,[56] she cannot but have the guilty conscience of a traitor to her own cause.

Mark, for his part, does not hesitate to leap at this chance. He immediately changes his tone and commands her into the bathroom to shave herself, which she eventually does. When he claims, "I'm giving you what you want" (64) he is right because the ambivalent role he plays satisfies two of Cassie's needs at once. As the self-proclaimed feminist, he tells her what she rationally wants to hear and recognises her self-determined identity; as the self-assured, aggressive lover, he provides the domination which she secretly desires in her sex life. Deceived by the role he plays, Cassie allows Mark to mediate between her demand for social and political gender equality, which she believes to be their common cause, and her private sexual desire, which is rooted in the "very long cultural tradition in which what was eroticized was precisely men's power and women's lack of it" (Illouz, *Why Love Hurts* 192). Her eroticism interferes with her political activism, causing the same problem Mark expresses with regard to masculinity: a

---

[56] In the intriguingly titled epilogue "Sadomasochism as a Romantic Utopia" of her short analysis of the *Fifty Shades* phenomenon, Illouz reminds us that this association is by no means necessary. What makes BDSM so enticing, she suggests, is that "roles are reestablished, but in a way that does not necessarily overlap with gender. [...] Thus BDSM avoids the confusion and ambivalence inherent in gender equality and reaffirms sharply defined and stylized sexual roles, yet without predicating them on 'hard' gender identities. In fixing clear roles detached from identities, BDSM provides the certainty that comes with scripted roles without returning to traditional gender inequality. This is because whatever inequality is enacted in BDSM is playful rather than inscribed in a social ontology of the sexes" (69).

keen sense of uncertainty resulting from the tension between the allure of traditional gender roles and the ideal of equality and political correctness. What Cassie does not yet see is Mark's insincerity regarding the principles to which he allegedly subscribes. While Cassie's sexual fantasies include the suspension of gender equality in her private bedroom, Mark's misogyny does not stop there. The way he plays with and betrays women shows his unwillingness to grant them the equality and respect which Cassie demands, if not for the particular context of her own sexuality, for the general treatment of women.

Meanwhile Rose, ignorant about this development and unwilling to simply accept the breakup of her relationship with Mark, traces his flatmate Tim and urges him to help her win back Mark. Her decision to fight for Mark, she tells him, was confirmed when she "saw a *cornflower-blue dress* in the window of Selfridges" (48), which she understood immediately as a sign of the universe, blue being Mark's favourite colour. Although unable to afford it, she plans to surprise Mark in this dress and so to rekindle his love. Tim, discernibly falling in love with Rose during their conversation, agrees to help her and later we see him stealing money out of the cash drawer of the chicken grill he is working in.

Shortly after, Rose appears in her and Cassie's flat. "*She looks fabulous in a cornflower-blue dress and red stiletto heels*" (68) and explains to a bewildered Cassie that "a miracle happened" and that she "got this envelope with [her] name on it and no listen to this in it was contained: the exact amount of the price of the dress" (69). Cassie is furious as Rose has not paid any rent so far, but Rose explains herself – explaining above all the extent to which her concept of love is shaped by romantic media productions:

Rose: Well I needed this dress! Right? It's the dress for the final scene! The dress to walk down stairs in so he sees me across a crowded room and realises he's in / love with me.
Cas: What are you talking about?
Rose: Cassie, what's rule number one when someone dumps you? You lose weight. Get a fabulous dress and then bump into them unexpectedly so they know they fucked up (70).

Cassie tries to talk Rose out of her plan, arguing that theirs was only a meaningless sexual affair that started in a pub with both of them drunk. When Rose corrects her, it is revealed that Mark had seduced Rose with the very same method of feigning interest in a woman's firmest beliefs.

Rose: Well you're wrong. A pub! Ha! Me and Mark have the most amazing *how we met* story ever. I was having a falafel and he was on a nearby table and I started telling him about numerology and he was really interested. [...]
During the following Cassie crumples, slowly. As though something has just collapsed inside her.
Rose: and he invited me round to his flat. Amazing. And he'd looked up our numbers? He'd actually looked up our numbers online which I'd already done of course but hadn't

> told him and he said we're eighty-eight per cent compatible. And
> he said
> listen he said
> I've been falling in love with you ever since I first saw you. And I just keep falling. And I know we've both been hurt before but I was wondering if tonight we could play a game. Pretend it's just you and me and we're two human beings who think they might know they might
> like each other.
> *Beat.*
> And then he kissed me. Isn't that the most beautiful thing you've ever heard? (72–73)

Rose's last question bears painful dramatic irony given that only a few scenes earlier Mark has not only seduced Cassie with the same trick but has also used almost exactly the same words (cf. 58–59). Devastated and infuriated, Cassie cannot bring herself to tell Rose what she knows about Mark. They end up arguing and Rose leaves for Mark's flat to give him what she thinks he really wants (cf. 85), which is oral sex that *"seems to go on for an uncomfortably long time"* (86). Having finished, she thanks him for giving her a second chance and apologises for not meeting his expectations before: "For some reason I wasn't exactly what you wanted or I let you down or / I did something anyway that wasn't quite right. But whatever it is you want, I can be that Mark OK? Just tell me what you want" (87). When Mark gives her to understand that he is seeing someone else and does not want or need her, she tries the reverse direction, stressing her own neediness: "I need you. [...] I have nothing without you" (89). Mark's reply – "I don't want you to need me" (90) – is the ultimate refusal of love, the refusal to desire the other's desire. When Tim tries to seize this definite break-up for his confession of love, he unfortunately and unconsciously reproduces Mark's hackneyed line again: "Ever since I first saw you / [...] I can't stop thinking about you" (92). The problem of authenticity and originality in love is carried even further when an emotionally hurt and confused Rose runs out and *"one of her shoes comes off"* (93), completing the Cinderella motif which had started with her encounter with a Prince Charming who is beyond her reach and the miraculous gift of the blue dress. The scene ends with a not too subtle suggestion of the transferability of emotions, when Tim cries "Why did you leave me?" (93), hugging the urn of his grandmother. Immediately after that, we see Rose in her karaoke bar, singing *"a power ballad"* (94) at the climax of which she gouges her eyes out with the heel of her remaining stiletto.

The ending brings together all four characters in Rose's room in the recovery ward of a hospital. Rose is still unaware of Mark and Cassie's by now terminated affair, and Mark's claim to be in a relationship with a dancer might only serve the purpose of sparing her the realisation that she has been betrayed not only by Mark but also by her flatmate Cassie. Cassie is having a baby, allegedly the result of a drunken one-night stand that had started in a pub, which for Rose is nothing less than "a miracle" (107) since she can hardly believe that "out of something so / Meaningless / you can get / Actual / Life?!" (108). Mark seems willing to believe this story, but it remains undeterminable as to whether Cassie really had this one-night stand during their relationship or whether she instead uses the same crooked 'how we met story' Mark had told her about him and Rose so as to deny his paternity and keep him out of her future life. The biggest surprise, however, is the news that Rose and Tim are now a couple and are moving in together somewhere by the sea.

It is the combination of climax and final resolution that has attracted most criticism. Henry Hitchings expresses irritation about Rose's act of self-blinding, complaining that "the ending is misjudged, marred by lumpen symbolism and an effortful striving after Sophoclean grandeur" (285), an accusation mirrored by Robert Shore, who is annoyed that instead of complex character studies "we get that unwarranted Sophocles-meets-Betty Blue plunge into the abyss and tiresome clichés about boarding schools and the like" (286). Quentin Letts' condescending critique culminates in his astonishment that, although the play went through "so many rehearsals and readings, [...] no one in that development persuaded little Miss Skinner to find a less ostentatiously immature climax, if that be the term" (285). For Charles Spencer, "it's the shocking and dramatically unearned act of self-mutilation that really makes the flesh creep, along with the grotesque suggestion that it actually provides a happy ending for the play's two most vulnerable characters" (286). Even Benedict Nightingale and Sam Marlowe in their mostly favourable reviews find the ending not convincing, criticising that the "denouement [...] could be more credible" (Nightingale 285) and that "the play declines into heavy-handed symbolism and narrative improbability in its later scenes" (Marlowe 287). Most of this criticism, I think, could be justifiably levelled at a play conceived as a naturalistic imitation of reality and a profound analysis of psychologically plausible characters. However, the fact that, as Nightingale observes, the occasionally almost caricature-like figures "sometimes behave in ways dictated more by plot than character" (285) and that, as the unfavourable reviews just quoted emphasise, the characters lack roundness and complexity and that their actions are not motivated logically by their psychological structure, is not an indicator for poor writing but as much part of the play's aesthetics as the climactic blinding scene. By stripping the characters of too much psychological depth and individuality, they can serve even better as vessels and mouthpieces of recurrent and recognisable fragments of the discursive field of love. Neediness, obsessiveness, blindness, egoism, altruism, instability, corporeality, spirituality – all these characteristics of love are thematised and, variously, embodied in the play by characters who lend themselves to unilateral exaggeration of one or a few of these characteristics. Consequently, the two overarching discourses about the compensatory and precarious nature of love become visible with special clarity.

8.2 Urban Loneliness and the Desire for Meaning

All figures at least implicitly indicate love's compensatory function, but it finds its most obvious representation in Tim's emotional development, for whom love – and the belief in love – come to compensate for the death of his grandmother. Already the very first scene shows Tim, standing on Eastbourne Pier and talking to the urn containing her ashes, and Mark's later description of Tim's day-to-day existence spells out his inability to cope with the situation: "And I know it's hard at the moment because obviously you're sad and you've had a 'loss' and of course you need time to get over it but Tim / it's January. It's been what / four? five? months? And every night I get home and you're still here and *she's* still here and I just think when is it going to change?" (24). It changes when Rose steps into his life: all of a sudden, he summons up new courage, finds a new objective in life. The loss of his one object of affection is compensated with the acquisition of a new one. Moreover, love also replaces a more general loss of faith Tim testifies to in his first conversation with Rose. When she tells him what she believes in (fairies, unicorns, pixies, true love etc.), how she once asked for (and received) a sign by the fairies to prove their existence, and that "it's important to try and believe in something [...]. Especially these days" (46), he replies: "I think I stopped believing because / when Nan died I realised for the first time that when you die / that's it. You die. You're gone. And you aren't ever / ever / coming / back" (47). But in contrast to this dismissal of metaphysical beliefs, in the last scene of the play Tim is seen on Eastbourne Pier with the urn again, asking his Nan for a sign that she is "there" (117). To his astonishment, he receives an answer in the form of *"loads of cigarette ends"* (117) he discovers in the ashes of his grandmother, who was a passionate smoker and probably died of lung cancer (cf. 44). Of course, the sign is not real – the cigarette ends have been put there by Cassie when she stayed at Tim and Mark's flat and could not find an ashtray – and the play also does not propagate the necessity to return to naïve metaphysical beliefs. But it suggests that love may have the force to replace a mixture of lethargy and nihilism such as Tim's with a more hopeful outlook on life. His love for Rose restores in Tim the capability of believing in such improbabilities as signs from the dead or true love.

This may well be called a form of illusion, but it is not thus something to be wholly renounced as it is also a source of pleasure and, here in particular, of orientation, meaning, and security.

This idea of love as possible compensation for a loss of orientation in modern societies is further supported by the characterisation of the remaining three figures. Most obviously, Rose tries to satisfy her need for meaning, security, and 'ontological rootedness' with a curious but firm system of beliefs. That 'true love' is equated in this system with pixies, dwarves, Atlantis, and other myths is not so much a statement of the play about love's non-existence than about the similar function love and these myths fulfil for Rose, which is to supply her life with sense and meaning. Her painful neediness and obsessiveness in matters of love indicate how desperately she strives for such meaning, while her desperation can be explained with her saddening loneliness, revealed by Tim in the waiting room of the hospital: "I went through her phone / when I got here / there was only four numbers in it: yours / mine / Mark's / and the *Daily Mirror* Astrological Helpline" (101). Possibly an intertextual reference to Marber's *Closer*, in which Alice's loneliness is illustrated by a sentence in Dan's novel that describes her as having only "one address in her address book; ours ... under 'H' for home" (Marber 19), this is the moment in Skinner's play that most emphatically draws attention to the striking parallels to *Closer*. With its figure constellation of two men and two women swapping partners and its figure conception of rather type-like characters identified as social representatives in the list of dramatis personae, *Eigengrau* shares much common ground with its nineties' forerunner. Parallels appear also in details such as the (faulty) narrativisation of first encounters into memorable 'how we met stories' or references to cultural templates like courtly love (*Closer*) or fairy-tales (*Eigengrau*). What connects both plays most intimately on an overarching level, however, is this very theme of love in an environment of urban isolation that is symbolised by the almost blank address or phone books. Like in *Closer*, the four characters seem to lack any significant social or family contact apart from their flatmates and their short-lived romantic relationships, and the belief systems to which they commit – Rose's and Tim's metaphysical beliefs, Cassie's somewhat desultory feminism, and Mark's belief in marketing – serve as their self-chosen sources of direction meant to overcome or prevent a sense of being lost. The motif of loneliness is further highlighted by an aesthetic device that seems to signify the drowning of the individual in the faceless, amorphous, anonymous crowd inhabiting the city of London. Six scenes of the play consist only of words spoken by impersonal voices expressing, as Nightingale puts it, "Skinner's cynicism about life in London – a city evoked by a loud ugly babble regularly heard inside the big ugly box serving as an auditorium" (285). For all four characters, then, love can be assumed to fulfil – or at least to promise – a compensatory function in the form of granting recognition, generating a sense of self-worth, and alleviating the ills of life in a society marked by isolation and a loss of meaning or groundedness. Loaded thus with downright existential significance, love becomes an incalculable risk in the face of its uncontrollability, especially for the play's two 'weaker' characters.

8.3 "I don't want you to need me": Love's Precariousness and the Desire for Control

The precariousness of love takes various shapes in the play. First, there is the obvious threat of losing the beloved person (Tim's Nan) or the beloved person's love (Mark's love for Rose). If it is questionable whether Mark ever felt anything like love for Rose, the only difference this makes is that the precariousness of Rose's situation consists not in the loss but in the unattainability of love. In any case, both Rose and Tim are painfully aware that their objects of love are beyond control. In Cassie's and Mark's concepts of love, the struggle for control turns into a struggle for power. For Cassie, love's precariousness is first of all the result of gender inequality and male domination, a specifically female vulnerability against which she has equipped herself with an armour of feminist theory which, however, she wears only half-heartedly. Mark, on the other hand, has chosen an aggressive instead of a defensive strategy against the threats of vulnerability and loss of control. In an Ovidian fashion, he has trained himself to mastery of the game of love, which he plays cunningly and successfully. As Singer notes about the *Ars Amatoria*, Ovid provides his readers with instructions on how to find a new beloved, how to flatter, seduce and deceive, "but he never offers instruction in methods of attaining love that can be both passionate and lasting" (Singer, *Plato to Luther* 144). The Ovidian lover knows how to "feign a passion for the other while nevertheless retaining independence. Apparently the goal is to get someone to love you without your having to give love in return" (145). Similarly, Mark seduces, deceives, manipulates and, as the 'marketing guy,' knows how to advertise and create demand for his person. He is not after permanence but after the joy of short-lived passion and he has steeled himself against the precariousness of love by learning how to control the game and to keep his own emotional investment at bay.

Rose is the character who suffers most from her powerlessness and she inflicts horrible pain and misery upon herself in her attempts to gain control. A still harmless but also inefficient attempt at overcoming love's contingency can be seen in her reliance on numerology and other esoteric methods for divining the compatibility between Mark and herself, through which she hopes to derive a kind of love-determinism that is more reassuring than complete randomness. Moreover, such determinism promises to relieve the pressure to choose well caused by the abundance of choice in the digitalised dating market. One connecting feature of the otherwise incoherent list of words and phrases uttered by the impersonal voices is their affiliation to the semantic field of marketing – of advertising, supply and demand –, with particular references to internet based housing and dating markets.[57] The way these wants and offers rain down on characters and audience reflects the duress enforced upon the inhabitants of digitalised modernity to meet the abundance of choice in love with skills in both 'selling' and 'buying,' that is, with the abilities to advertise the self and to determine one's needs and demands in order to choose properly. Rose's numerology is an attempt at avoiding both love's uncontrollability and the demand to partake in this business of 'selling' and 'buying.' However, when it proves to be ineffective, she decides to actively take control and turn herself into Mark's object of desire which, for her, means to reduce herself to a sex object. The words with which she advertises her performance of oral sex indicate that she hopes to embody the epitome of male sexual fantasy, an irresistible object of desire:

> What's wrong with you Mark?
> Don't I look sexy?
> On my knees in front of you?
> Isn't this what you want?
> Oh! I think it is ...
> [...]
> *Hard* day at the office
> get home
> blonde girl comes round offers to *suck* your *cock* for you and what?
> You're going to say no? (85)

---

[57] The peculiarities of internet-based communication might provide a further connecting feature within the seemingly unrelated series of words and phrases. Successions such as "large kitchen / well hung / ample storage / spacious / ladies only / huge erection / bright and / light filled / Sky / Sky Plus" (66) or "I want / someone / to do / to let / to buy / to get / pants / unseen / socks / unheard / a wig / last known / a dress / underground / over / ground almonds / two eggs / white wine" (67) not only make use of the ambiguity of words that have different meanings in different semantic fields (e.g. "huge erection" in the semantic fields 'construction/architecture' and 'sexuality') but also imitate the directionless progress of internet surfing that moves along hyperlinks which connect all kinds of more or less related contents and where 'housing,' 'dating,' 'pornography,' 'clothing,' 'public transport,' and 'cooking' are always only a few clicks apart.

Believing to have found what Mark really desires, she offers to provide it. After the fellatio, she further promises to be his object of desire no matter what form his desire may take: "But whatever it is you want, I can be that Mark OK?" (87). When she vomits on the floor after Mark's revelation that he is seeing someone else, she cleans it up in front of him, desperately offering an affair, an open relationship, and a threesome (cf. 87–89). The characterisation of Rose as a type-figure representing misguided female submissiveness, which has been pushed ahead so far in the presentation of her relentless pursuit of the dismissive Mark, culminates in this scene. As if to conform to the stereotype of female love postulated by Nietzsche and lamented by de Beauvoir and Firestone, Rose wilfully agrees to her objectification and offers herself as an item of possession. Having idealised Mark into her own romantic demigod, she sacrifices her dignity and self-worth in an act of total submission. In its imperturbable fixation on this illusion of a perfect love object, her blind infatuation fully lives up to José Ortega y Gasset's scathing criticism of 'falling in love' as "a state of mental misery which has a restricting, impoverishing, and paralyzing effect upon the development of our consciousness" (44). Not to be confused with love proper, Ortega calls this phase a form of "mania" (50) that manifests as "attention abnormally fastened upon another person" (52), a "transitory imbecility" (55) which suspends sober judgement and reasoning. Moreover, in her desire to gratify her beloved and to become his object of desire, Rose also strongly resembles Alcibiades in his devotion to Socrates. In Lacan's reading, what Alcibiades desires is both Socrates and Socrates' desire. Rose, like Alcibiades, desires and wants to be desired and is refused by the beloved, although Mark lacks the Socratic consistency to deny sexual gratification before denying the lover. Like Alcibiades, Rose has humiliated herself, has opened up and made herself vulnerable in the attempt to become the object of the other's desire, only to learn that this goal is beyond anyone's control.

The most graphic expression of love's precariousness in the play is certainly Rose's act of self-blinding, which has received so much criticism and for which I would like to suggest a threefold interpretation which shows that there is much more to it than ungainly symbolism. First of all, there is of course the reference to Sophocles' *Oedipus Rex* which some of the critical reviews have pointed out. Analogous to Oedipus, who punishes himself with physical blindness for his previous lack of insight, Rose's self-mutilation can be understood as a form of punishment following the *hamartia* of 'spiritual' or 'mental' blindness that had prevented her from seeing Mark's insincerity and from recognising Tim's genuine affection.

Second, the fact that it is Rose's self-inflicted blindness that leads to her relationship with Tim points towards an ambiguous concept of love and the role of literal and metaphorical blindness or sight. On the one hand, by sacrificing her eyesight Rose seems to win the ability to see more clearly Tim's loveable character. On the other hand, optical blindness seems to be a presupposition for her developing of loving feelings for Tim in the first place. The obvious compatibility of their characters – something Rose articulates already during their very first encounter when she calls him a "kindred spirit" (47), even though she might not be fully reliable in this situation as she needs Tim for her plan to win back Mark – is not enough to make her fall in love with him. The play thus refutes the idea of love or attraction altogether unaffected by physical appearance. As long as the beautiful Rose is able (or forced) to see the unattractive Tim and hence inevitably compares him to the handsome Mark, no love emerges. In other words, while Tim's conversion from disillusioned sceptic to romantic believer due to his encounter with Rose follows the familiar pattern ("Then I saw her face, now I'm a believer"),[58] for Rose, who starts out as a believer, it becomes necessary to avoid the sight of Tim to keep her faith. This might be read as an unpleasant character trait in Rose, a contemptible superficiality that she then bitterly regrets. I think, however, her self-blinding is better understood as a hyperbolic *mise en scène* of a social-realist conviction – a criticism of the dewy-eyed idea that physical appearance and mutual bodily attraction play no role in true love. It is useful to remember here that even for Ortega, who is at pains to distinguish the ephemeral phase of 'falling in love' from genuine love, the former always precedes the latter. Without the fixation of increased attention on a particular person as a result of spontaneous and irresistible physical attraction, he insists, no true love can ensue (cf. 81). The importance of physical attraction for infatuation is confirmed by Elaine Hatfield and Susan Sprecher's empirically verified "Passionate Love Scale" (1986), in which "[a]ttraction to other, especially sexual attraction" (Hatfield and Rapson 110), figures as a central indicant for passionate love (which is used synonymously for infatuation). This physical and sexual attraction, however, is absent from Rose's relation to Tim, and gouging her eyes out is her way of eliminating it as a necessity for love.

That she is prepared to go to such extremes demonstrates the degree to which she has internalised the socially and medially transmitted compulsion to find love and build a functioning relationship that is evoked by the impersonal voices immediately before she blinds herself ("fall in love [...] get married [...] have kids [...] move to the country [...] live happily ever after [...] just someone /

---

58 Written by another artist (Neil Diamond) and performed by a casting band surrounded by controversies and doubts concerning their ability to sing and play their own instruments, the Monkees' "I'm a Believer" is a fitting intertext for a play that repeatedly questions the possibility of authenticity and originality in the discursive field of love.

please someone / is anyone / anyone / hello? / hello? / is anyone there?" [96–97]). The desperate measures she takes epitomise the pressure exerted on her by a collective mindset that castigates the inability to be in a romantic relationship, and the suffering resulting from it, as symptoms of a "flawed self" (Illouz, *Why Love Hurts* 130). The act of self-blinding is thus a lurid metaphor for the precariousness of conceiving of love as the primary generator and indicator of self-worth, as well as a cruel reversal of the pattern in which distortion and loss of clear sight are usually treated in discourses of love. While love is customarily blamed for distorting the lover's view of the beloved, idealising (non-existing) virtues and blinding the lover to faults in the beloved, here blindness works as the prerequisite for love. In fact, Rose undergoes both kinds of blindness: her infatuation with Mark renders her blind to his insincerity, to their incompatibility, and to Tim's genuine love for her; but in order to appreciate Tim's affection, she has to blind herself physically first. In German, the proverbial blindness of love usually translates as *Liebe macht blind* ('love makes blind'), alluding not so much to love's contingency but to its impact on the lover's ability to see clearly. It is precisely this form of love-blindness the play thematises, not least with its German title *Eigengrau*, denoting the colour the human eye sees in absolute darkness, which is perceived as slightly lighter than black due to the lack of contrast with lighter colours. As a colour the human mind produces on its own, independent, as it were, of the object of perception, it points towards the unreliability of human sight, both visual and spiritual. In the darkness of love, vision loses its grip on reality and tends to see not what is there but what it produces. In a sense, it is Skinner's darker version of Stendhal's concept of 'crystallisation,' the imaginary process the French writer saw at work in all instances of passionate love and which discovers – or rather creates – desirable qualities in the beloved person which he or she does not really possess (cf. Singer, *Courtly and Romantic* 360–61). At the same time, *Eigengrau* is what Rose sees after blinding herself, a vision void of contrast in which 'all cats are grey' and in which Tim's beauty of character is no longer overshadowed by his physical unattractiveness. Blinding herself is her last resort in the attempt to gain control over love, taking it in her own hands to bring about the delusionary power customarily ascribed to love. Rather than presenting a "happy ending" enabled by "grotesque" means, as Spencer argues, Rose's love for Tim is rendered highly ambiguous and doubtful by the fact that she literally had to force it upon herself, deliberately producing the delusion that is supposed to be love's effect. It is the biting irony of the play that no such force was necessary for her infatuation with Mark, who engendered enough delusion in Rose by his mere attractiveness. Yet both her eagerness to be so easily deceived by Mark, her self-humiliating perseverance in her devotion to him, and her willingness to inflict bodily harm on herself for the sake of loving Tim may be traced to her dependence on a romantic relationship as the most salient source of self-worth.

The third meaning of Rose's blindness connects the play with the theories discussing love as 'Gift-love,' as a wish or even need to care for others. Spaemann, Pieper, and Lewis all agree on the existence of a human need to bestow love, to care for the well-being of another person. Frankfurt, who defines love as caring for others, is the one who most clearly ascribes to it the function of investing human lives with meaning, even though this impairs his own idea of love's disinterestedness (see 2.1). The figure who embodies this side of love, the joy and sense of meaning to be derived from caring for others, is Tim. Tim is certainly no self-sufficient fountainhead of disinterested love. Throughout the play, his wish to be a carer is marked as a need rather than an expression of voluntary altruism. As Lyn Gardner observes, "all he really wants is somebody to love and look after" (286), and it is plainly obvious how much his own well-being depends on the opportunity to be responsible for that of somebody else. The main function of Rose's self-blinding is precisely to lift Tim into a position where he can do that, where he is needed. When Mark reacts unenthusiastically to Tim and Rose's plan of moving in together, Tim's final argument – "Rose needs me" (112) – eventually convinces even Mark, who knows Tim well enough to recognise that caring for the blind Rose can be the source of meaning he had been waiting for since his grandmother died. As far as Rose is concerned, her act of self-mutilation can thus be construed as an act of deep love: for Spaemann, to love a person wholeheartedly means to let him or her feel that he or she is needed; for Lacan, "love is giving what you don't have" (*Seminar VIII* 34), that is, giving your 'lack' or 'need.' Rose's blindness, her lack of sight, is a gift for Tim, a gift that allows him to be needed, that makes him 'desirable' as a carer. This gift is an act of love – but it is no more disinterested than the love it enables Tim to bestow. In her relationship with Mark we have seen that Rose, too, is not simply a needy person but that her need is to be needed, or in Lacan's words, her desire is the other's desire. In Mark, however, she could not find this. For whatever reason, she could not turn herself into that which Mark needs or desires; nor did Mark want to be that which she desires. In Tim, on the other hand, she finds the required neediness of which she can make herself the object. By making herself dependent on a carer, she turns herself into the object of Tim's desire for someone who needs him.

8.4 Conclusion

The foregrounding of the dominating discourses of love in Skinner's *Eigengrau* is facilitated by the choice of somewhat monodimensional characters and the predominance of plot over characters and minute psychological motivation of their actions. In particular, the two shocking scenes, which seem not to be in line with Rose's rather timid character, are not to be seen as inept deviations from an attempt at dramatic realism but as the most powerful images in a play that is, despite its often brilliantly realistic dialogues, at its core profoundly symbolic. The discourse of compensation finds expression in the characters' search for meaning in a world without fixed points of orientation, an existential condition the play tries to mould into a theatrical experience by wrapping the audience in absolute darkness whenever the impersonal voices expose them to a plethora of wants and demands.[59] As Gardner writes, the play is "an urban fairytale about a generation with nothing left to believe in while adrift in the big city" (286). Cassie, Mark, Rose, and Tim all struggle against isolation and disorientation; they all hope to find meaning through their self-chosen belief systems; and for all of them, if only briefly for Mark and Cassie, love appears as the most promising way to the meaningful lives to which their self-chosen paths could not guide them – even if the play's demystification of love in the plotline and its hybrid aesthetics comprising fairy-tale and romcom set pieces maintain the uneasy suspicion that love may be no more than a reassuring illusion.

It is probably precisely because love is not simply self-chosen that it is a more promising purveyor of meaning. Its uncontrollability and mysteriousness lend it the powerful aura of fate and determination, not only disempowering the human will but liberating it from the task of making insecure choices. But love's uncontrollability, which makes it, if successful, a source of meaning and security, is also what makes it precarious and, if unsuccessful, a source of pain. In *Eigengrau*, the precariousness of love is represented in various shades. The way Cassie and Rose open themselves up to and are hurt by Mark bespeaks the intense vulnerability in love. Cassie, almost ironically, puts aside her armour against all men and discloses to Mark the most intimate depths of her personality, only to learn that Mark fully lives up to her former prejudices about the insincerity of men and their unwillingness to be emotionally honest. Similarly, Rose, in her attempt to win back Mark, humiliates herself in front of him and

---

[59] The only exception to this is Scene Fifteen, where the voices are accompanied by pink light radiating from a disco ball in Rose's karaoke bar and where the blackout follows the act of physical self-blinding (cf. 94–97).

offers herself up for sexual possession. Where Cassie sacrifices her feminism out of (blind) love, Rose sacrifices her fairy-tale belief in true love to a debased version of unreciprocated and submissive infatuation. Both open themselves to the other, expose themselves in their full vulnerability, and are hurt. Rose's unsuccessful self-sacrifice also points towards the precariousness that results from the uncontrollability of the other – and of the other's desire in particular. Nothing she does enables her to control Mark's desire. She is more successful, eventually, with Tim. She indeed turns herself into his object of need, thus gaining control, as it were, of Tim's desire. Paradoxically, however, the price she pays for this control is complete dependence. The shocking act of self-mutilation provides a forceful image of what it means to love in the Lacanian mode, that is, to give what you don't have: it means to give insufficiency, neediness, dependence in the hope that this is what the other desires and in the hope of receiving the same in return. It means giving up all control in the hope of overcoming love's uncontrollability. Self-blinding is a fitting image for the precariousness of this risky undertaking.

9 "We don't need ties": Rebellious Love in Mike Bartlett's *Love, Love, Love*

9.1 Synopsis

Although its title seems to suggest otherwise, love is not the main topic in Mike Bartlett's *Love, Love, Love* (2010). Only on closer inspection the play reveals its tight links to the sociological positions outlined in Chapter 3 in that it foregrounds the specifically contemporary sources of love's precariousness and, as it were, transforms the schematic story of love's historical development from a promise of absolute freedom to a threat of limited self-realisation into the concrete story of one couple's love and marriage. Its principal subject, however, is a different sociological thesis. In three acts, the play follows Sandra and Kenneth, representatives of the baby boomer generation, from falling in love to the Beatle's "All You Need Is Love" (hence the title) in 1967 to marriage crisis and divorce in 1990 and into their retirement age and an awkward family reunion with their now grown-up children in 2011. The play's focus, as the critics have unanimously observed, is on the thesis that the baby boomers are a generation of selfish and self-indulgent individualists, irresponsibly wasting the nation's resources and living at the expense of their children's generation. Starting as easy-going, pot smoking Oxford students living off grants as high as their parents' income (cf. 27), the sky is the limit for nineteen-year-old Kenneth and Sandra. They feel that there is a momentum in history and that the future belongs to them. As Kenneth tells his hard-working brother Henry, who is only four years older but appears like a relic from a bygone historical era:

> Nothing like this has ever happened before. The laws are constantly being overthrown, the boundaries what's possible, the music's exploding, the walls collapsing. That's what's going on. That's what's changing. We travel, do what we want, wear what we like. Enjoy it. Experiment.
> We're breaking free. (31)

Kenneth and Sandra know they are "the future of this country" (27), and their recipe for success is that "[y]ou've just got to want it enough" (39). Twenty-three years later, they live in a "*medium-sized terraced house on the outskirts of Reading*" (60) with their fourteen-year-old son Jamie and their daughter Rose, who just turned sixteen. Their dreams of unbridled freedom have not come true, obviously, as they both feel trapped in the life they are currently leading. Kenneth sums it up: "We live in Reading. / Something's gone wrong. [...] It's all house. Children. Work. We never wanted it like this. I'm not happy you're not happy so ..." (90–91). So they get a divorce. They were unable to preserve the freedom and carelessness they cherished as teenagers in their marriage and working life, but this does not prevent them from taking the liberty to pursue their own individual happiness with utmost carelessness towards their children. They are presented as terrible parents throughout, so busy with work or their extramarital affairs that they forget to attend their daughter's violin concert and cannot remember her A level subjects or her precise age. Towards the end of Act Two, the family sits around Rose's birthday cake (inaccurately decorated with fifteen candles) while Sandra, having offered her children wine and cigarettes, nonchalantly informs them that their parents are going to separate.

Act Three is set in 2011 in Kenneth's new residence that unambiguously indicates further social and financial advancement. The once bright and talented Jamie, now thirty-five, is living with his father and seems to have little control over his life which he spends sunbathing or apathetically watching videos on his iPad. Rose, who is sharing a flat in London and earns about a third of what her retired father is making with pensions, payments, and income from a house in Birmingham, has asked her parents for a family gathering to confront them with deep felt resentment. "It's all your fault. All of it" (118), she says, blaming her parents and their entire generation for her financially and emotionally wrecked life. "Everyone I know has less than their parents did at their age," she claims. "They're bringing their children up in these little houses, these tiny flats, the best they can afford, while their parents sit on all the money, in huge houses, with big empty rooms. It's disgusting" (124). Regarding the youth movements of the sixties, her judgement is scathing: "You didn't change the world, you bought it. Privatised it. What did you stand for? Peace? Love? Nothing except being able to do whatever the fuck you wanted" (121). The condemnation of her parents' generation ties in with the reproach Rose directs at Kenneth and Sandra as particular individuals who, as parents, displayed the same self-centred and irresponsible behaviour and ruined their children's lives with their sloppy parenting, the encouragement of fruitless dreams such as becoming a violinist, the instilment of the to-be-frustrated conviction "that a woman can have it all" (119), and their heedless and egoistic divorce. For all this, Rose now demands financial compensation, but her request to buy her a house is never taken seriously by Kenneth and Sandra, who feel not so much guilty about their roles as parents as they do disappointed with the performance of their children. "I thought our children would be heroes" (125) Sandra wonders after Rose leaves the room.

> I imagined they would soar. Standing on our shoulders I assumed that our kids would reach heights we never imagined, change the world entirely. But look at them. They sit

> on computers, not living, typing messages about nothing. Watching meaningless videos, and waiting for Friday night, they want to be rich and famous, in fact that's all they want to be, but they never lift a fucking finger.
> Do they?
> They don't read, they don't work and they don't think. They want it all on a plate.
> And then strangely when nothing arrives, it's our fault. (126)

There is a short moment when Sandra, usually the more negligent and self-absorbed of the two parents, gives Rose's outburst a second thought and, remembering their daughter's suicide attempt on the night they decided to divorce, concedes that "maybe she's got a point" and "[p]erhaps we just got lucky" (127). But this is the closest either of them ever gets to an admission of guilt, and the thought is quickly wiped away when Kenneth brings up the idea of rekindling their romance and they dance and kiss to the sound of "All You Need Is Love" until the final curtain, while an incredulous Rose, who needs a lift to the station, tries in vain to win their attention.

* * *

The final act allows for an open reading that does not side with one of the two generations and does not take Rose's litany of charges as representative of the play's dominant perspective. Many critics, however, have focused in their reviews on the sociological thesis about the baby boomers that is expressed in Rose's accusation and have elevated the question concerning its accuracy and justifiability to the question deciding over the play's value. Quentin Letts' favourable review rests on his assumption that Bartlett's depiction of "the Fifties babyboomer generation which 'hung out' in the flower-power era, smoked weed, received full student grants, divorced like Tudors and has now taken early retirement on large pensions" is quite accurate, that the playwright draws a "dart-sharp, horribly true" picture of "those ghastly perpetual groovers with their sub-Paul McCartney ways, their contempt for family loyalty, their insistence on doing their own thing" (463). Similarly unrestrained consent shines through the lines of Charles Spencer's praise for Bartlett's "knock-out blows on the complacency and selfishness of the have-it-all baby-boomer generation" and his insistence "that the soppy, sloppy self-indulgent values of the Sixties were often deeply selfish" (464). Susannah Clapp chooses a more neutral assessment of the play, pointing out the currency of the position the play seems to take up: "Baby boomers are declared to be busted flushes. They've always had it easy; they have been spoiled, indulged, and pretty much ruined things for everyone else. [...] There is nothing very startling about this as a thesis: it's the received wisdom of many newspaper features" (463–64). Sarah Hemming sounds still more sceptical when she refers to "the oft-repeated charge that the baby-boomers pulled up the ladder behind them" and argues that "while Bartlett might simplify issues himself, what he demonstrates with great flair is how every generation simplifies the faults of the previous one" (464). Michael Billington is torn between consent and objection to Bartlett's verdict on the baby boomers:

> As a survivor of the 60s, I think Bartlett is unfair to a decade that saw Britain become a better, more tolerant place: capital punishment was abolished, homosexuality decriminalised and racial discrimination outlawed. But he offers a wholly persuasive portrait of a couple who typify some of the less attractive aspects of the period, including its naivety and narcissism. (463)

In his introduction to the play, director James Grieve suggests a perspective that differs from most reviews in that he does not take it for granted that the play is a scathing condemnation of the baby boomer generation on behalf of their less fortunate children. Not that he is uncritical of "the sense of entitlement of the baby-boomers, who for the first time in human history firmly believed they could do anything they wanted to do" (6), but he also displays sympathy for the disappointment of parents who sacrificed their dreams of freedom to their careers and now "look to their children to fulfil the promise of the sixties" (11) while these children, instead of enjoying and making the most of the freedom and possibilities their parents had won for them, regret that they do not have the same exhausting but well-paid careers. "She has had more opportunities than any generation before her, more freedom," Grieve writes about Rose. "Now she is left only with excuses. 'Perhaps we just got lucky,' concedes Sandra. But perhaps she had to fight harder for it, and perhaps she just wanted it more" (19). In the same vein, Peter Brown argues that next to the "literal view" that sees the play as an "unbridled condemnation of a spoilt-brat generation who had and have everything – nice things, houses worth a small fortune, great pensions – and whose self-indulgence rides rough-shod over the fate of their offspring" (P. Brown), Bartlett offers a view that is also critical of this offspring generation who feel no less entitled to material possessions but are less prepared to work and make sacrifices for it. Rose's notion that her parents owe her a house since they have ruined her prospects by encouraging an unprofitable career as a musician, Brown writes, "suggests a generation who depend on their parents instead of carving-out a life of their own" (P. Brown). Brown further notices that "the whole piece has an exaggerated feel to it," an observation that is also expressed in other reviews which describe Sandra and Kenneth as rather "archetypal" and close to "caricature" (Brantley), hold that "Bartlett exaggerates the damage they do to their children" (McGinn 465), find fault with the directness and lack of subtlety with which Rose brings forward the play's socio-political argument (Soloski; Teachout; McGinn 465), or dismiss the entire play as "ham-fisted," as does Neil Norman who writes that "[t]he theme of Bartlett's sermon is irresponsibility and he expresses it with the finesse of a hod carrier with a headache" (465). Hardly anyone of his fellow critics followed Norman in his total disapproval, but several felt uncomfortable about the directness and blatancy with which Bartlett made some of his arguments. The depiction of Sandra and Kenneth's parenting is a case in point. While forgetting to attend Rose's violin performance is still credible, forgetting how old she is or inviting fourteen-year-old Jamie to smoke and drink with his parents seems over the top. Additionally, the playtext is conspicuously clear in its demand that "*[t]he play should take place in a proscenium arch theatre. A red curtain should close between scenes*" (21) – an aesthetics that, as Brown argues, "has more in keeping with a game show or variety than serious drama" (P. Brown). In sum, the exaggerated style of the play made Brown "wonder if Mike Bartlett thinks the attacks on the baby boomers are overblown" – a thought that is well worth considering. It seems unlikely that Rose is the mouthpiece of the author or fully represents the intended reception perspective on the play and the sociological thesis around which it evolves. The vitriolic condemnation of the baby boomers may be expressed more directly and more extensively in the play, but the Generation X of Rose and Jamie does not go uncriticised. Similarly, the play may call louder for sympathy with the neglected and traumatised offspring generation, but the love between Sandra and Kenneth feels genuine, and so does the hope they put first on their own and then on their children's freedom, just to be disappointed twice. Grieve writes in the introduction that in the conversations and heated debates among theatre goers after the performances "[n]o-one could agree who Mike Bartlett sided with. And in that binary lies the lasting force of the play" (20) – an assessment which appears to be confirmed by its successful revival at the Lyric Hammersmith in 2020. As far as the play's central topic is concerned, I will leave it at that. After all, the focus of my interest lies elsewhere.

Love is not the central theme of the play. It is neither discussed as a topic by the characters nor presented as an object of scrutiny to the audience, and consequently has received little attention both in the reviews quoted above and in those of its revival (cf. Akbar; Cavendish 2020; Lukowski; Waugh). And yet, throughout the play the protagonists' conception of love is evident in its impact on their weighty decisions. It surfaces, stealthily but discernibly, in the pivotal moments of all three acts, and it is congruent with and offers an explanation for their behaviour. In a sense, the large-scale historical-cultural development of romantic love in the modern period that I have delineated in Chapter 3 is mirrored here in the micro-scope of forty-four years and a single couple, complete with the promises, frustrations, and paradoxes this concept entails. Sandra and Kenneth, the rebellious youth of the sixties, renouncing all that is old and rushing towards a glorious, prosperous future, share a vision of passionate romantic love that is in accordance with their championship of absolute freedom, a love that is free from and liberates them from social responsibility and encrusted structures and traditions. The story of their marriage, however, exemplifies the difficulty of integrating all features of the programme of romantic love, which includes not only the right to unreservedly follow one's passion but also the promise of durability. Notably, what terminates their relationship is not a lack of love or the death of passion but their unwillingness to turn in their goals of maximum individual happiness and self-realisation for the rewards of a shared life. The fundamental selfishness and rationality underlying this attitude is also what prevents them from establishing their family as a countermodel to and refuge from the world of capitalism, consumerism, and cold rationality. Instead of taking up this function, which is ascribed to matrimony and family in the sociological narrative of the development of modern love, the family created by Sandra and Kenneth is suffused with the selfish rationality of capitalism, resulting in a bizarre and eventually cataclysmic symbiosis of cold rationality and the flower power ideas of love and freedom. In this atmosphere, they raise children who adopt their parents' materialism and desire for autonomy and self-realisation but at the same time desperately long for the security and stability once provided by those encrusted structures and traditions their parents have torn down so fervently.

In my analysis, therefore, Sandra and Kenneth are not primarily representatives of the sociological cohort of the baby boomers. Primarily, they are individuals characterised by rationality, selfishness, and individualism who share a conception of romantic love that befits their lifestyle and philosophy. As such, however, they become sociological types again to some degree, since the vicissitudes of modern love described by sociologists such as Luhmann, Illouz, Bauman, or Beck/Beck-Gernsheim are bound up with a notion of the modern individual who is characterised chiefly by self-centred rationality and unbridled individualism. If, that is to say, Sandra and Kenneth sometimes appear like types or caricatures, the reason is not that they are conceived as monodimensional characters lacking complexity and individuality but that, their skilfully crafted individuality notwithstanding, at the crossroads of their lives they act and decide in line with the sociological type of the self-centred modern individual. In the sociological history of love, theirs is the type that inherited the romantic ideas of love as passion and freedom from social restraints, added the demand for absolute self-realisation, and was succeeded by the postmodern type who oscillates between these individualist ideals and a nostalgia for the security of structures. Bartlett stages this development in three acts.

9.2 Act One: Breaking Free

In Act One, we witness the inciting moment of Kenneth and Sandra's drama of love, and we catch a glimpse of the concept of love on which they base their shared future. Already here, the seed for future disaster is planted as their love is from the start marked by the malign combination of a rebellious urge for freedom and selfish rationality.

Their love epitomises their view of the world. It is rebellion against all that is old, traditional, and restrictive. It opposes all forms of limitation or obligation that might be imposed from the outside on the loving individual. In accordance with one of the central features of romantic love, their only obligation is towards their own passion: to follow only their hearts and accept the legitimating force of love at first sight, which sanctifies their infatuation as good and right and liberates them from social and moral responsibility. The first victim of this concept of love is Kenneth's brother Henry, whose messy north London flat, in which Kenneth has lodged himself for the term break, provides the setting for the first act. Henry ineffectively tries to get rid of Kenneth for the evening since he has invited Sandra, whom he has been dating for two months. When she finally arrives, she immediately loses interest in the somewhat stiff and old-fashioned workman Henry and is drawn to his younger happy-go-lucky brother and fellow Oxford student Kenneth. The attraction is reciprocal. As Grieve writes, "Sandra is everything Kenneth wants to be. Self-assured, intrepid, uninhibited. She smokes weed openly and thinks nothing of sleeping in the park and calling a policeman a 'cunt.' They click immediately" (7). At the same time that we are made to sense the budding romance between Sandra and Kenneth, we also feel the growing distance between Sandra and Henry – a distance that issues from their different outlooks on life and relationships and from the small but significant age difference (cf. Grieve 7). When Henry wonders why Sandra did not tell him that she lost her job recently, she makes this obvious:

Henry: You didn't tell me you'd lost your job.
Sandra: No
Henry: Why not?
Sandra: We're not married Henry.
Henry: Didn't say we were but –
Sandra: Not even close. Don't want to get married. Probably never will. There's no loyalty to you Henry, I never promised anything.
Henry: So you say.
Sandra: So I do say that's right I say, I propose quite the opposite of marriage, I think you're very attractive and very nice, and all that but I don't feel the need to make any kind of commitment, especially one that restricts the woman in the arrangement. I'm not

ready to obey anyone. I mean really, I'm not twenty yet, neither's Kenneth, how old are you?
Henry: Twenty-three.
Sandra: Yes, well that's a bit older but we're not our parents. Things are different now. There's freedom. (48)

This is her programme. No commitment, no restrictions, none of the obligations that burdened the marriages of their parents' generation, nothing to limit her personal freedom – and it is what Kenneth is looking for, too. Theirs is a "meeting of minds and a shared appetite for life," as Grieve puts it. "The perfect storm of opportunity. They will facilitate each other" (8). Before long, Sandra provides some privacy for them by sending Henry away to fetch some fish and chips (to make up for the dinner she had offered but forgotten to bring) and effortlessly seduces a more than willing Kenneth. When Henry comes back, they dance and kiss to the sound of "All You Need Is Love." He storms off (never to be seen again in the play) and while Kenneth seems to feel some qualms of conscience Sandra reassures him that Henry will understand that they are "better suited" (59) and that "when he's found a new girl to go out with, one that's really suited to him, a slightly duller traditional kind of girl [...], he'll be really pleased this happened." And after all, "Sometimes you have to do what feels right" (59).

Sandra's justification of their joint betrayal of Henry reveals how her concept of love mingles the idea of romantic passion that overwhelms and can only be passively endured with the rational approach of assessing a potential partner's compatibility. Their love is justified because it 'feels right,' and it feels right because they are well 'suited,' which provides the basis for the kind of relationship Sandra is looking for. To be sure, she is not after a meaningless "one night thing" (58). What she wants is to make the most of the opportunity provided by the energy of her youth and the freedom of the time, together with a kindred spirit in a relationship of love that is free from the stain of commitment and finality. She wants to be with someone sharing her love of life and freedom, but she is not thinking in the traditional long-term categories that, as she obviously supposes, determine Henry's conception of love. Rather, her plans extend only to the next half year or so. "The whole of the summer, we can do whatever we want" (59) she says to Kenneth, and then they are "going to live together in Oxford next term" (58). And Kenneth is the kind of man she has recognised immediately as the kindred spirit who shares this longing for open-ended, commitment-free togetherness. In contrast to Henry and his preference for classical music, Kenneth likes rock and roll, which for Sandra is "simply a better sort of music" because "[y]ou can dance to it" (52), and her philosophy of dancing mirrors her idea of the sort of relationship she seeks: "I love dancing. It's form and chaos all at the same time. Freedom and restriction combined. Anarchy and fascism. I love it" (52). Her vision of a love relationship is expressed in this image, in the precarious and paradoxical combination of freedom and anarchy with the restrictive form of the couple, and she has determined that Kenneth is 'suited' for this kind of dance. Passion and rationality are already intertwined in these first moments of their relationship. The desire to be free within a passionate, exciting relationship is supported by her rational calculation of the compatibility of the potential partner. Quite as Illouz describes the modus operandi of modern lovers, Sandra employs acts of 'romantic rationality' at the early stages of a relationship in order to ascertain the compatibility of the partner and thus her own chances for self-realisation within this relationship. It is not – or at least not only – irrational infatuation that sparks the love between Sandra and Kenneth, but the justified assumption that they are suitable partners, able and willing to 'facilitate each other,' that is, to enable the other's self-realisation.

As Illouz and Kaufmann argue, this fundamental and ultimately selfish rationality that characterises the beginning of many love relationships is not necessarily detrimental to them, provided it is gradually replaced by more irrational and selfless forms of love that prioritise the flourishing of the relationship rather than individual self-fulfilment. This does not happen, however, with Kenneth and Sandra. As the analysis of Act Two will show, their love does not successfully make the step from rebellion against the world to the creation of an enjoyable world of their own, of what Kaufmann calls the "house of little pleasures" (151). Like their selfish rationality and irresponsibility towards others, which is indicated in Act One and later breaks out in full force in Act Two, their inability to transform their love is already foreshadowed by the special emphasis that is put on the oppositional nature of their love in the first act. Despite all its rationality, their love is doubtlessly passionate and, as Kaufmann writes, "passion is always a form of rebellion that rejects this world" (135). From the beginning, their love appears as a driving force *against* a common enemy rather than as the motivation *for* a common project. It is the emotional continuation of their rejection of all that is old, traditional, and restrictive and thus unites them in their struggle against the world inhabited by their parents or people like Henry. "We don't need ties, we don't need jobs. We don't need these *structures*" (54), they both agree, and their love accordingly emerges in an atmosphere that is free from these impediments: they don't need jobs (thanks to their generous grants); they neither wear ties (Kenneth is wearing a dressing gown throughout Act One) nor do they feel obliged by family ties (they both have effectively broken off communication with their parents, and Kenneth seems to think nothing of pinching his brother's girlfriend); and they are up for 'the opposite of marriage,' rejecting the idea of love limited by traditional structures. In line with this oppositional stance, they conceive of their love and their relationship-to-be as a continued rejection of the ordinary and, like in Illouz' analysis of the commodification of romance, this conception finds its clearest expression in the motif of travelling. When Sandra suggests that they "travel around together this summer" (59), she evokes the most vibrant image of the romantic escape from ordinariness. They are going to be free from the routines, the boredom, and the rules of society. The final dialogue stresses this essential ingredient of their love once again:

Sandra: [...] Are you ready?
Kenneth: For?
Sandra: Adventures. (59)

Before their relationship has even started, they have already defined it *ex negativo* as something that is extraordinary and unconventional, a countermodel to the traditional concepts of previous generations that is characterised not so much by concrete positive features but by its newness and the *absence* of old rules and limitations. It is profoundly adventurous in the complete uncertainty of its duration and direction and it is fundamentally rebellious in its rejection of ties and structures. Unsurprisingly, adventure and rebellion prove to be impractical as a permanent condition.

9.3 Act Two: Trapped

It takes only a few moments into Act Two to realise that Kenneth and Sandra were not able keep up their complete rejection of all social norms and structures. Twenty-three years after Act One they live what seems to be the most ordinary family life. The "[r]*easonably tidy*" dining room of their terraced house in Reading, adorned with family photos, contrasts as sharply with the messy setting of Act One as does Kenneth's neat appearance in "*jacket, shirt and trousers*" with the casual look of his younger version. It is immediately clear that they have moved up the social ladder and that 'jobs' and the traditional 'structure' of the family are no longer excluded from their lives. Details are scarce and we never learn how all this came about and how exactly Sandra and Kenneth are making money, but it becomes clear that they are both working hard (and often late) in well-paid jobs that allow them to send their children to a public school and make plans for their future studies at Oxford. Obviously, at some point in their relationship and for reasons never specified, they turned in freedom, adventure, and rebellion for the restrictions, routines, and conformity of work and family. The consequences are dire, for them and for their children.

The first of two fateful results of their changed lifestyle are their ill-guided attempts to preserve something of their rebelliousness within these self-chosen ties. These attempts are visible above all in the treatment of their children, whom they raise in a curiously mixed atmosphere of Thatcherite work ethics and sixties idealism (cf. Grieve 10 – 11). They are under immense pressure to perform well at the expensive public school and in their extracurricular activities such as music lessons or maths club, not least because they are duly reminded by their parents how favourable their opportunities are. As Kenneth tells his daughter Rose, "Wish I'd learnt an instrument. But we didn't have the facilities you do these days" (66). At the same time, Kenneth and especially Sandra are wondering at the absence of a rebellious spirit and the urge to break free and reach self-fulfilment in their children. Sandra does not remember what grade Rose is in with her violin lessons or that she is going to do music in her A levels ("Don't get offended Rosie. I've got a life of my own. I can't remember every detail." [75]), but she encourages her to pursue the dream of becoming a professional musician. Rose's pragmatic argument is swept aside:

Sandra: Grade six that's really something. You could keep it going Rosie – a professional musician. Touring the world.
Rose: Pays really bad –
Sandra: It's not about the money, it's the passion, the audiences. You enjoy playing don't you? I can tell. [...] It's important to do something you love. (73)

As Grieve points out, Sandra tries to convince her daughter to do what she denied herself: to follow her passion instead of sacrificing it to financial security and material conveniences (cf. 11–12). The life Sandra and Kenneth have chosen has nothing to do with the life they dreamt of in their youth and their children are supposed to fare better. Expensive education and the encouragement of their dreams are supposed to make them happy, happier than their parents. But the plan fails, both in the long run, as we will see in Act Three, and in the present, where Rose is craving more than financial and ideological support. She wants recognition of her present self rather than an investment in her future; she longs for her parents' attention *here and now*; she needs immediate support rather than the provision of a framework within which she is allowed (or forced) to develop freely and independently. "Rosie don't be stupid you're looked after you get everything you want" (65) says Kenneth when Rose indicates that she feels neglected, and he has already left the room when she sighs that "He never *listens*" (65). Sandra is equally unsympathetic when Rose tries to explain why she is still sulky after her parents had nearly missed her violin concert:

Rose: It was just before, when all the other parents were all there. Before it began.
Sandra: Oh – no. We're still / on this.
Rose: It was embarrassing. All the other parents were there. Dad only got there at the last minute, and you –
Sandra: I'm very busy Rose.
Rose: You're a modern mother a working woman / I know I know
Sandra: These other mums are probably at home all day, probably don't work the hours I do.
Rose: I think they do actually. Most of them do work. (71)

Sandra's explanation must certainly seem contradictory to Rose. If, as her parents keep on assuring her, life is not all about money but about doing what feels right, why does their working life keep them from spending time with their children? Why are other parents capable of balancing work and family life while Rose and Jamie are left alone within the framework of material security and top education supplied by Kenneth and Sandra? And it gets even more contradictory when Sandra seems to rebuke her daughter for not living the sort of wild life which she and Kenneth gave up at some point, ignoring entirely that it was them who put Rose and Jamie on the tracks that are supposed to lead from public school to Oxford on a journey that is characterised by an enormous pressure to perform rather than by the carefree levity that dominates the melancholy memories of Kenneth and Sandra's adolescent years. The moment this transpires most visibly is when Sandra warns Rose not to "get too attached" (76) to her boyfriend Mark, not only because he is an unreliable flirt but also because she would like her daughter to reject ties and obligations as much as she did when she was young:

Sandra: Oh look at you, all little and upset and red – you're young is what I'm saying, and you're pretty, when you make the effort. You look miserable. You should be having fun.
Rose: Like you had fun you mean.
Sandra: Well yes it's true we did have *fun* when we were your age we did.
Rose: Fucked around a lot did you?
Sandra: We certainly weren't hung up on sex no.
Rose: Now you think I'm *hung up on sex*.
Sandra: We didn't put all our eggs in one basket, it all seems so important for you, and you've got all these exams, I'm worried, you look depressed.
Rose: Not my fault about the exams. (77)

In the advice to her daughter to have more fun we see Sandra's attempt to bequeath something of the spirit of the sixties, which she and Kenneth sacrificed years ago to the world of work and money, to their children. If they, for some reason, were unable to live the dream of total freedom, at least their children should follow this path. What she does not see in this moment, however, is the pressure they assert on them whenever they remind them of their tremendous opportunities for which they ought to be "grateful" (76) or consider whether doing the thing she loves, that is playing music, will eventually equip Rose with an Oxford degree (cf. 74).

The most obvious attempt of Sandra and Kenneth to preserve the element of rebelliousness that had characterised their young love are the affairs they are both having. Not surprisingly, their conception of romantic love as breaking free from the ordinary and the everyday was bound to be frustrated once their relationship became permanent. Adventurous moments become less frequent the more the relationship turns into the 'new normal' or the ordinary. To repeat the excitement, impulsiveness, and passion of fresh romantic encounters becomes increasingly difficult within the relationship, while they are still available outside of it. It seems plausible that the dramatic structure of the amorous adventure of an affair is particularly alluring for figures like Sandra and Kenneth, whose love and mutual attraction is (or was) based so firmly on a longing for adventure and, moreover, whose concept of love emphasised freedom and self-fulfilment over loyalty and obligations. None of them has any deeper feelings for the person they slept with, there was not even an irresistible bodily attraction. Kenneth's affair Frankie was considerably younger than Sandra but "she wasn't a patch on [her]" (88) otherwise, and Sandra's affair Chris "is not the sort of man you would expect me to be sleeping with" (95), as Sandra tells the assembled family in the final moments of Act Two, after she has decided to use the awkward midnight birthday celebration she has planned for Rose for the announcement of their divorce. For both, their affairs are meaningless adventures motivated by their longing for precisely this: an adventure, a breaking free from the ordinariness of daily life. And they are not the reason for their divorce either, as Sandra's further explications make plain. In fact, they are merely symptoms of the deep frustration that burdens their love and that has its sources elsewhere.

Their frustration is the second major consequence of Kenneth and Sandra's betrayal of their former ideals and their inability to fully abandon and replace them with new ones. In the second act, Grieve writes, "[o]ur heroes are no longer adventurers, but hard-working suburbanites, tired and drawn" (8). This is not the life they have imagined for themselves, and it does not make them happy. The stark contrast in which it stands to their former dream of living a life of permanent extraordinariness and rebellion makes it all the more unbearable. They always agreed that "there's nothing worse than being stuck" (81) but they feel that this is precisely what happened. Their affairs are nothing but temporary jailbreaks after which they return to the prison their marriage has become. As Grieve observes, they are "feeling trapped" in their situation. "Their dreams of freedom have congealed into the harsh reality of incarceration. They have to break free" (13). In the situation presented on stage, Sandra is the driving force behind their move to 'break free' by means of a divorce, but Kenneth, although he is uneasy about discussing the matter in front of their children, largely agrees with her. Their decision is totally in accordance with the mixture of rebelliousness, selfish rationality, the feeling of entitlement to the reckless pursuit of individual happiness and self-realisation, and the thirst for adventure which have been indicated as determining character traits already in Act One, and Sandra has no difficulty convincing her initially hesitant husband or justifying her resolution:

Sandra: I think things are different now I think things have changed, we're entitled to do our own thing follow our own path, no one can tell us what's *right*, not church not the government, not even our children, it's no one's business but our own.
We've got our lives to live Ken, you, me, Rosie, Jamie, when it comes down to it we're all separate people these days. On our own paths.
There's a bit of you that's excited about this. Isn't there?
There's a little bit of you excited about the possibilities of striking out, on your own?
Kenneth: Yes.
A Pause.
Sandra: So there we are children. Mum and Dad are going to be happier.
And trust me. You'll be happier too. (97)

All the constituents of Sandra and Kenneth's concept of love reappear in this statement. In contrast to previous generations, theirs is a love that rebels against traditional structures such as religion, the state, or the family. Their only obligation is towards themselves – as lovers who enable each other to experience the joys of romantic passion and, even more so, as individuals who are solely responsible for their own personal happiness. And if the relationship they find themselves in turns out to be impedimental to the pursuit of happiness, their concept of love demands they break with their family rather than with the ideal of self-realisation. Retrospectively, thus, their marriage is revealed to be the kind of 'pure relationship' described by Giddens, a purely profit oriented temporary coalition, free from romantic figments like uniqueness, unconditionality, and eternity, which is terminated as soon as the effort to maintain it exceeds the pleasure it yields. That for Sandra and Kenneth the expenses exceed the revenue has to do with the stubbornness with which they cling to their concept of selfish and adventurous love. As I have indicated above, they are unable to transform their rational, profit-oriented love that seeks pleasure in the raptures of passion and freedom into an economically irrational form of love that derives pleasure from acts of *giving* within a self-imposed structure the maintenance of which is a shared goal. According to Jean-Claude Kaufmann, there are two forms of love that have come down to us from two very different traditions. Passion is a utopia, and, for a moment, our emotions turn it into a reality that is not of this world. It is inevitably doomed to die, at least in its original radical form, as it can survive only if it rebels against the world. Couples very quickly set about building a new world – often with considerable pleasure – and elaborate a shared culture that takes the material form of a style, rhythms, ways of doing things, and objects. The couple's second life has already begun, and this is where the other love – which has yet to be invented – becomes crucial. It will not grow of its own accord. (131)

This other love, for which Kaufmann uses the term *agape*, is markedly different from the love on which Sandra and Kenneth have based their relationship. It does not rebel against the world but tries to construct its own; it does not seek individual profit but the flourishing of the couple; and it prefers stability to adventure. "*Agape*," he writes,

> puts life on a very different footing, especially when it is enriched by multiple expressions of a shared well-being. Too many analyses insist that there is an emotional pressure-drop once the heady days of passion are over. They fail to see that a very different love is emerging (or should be emerging). It is less demonstrative and less exciting, but it is deep, magnanimous and subtle. (131)

For this love to emerge, however, the couple must be prepared to change, to replace the demand for absolute self-fulfilment with commitment to a shared project and to abandon the rational and egocentric profit-orientation of the "calculating individual" (52). As Kaufmann argues,

> We simply cannot commit ourselves unless we are in love, and unless we undergo a change of identity. Rational thought cuts us off from the world and makes us fall back on our own points of reference. In our dreams, we will always be the way we are now. We have an easy conscience and assume that we can simply introduce someone else into our lives. Now, that is quite impossible. There can be no commitment unless we change. (50)

Neither Sandra nor Kenneth, however, are prepared for this kind of personality transformation. To relinquish their dream of happiness through absolute self-realisation is never an option for them. The 'little pleasures' offered by *agape* and the world they have constructed for themselves cannot compete with the excitement and freedom promised by their dominant concept of love. Their divorce does not follow from a revelation that they have stopped loving each other but from the recognition that they cannot love each other the way they would like to within the structure they have constructed, which they experience as confinement rather than a shared project to which they feel committed. "Relationships that peter out do not die because there is no passion left," Kaufmann contends; "they do so because there is not enough *agape*, because the couple concerned cannot establish the preconditions for a well-being that they can share" (132). This is true for Sandra and Kenneth. When they rekindle their love in Act Three we see clearly what is already perceptible throughout Act Two: that despite their constant quarrelling, their affairs, and their final decision to divorce it is not a lack of love *per se* that terminates their relationship but the lack of the kind of love that can survive within the boundaries of a family. In prioritising their autonomy, freedom, and individual happiness they rule out any form of love that is so irrational as to sacrifice absolute self-realisation to a form of 'compromising' togetherness.

9.4 Act Three: The Lost Generation

At the climax of Act Two, Sandra reassured her children that "Mum and Dad are going to be happier. And trust me. You'll be happier too" (97). Act Three reveals that twenty-one years later only the first half of the promise has come true. Everything went well for Kenneth and Sandra. Throughout the scene, they put forth an impression of light-heartedness and youthfulness that puts them into stark contrast with their children. They are both retired now, but Kenneth still makes three times as much money as his working daughter (cf. 110) and Sandra, "*dressed expensively and tastefully*" (111), is living in a house with garden, gym, and pool (cf. 127–28). They are both sixty-four years old but are healthy and in good shape (cf. 104, 111). They both have love lives but neither seems to bother too much about it. Kenneth has ended a relationship some time ago because she was not on his "intellectual level" (128) and Sandra is married to a man for whom she has no passionate feelings and who is not well medically so they "ply him with booze and it does the trick" (113). She also loves Facebook because "[i]t's good for flirting" (114) and it is obvious that she still feels little obligation towards her current partner. To some extent, Act Three mirrors the situation of Act One: Sandra is in a relationship with another man, but when she meets Kenneth she senses that there is a deeper connection to him than there will ever be to the other man. When he puts their Beatles song on, the mutual attraction is as irresistible as forty-four years ago. Unlike then, however, they are now in a position to really live their dream and their conception of love. When they were young, they dreamt of a life without 'ties,' 'jobs,' or 'structures' and filled with 'adventures' and 'travelling,' but they were soon knocked out of the skies by the demands of real life. The necessity to live within the confining structures of work and family killed off their love which rested on an ideal of unlimited freedom. "We never travelled [...]. We never saw the world together" (130), Kenneth remarks, reminding Sandra and the audience of the sacrifices they made when they settled down to family life. Now, retired and well-off financially and with the children long grown up, they are eventually as relaxed and untroubled again as they were at the end of Act One and their love can rise anew on its preferred foundation of freedom and adventurousness. "I'll sell this place, and off we go, you and me, world tour, whatever we want" (130), Kenneth proposes, and although Sandra initially feels uneasy about the idea in the face of Rose's precarious situation and her own marriage, the way they are absorbed by each other in the final tableau, kissing and dancing and utterly unperturbed by Rose's presence, makes the notion of an 'elopement' at least conceivable. With the tiresome life of hard work behind and the prospect of a carefree future before them, they can now lead the life they always desired and, consequently, they have become compatible partners for each other again. If they were no longer 'facilitating each other' but conceived of each other as obstacles in their projects of self-realisation in Act Two, now that they have reached financial independence and have no obligations but to themselves they can again perceive the other as an enrichment rather than a burden.

For Jamie and Rose, on the other hand, the promise of happiness remains unfulfilled. The once bright, talented, vibrant Jamie appears to have been ruined by his parents' laissez-fair education and divorce, vegetating in his father's garden, drinking wine, smoking weed, playing games on his mobile phone, working in jobs far beneath his capability, dropping out of further education courses, and displaying lapses of concentration in his repetitive and distrait conversations. When Rose confronts her father with her observation of "Jamie's stasis and intellectual depreciation" (Grieve 16), he plays it down:

Rose: He used to be bright.
Kenneth: He's his own person. He's very intelligent.
Rose: But he's not Dad. He really isn't. Not anymore.
[...]
I just wonder if he should ... get help or something.
Kenneth: Leave him alone. He's happy.
Rose: But that's not –
Kenneth: What?
Beat.
Rose: Nothing. (109)

They leave it at that and the topic is not brought up again, but the apathetic impression Jamie leaves in Act Three, especially if contrasted to the impassioned singing and dancing performance with which he opened Act Two, makes it hard to accept Kenneth's claim that he is happy. For the audience, in any case, the spectacle of the figure's development is a saddening one.

As far as Rose is concerned, there is no doubt about her unhappiness. For her, life is miserable. While her friends "who got proper boring jobs" (120) can afford to go on holidays and start families, she broke up with her boyfriend, is sharing a tiny flat in London, and is "still temping between gigs" (120) without any prospect of improving her situation. "No flat, no kids, no partner, no car, ten thousand in unsecured debt" (118) – this is her present situation, for which she blames her parents who encouraged her to study music, to follow her passion, and to invest into a career that never happened while they "just watched [her] entirely fuck up waste [her] life away" (120). I will not resume the question, however, whether Rose's accusations are justified and whether her parents, as individuals and as representatives of their generation, have ruined the world for their children when they "[c]limbed the ladder and broke it as [they] went" (121). I am more interested in the conspicuous change of dreams and ideals from one generation to the next.

Sandra and Kenneth have always defined themselves as rebels, and despite the long period during which they subscribed to the rules of society and submitted to ties, jobs, and structures, they never stopped seeing themselves as the generation who broke free and rejected and overthrew traditional norms and strictures. When Rose blames them for their education, Kenneth refers to this attitude of resistance which he misses in his daughter:

Kenneth: […] No one made you play the violin.
No one made you keep going.
Rose: But you always said what a shame it would be –
Kenneth: Why did you listen to us?
We're your parents.
Sandra and me, we never listened to a word our parents said.
Why the hell did you take any notice of what we told you?
You're supposed to rebel. (122–23)

But rebellion against her parents was never what Rose wanted. In fact, with their mixture of laissez-faire education and monetary pampering, allowing their children to drink and smoke, paying for their public schooling, and always encouraging rather than forcing them to do the 'right thing,' they gave them little reason to rebel. Moreover, the lack of dedication to and interest in their children's present life (as opposed to the financial investment into their future) evoked in them, and in Rose in particular, a desire quite contrary to Kenneth and Sandra's own teenage attitude. While Kenneth and Sandra sought liberty and could not wait to emancipate themselves from their parents, Rose always longed for a closer relationship. While Kenneth and Sandra would try everything to avoid contact with their parents during the term break, Rose is deeply hurt by her parents' near absence from her violin concert or their unwillingness to listen to her properly. She does not want her parents to leave her alone or let her do whatever she wants. There are no restrictions put up by her parents that she needs to tear down, no oppressive structures to revolt against. As Grieve puts it: "Unlike want-away Kenneth and Sandra, Rosie doesn't prize freedom and independence – she wants security and protection" (10). When she confronts her parents in Act Three, she is very articulate about what she feels was lacking in her childhood and adolescence: "I needed guidance, real honest guidance when I was young" (120). Against the backdrop of this longing for close parental guidance, the divorce was particularly disastrous for Rose (and, she assumes, even more so for Jamie) and she resents the levity with which it was decided: "[...] Granny and Granddad made an effort. To stay together. For You. But you just ... one night. Over. Done. Not even what happened to me made you think about it. You still said you didn't want to be trapped. It's not a trap, it's called responsibility" (122). To her parents, this sounds overly "*dramatic*" (122) and even though Sandra is, after all the years, still shocked by the memory of Rose slitting her wrists on the night they decided to separate, neither of them fully understands why she was and is making such a fuss about it. They do not understand it because Rose's anger and despair result from the frustration of ideals that are contrary to their own. Grieve indicates the bewilderment they must feel when he asks: "Is the world turning full circle? The pragmatic, frugal ethos of the fifties so despised and forcibly overthrown by the babyboomers is what Rosie now yearns for: certainty, structure, responsibility" (15). She embodies the flipside of the paradoxical figure of the modern lover as described by Illouz and Bauman, the side that craves security and stability in a fluid and individualistic world. Rose is the kind of character whose self-worth depends on the experience of love. Lacking a successful professional career, she longs for recognition as a beloved wife and mother – in contrast to her parents, who seemed hardly affected by the collapse of their marriage and family, presumably because their sense of self-worth depended on their ability to live up to the ideal image of an autonomous, successful individual rather than the experience of being loved. For them, partnership and family never assumed the function of a countermodel to capitalism and individualism. Their love was rebellion, not against the cold selfishness of a profit-oriented market society but against the traditional structures of the previous generation. And indeed, they are too successful in this brave new world of post-war capitalism as to long for a countermodel that propagates unconditional love, support, and loyalty. If anything, they have modelled their partnership and family after the logic of the market, where they are "all separate people" (97), where everyone is responsible for their own flourishing, and where alliances are dissolved if they are no longer profitable.

For Grieve, "their parenting is positively Thatcherite. Here it's self-determination; survival of the fittest. No-one's going to do anything for anyone else – you're on your own. If you're good, you'll make it. If you're weak, it's your fault" (10–11). This individualistic world has crushed Rose and she longs for the values disparaged in the concept of her parents: unconditionality, selflessness, responsibility, and the readiness to subordinate self-realisation to the needs of loved ones.

9.5 Conclusion

Critics Dominic Cavendish and Henry Hitchings, in their otherwise favourable reviews of the 2010 production, have described Bartlett's play as somewhat "schematic" in its plotting – a description that also applies to its figure conception. However, the play's overall aesthetic design suggests that these features are strategic devices rather than faults. If produced in accordance with the stage directions, a red curtain is raised three times to reveal views of three realistically designed box sets which, framed by a proscenium arch and supported by music and television programmes of the time, attempt a reconstruction of 'slices of life' from three different historical periods. What the audience effectively looks at is comparable to a series of life-size dioramas in a human history museum, meant to facilitate a combination of the stance of the historically interested, objective observer with the fascinating allure of realistic illusionism. Like figures in a diorama, the characters in the play bear individual features that raise an interest in them as singular persons, but they are at the same time types or representatives standing in for larger groups or generations. And just like dioramas inevitably steep what they depict in an atmosphere of theatricality, so the curtain and the proscenium arch mark the events on stage as theatre – not, however, to set it off from reality but to remind the audience of reality's theatricality. Members of a society inevitably play roles, scripted by the *zeitgeist* and structure of feeling in which they move, and all individuality they undoubtedly possess does not eliminate this element of theatricality. A playwright who tries to distil these collective roles into dramatic figures, necessarily neglecting individuality for the sake of representativeness in this process, hardly differs, in that, from the sociologist, who likewise constructs representative types meant to embody not individual traits but larger, more collective sets of behaviour or emotion.

In accordance with this figure conception, the characters in the play are designed in sharp contrast to each other. In Act One, Kenneth and Sandra are contrasted with Henry, who represents the mindset of the previous generation. Later, Jamie and Rose are contrasted with their parents, signalling another change of ideals and desires from one generation to the next. While in the case of Jamie the most important contrast is that between his appearance in Act Two, still resembling his parents in his liveliness and palpable intelligence, and his appearance in Act Three, where his apathy and dullness point towards the disastrous effects of their irresponsibility. Rose's character is designed in stark relief to her parents, her mother in particular, in order to highlight their divergent values and views on life. Whenever Rose is on stage, she appears vulnerable, needy, and dependent, whereas Sandra radiates self-confidence and autonomy throughout. Even in the moment of marital crisis, Sandra remains fully in charge of herself and the situation, while Rose suffers an emotional breakdown. Sandra and Kenneth preach self-reliance and the creed that you can reach anything if you "want it enough" (39), while Rose argues that parents are supposed to look after their children. Sandra and Kenneth, who dreamt of freedom and rebellion, betrayed this dream for the benefit of careers and financial security but followed it when they decided to break with a family life that hindered their autonomy and individual happiness. Rose, on the other hand, encouraged by her parents, followed her passion when choosing her profession but now completely lacks financial security. Most of all, the contrast between Rose and her parents is visible in their central goals and values: individualism, self-realisation, and freedom from structures and obligations on one side, security, guidance, and structures on the other. Thus, the play spreads across distinct, antagonistic characters what can be found within single characters in other plays and in the sociological descriptions I have quoted above (especially that of Bauman): the contradictory desire both for individual autonomy and emotional support and security provided by an intimate partnership. What is primarily an inner conflict troubling (some of) the characters in the plays of Marber, Ravenhill, Kane, Kelly, and Skinner is presented as a primarily generational conflict by Bartlett. Within Rose, the paradoxical internal striving for autonomy and embeddedness is only slightly indicated in that following her dream career proves detrimental to her wish to raise a family – her partner, we learn, has left her for someone who is "twenty-four and desperate for kids" and "doesn't want a career" (120). But due to the facts that the desire for autonomy and self-realisation appear so much stronger in her parents and that her own choice of career seems as much a matter of conscious self-realisation as of parental influence and encouragement, Rose hardly makes the impression of a figure characterised by an inner struggle between the equally strong contradictory desires for independence and security. In the overall picture, the latter desire clearly preponderates.

10 "this poetical ... *shit*": Coming to Terms with Love in debbie tucker green's *a profoundly affectionate, passionate devotion to someone (-noun)*

It seems appropriate to conclude this volume, which has sought to illustrate the uncontrollability and contradictory nature of love and its discourses, with the analyses of a play whose focus of interest is evenly distributed between the instability of the terms and conditions both of love's existence and its modes of expression or 'terminology.' The very title of debbie tucker green's *a profoundly affectionate, passionate devotion to someone (-noun)*, which premiered at the Royal Court Upstairs in 2017, is a reminder of the elusiveness of the (absent) signifier 'love' and of the difficulties inherent in any attempt to explain a heterogenous and abstract concept with the help of other similarly heterogenous and abstract signifiers. The dictionary-styled circumscription in place of the simple word 'love' in the title seems to acknowledge that no play or any other piece of art or writing can hope to capture the 'essence' of the phenomenon or the 'centre' of the word's meaning but can only try to revolve around it in narrowing and widening circles in an attempt to approximate it without letting slip one of its many denotations.

In a series of loosely connected episodes, the play casts light on various shades of living in a love relationship. What distinguishes the play aesthetically from all other plays examined in this volume is the extent to which its effectiveness relies on the recognisability of speech patterns and situations. Despite an emphasis on the daily power struggles in love relationships, the play is remarkably quiet. It is emotionally gripping and touching, but in a uniquely calm and subtle manner that makes for a distinguishing feature of the play. Throughout, the play gets along without extreme situations, without graphic stage images, and even without a proper plot and plot-related suspense, although the episodes do add up to a comprising story which, however, remains extremely fragmented and is of little relevance to the play's efficacy. In a way, tucker green's piece appears like Barthes' *Fragments of a Lover's Discourse* turned into a play, offering glimpses at various facets of love life in different episodes arranged more like entries in a reference book on love than like stages in a coherent plot. For some critics, this absence of a coherent narrative is disconcerting. Aleks Sierz writes that the play "fatally lacks a really satisfying story. It is the victory of the abstract over the dramatic" (*Sierz.co*), while Michael Billington, who was additionally put off by the stage design of the original Royal Court production where spectators were seated on swivel chairs and surrounded on three sides by a narrow strip of empty stage, was "frustrated both by the staging and by the lack of specificity" (*Guardian*). What most critics acknowledge, however, are the "odd moments of recognisability" (Billington) offered by many of the episodes. According to Dominic Cavendish, "Green's achievement is to stir a visceral recognition of life and love at its most coolly recriminatory" (*Telegraph*), and Joe Vesey-Byrne writes that "green's play is intelligent without resorting to pompous dialogue or histrionic characters, and the evening feels like a privileged peep inside the lives of three recognisable love stories" (*Independent*). Notably, this recognisability is never hampered by tucker green's idiosyncratic use of language. Although her dialogues more often than not assume a poetic quality that elevates them above everyday speech, they nevertheless remain strikingly naturalistic, retaining the awkwardness, clumsiness, and uncertainties of everyday speech. As Natasha Tripney puts it, "[i]n her hands the stop-start patterns of everyday speech, interrupted thoughts and broken sentences, hesitations and repetitions, becomes a kind of poetry" (*The Stage*). Similarly, Cavendish praises her "ability to craft the way people talk – unwieldy grammar, stunted vocabulary and all – into something that has its own music" (*Telegraph*), pointing to this peculiar quality of tucker green's plays to make a clearly poeticised language so strongly reminiscent of everyday speech.

In the first part of this chapter, I will combine a synopsis of the play with an analysis of a number of themes and motifs that it brings to recognition and which work as manifestations of love's precariousness in the daily lives of couples: jealousy, dwindling passion, the little nuisances and power struggles of a relationship, and the pain of loss. The second part will be dedicated to what I take to be the two dominating topics of the play: first, the problematisation of language, communication, and the ability both to understand the other and make oneself understood; and second, the play's presentation of 'need' and 'want' as the dominating principles of love, especially in its manifestation as a 'need to be needed' or, simply, 'devotion.'

10.1 Synopsis

The play consists of three loosely connected parts, each featuring a man and a woman in a crisis-ridden romantic relationship. Part One is certainly the richest and most memorable, presenting in episodic scenes the ups and downs of love in a relationship that stretches over an unspecified but considerable period of time, while the shorter Parts Two and Three are dramatic presentations of single moments of crisis, taking up and emphasising some of the themes already addressed in Part One.

Scene One plunges the audience right into one of the many quarrels between the characters of Part One, simply called A and B and described only as *"female, Black"* and *"male, Black"* in the stage directions (2). Their dialogue sets the tone for the entire play, which is dominated by mutual accusations of inattentiveness, negligence, and incomprehension such as "It's not always all about you" (3) or "You ent never said something just to make me feel better" (4) or "You didn't understand my complexity" (6). In all three parts of the play, the couples accuse each other in this manner, imposing upon the spectator, who is often positioned in the middle space between the two arguing parties, the embarrassing yet fascinating role of an eavesdropper to private situations in which miscommunication and long harboured resentments at the other's faults and character traits are revealed as main subjects of controversy. In Scene Two, for instance, when B tries to express the intensity of his love, A complains that he has never told her this before. In Scene Three, A discloses to B that "[w]hen you do that-did that (thing), that you do – did. With your thing. To me – /[...] Thinking I liked it. / [...] I didn't. /[...] I never liked it" (13), referring most probably to a sexual practice that B was convinced his wife enjoyed. Scene Four is about B's jealousy which, however, is not caused by another man but by A and B's daughter. B senses that the child loves her mother more than her father ("She looks at me and sees it's not you and looks for you") and, moreover, perceives the baby as a rival for A's love ("I look at you./ [...] But you're busy looking at her" [17]). Scenes Five and Eleven mirror each other as each time one partner tells the other in almost the same words that their sex life has become boring for them. Both scenes end with the nervous question "Are you tired of me?" (22; 39), which is answered only inconclusively. In Scene Six, A accuses B of providing little assistance during her depressive period following the death of her mother, whereas B protests that she did not give him the opportunity to help or care for her but stopped talking completely for ten months instead. Scenes Seven and Ten offer some comic relief in a play that otherwise, despite the often sharply humorous verbal duels, mostly stifles laughter because the vulnerability of the characters and the emotional depth of their quarrels is always discernible behind their pointed attacks. In these two scenes, however, the object of dispute is trivial enough to allow for laughter, even though it is clear that when A complains about B's annoying behaviour when she tries to watch a TV programme or when B expresses his desire to "have a solitary shit" (35) without being disturbed in the bathroom by his wife or kids, they are addressing daily little nuisances which are not blocked out or made irrelevant by their mutual affection but are, instead, tiresome enough to burden their relationship. Scenes

Eight and Nine, which show them deeply in love and full of joy after the birth of a son, contrast sharply with Scene Twelve, where we witness the verbally harshest of their arguments caused by A's rejection of B's wish to have a third child. Scene Thirteen, then, stresses the strength of their love again, which even lasts beyond death. Although A and B are talking (and quarrelling) like they did in all the scenes before, it transpires that A has died.[60] She reproaches B for not talking with their traumatised daughter, who "cries most nights" (44) but has stopped talking – just as A had done years ago in the same situation. But B is troubled too much with his own pain of loss: "Nights are fuckin endless and days are a disaster. That soft neck-back piece a me yearnin for that touch by you has given up waitin and gone hard and inside I gotta gap left by you bigger than my outsides" (47). The pain of loss prevents him from comforting his children (**A:** "they need their dad." / **B:** "I need my wife. I want my wife" [48]). The death of his wife, it appears, is his worst-case scenario: "I'd swap my place for your place – / [...] I'd swap her place for you / [...] Both of them for one a you –" (48). To him, both his own death and the death of his children seem preferable to being left alone without his beloved wife. In view of this profound love, Scene Fourteen comes as a surprise as it presents another heavy argument between B and his deceased wife, the reason being this time that B has not observed A's last wish to be buried but had her "burnt and scattered" (52) instead. However, Part One ends not with the mutual "Fuck you" (53) of this scene but with a wordless Scene Fifteen, where A and B *"watch"* and *"are tender with each other"* (54).

Despite the conciliatory tone of its ending, the prevalent mood of Part One is tense. The relationship between A and B is presented as an incessant struggle for power and control that is only made bearable by recurring moments of love and tenderness. Mutual accusations, the rejection of the other's wishes, threatening the other's status as an object of (sexual) desire – these are all strategies of securing a position of power. The play does not create the impression, however, that the characters seek to manipulate or subordinate the other, as do the lovers in Ovid's or Nietzsche's conceptions. Nor are they threatened by a loss of individuality, freedom, or alterity, the dangers of love in the theories, for example, of Sartre, de Beauvoir, and Levinas. In fact, many of the possible sources of precariousness bound up with a 'loss of self' which I have discussed in Chapter 2.2.2 are treated as insignificant in the play: The characters' sex and gender does not surface as a cause of unequal power relations. The wish to possess the other is

---

60 The characters' use of past tense when talking about their relationship in many of the preceding scenes allows for the interpretation that these, too, take place after A's death and are imaginary conversations between B and his late wife. This reading, however, is only possible retrospectively since there are no further indications of A's death previously to Scene Thirteen.

not presented as a threat to freedom but as a desirable sign of affection, especially in Scene Eight, when A and B claim possession of their favourite part of the other's body, the very parts they caress in the wordless final scene of Part One. Similarly, the wish to know and understand the other, an impermissible threat to alterity for Levinas, is demanded by B in the first scene, when he confronts A that "knowing you didn't understand want to understand never did understand me was quite an underwhelming experience" (7). In short, the characters are not threatening each other's freedom and individuality, and their power struggles are neither attempts to permanently subdue the other nor acts of self-defence against suppression or absorption. Rather, Part One presents two characters that appear 'shattered' by love in Nancy's sense of the word. A and B are broken into by love and opened to and for the other. At moments when they indulge in love, they relish this mutual openness, the profound knowledge of the other, and the unconditional availability to and of the other. At other moments, however, they experience the enormous precariousness of love – the inevitable lack of control on which genuine love between equals necessarily rests. B's jealousy of his daughter as A's primary object of affection or A's refusal to have another child are examples for the uncontrollability of the other and her feelings that are hard to accept for B. In a way, all their daily power struggles can be seen as attempts to compensate for the lack of control that comes along with love. Gaining a position of power and control in a concrete situation, discussion, or argument is supposed to make up for the absence of control over the other's love. It is the characters' inability to accept the utter precariousness of love which fuels their daily squabbles. Scenes Thirteen and Fourteen offer the most touching rendering of this nexus, after the precariousness of love has manifested itself in its most disastrous form, the death of the beloved. Grief-stricken, B blames A for his suffering:

B: You shouldn't have left me.
A: I didn't leave / you.
B: You shouldn't have left us
A: I didn't leave / you.
B: you left / us.
A: I died, I died.
B: ... You left me. (49)

Confronted with the total lack of control over his wife, his love and pain turn into anger. The reaction is in line with Martha Nussbaum's explanation of adult love as rooted in the child's needy love for the mother (see 2.1.3 and 2.2.2). Just as the recognition of the independence, separateness, and uncontrollability of the mother gives rise to anger in the child, B's anger is a result of his insight into his lack of control demonstrated by the death of his love object. According to Nussbaum, mature love requires the abandonment of the infantile wish for omnipotence and the acceptance of the precariousness of love, but of course this is easier said than done, and B is not able to acquiesce silently in his lack of control. Consequently, he not only blames her for leaving him and the kids but finds additional ways to vent his anger by ignoring her instructions concerning her funeral service and burial method. But the misrecognition of her last wish serves not only as an outlet for his anger or a way of taking revenge for being left alone. At the same time, it is an opportunity for B to exert control over the person whose uncontrollability caused him so much pain. Botching up her funeral is his chance of gaining at least a small portion of the control which is always necessarily lacking in love and which he missed most dreadfully when he had to realise the insecurity of her presence.

Part Two, featuring "WOMAN, *Black or Asian*" and "MAN, *Black*" (2), takes up the struggle for power and control within a romantic relationship as the central topic. The single, coherent scene again pitches the audience into the middle of a heated argument during which the partners accuse each other of inattentiveness, lacking communication skills, selfishness, and ignorance regarding the other's true needs. Knowing each other well, their mutual accusations have an especially acerbic quality as they both are in a position to formulate their reproaches as attacks on the other's individual personality and character traits. When WOMAN complains about MAN's habit of pretending to be attentive while not listening properly, for instance, she grumbles "you do it on – doin it on purpose to make me look bad when it aint me at all it's you, you and your passively, pathologically, aggressively fuckingly not-right-in-the-head bullshit version of 'listening' thatchu do. Are doin. Are doin badly" (55–56). MAN, on his part, blames her for "always sayin so much but not sayin nuthin" which makes listening to her "like tryin to separate the shit from the shovel" (57). After a few lines into their argument it becomes clear that what they are both fighting for is an apology. In other words, their argument is a power struggle about the inferior position of the one who offers the apology and the superior position of the one who receives it. The stakes are high for both, and they repeatedly draw back from fighting and "*are busy,*" as the stage directions frequently stipulate, as if to recover from the increasingly acrimonious duels only to return with refreshed vigour. During the first half of the scene, WOMAN is aggressively but unsuccessfully demanding an apology from MAN, but it remains unclear what he is supposed to apologise for. It gradually unfolds that, in addition to a general unwillingness to apologise, which she even pathologises as "something psychological – a blockage or summink – some mentally constipated something that's stopped it comin out" (58), what WOMAN resents is MAN's behaviour during a period when she was ill and he "was there in body not in mind and sat there resentful" while she "[f]elt bad about feelin bad havin you sittin there with your awful aura in the room. Bad airs, bad air, bad aura. [...] You didn't even ask how I was feeling" (67–68). MAN's response, however, throws a different light on his role during her illness:

> I *knew* how you was 'feelin'. Fuck. I could *see* it.
> ... And a thank you for changing your sheets woulda been nice
> and a thank you for spoonfeeding you
> and a thank you for lifting fresh water to your lips
> and a thank you for cooling your fever
> and a thank you for easing your shakes,
> a thank you for takin out the sick bucket – for takin out the shit bucket,
> a thank you was it for cleaning you gently and
> thank you again for redressing your dressings
> gentler than they did and a thank you for – I could go on.
> [...]
> ... And a final thank you, for not sayin how fuckin frightened I was. (68)

Instead of an obligation to apologise, MAN feels a right to demand an expression of gratitude from WOMAN. But this is not the only twist which the information about the act of caring for his partner gives to their power struggle. His admission of being frightened reveals his insight into the precariousness of love. Like B in Part One, who experienced his lack of power and control most painfully when his wife died, MAN was faced with the precariousness of love through the illness of his beloved. He was frightened because it made him recognise his ultimate powerlessness and lack of control concerning the presence and existence of the person on whom his happiness and his ability to love and be loved depend. In analogy to B's behaviour in Part One, his refusal to apologise might hence be understood as an attempt at compensating the loss of control that entering a love relationship necessarily entails. Winning the power struggle against his partner is supposed to alleviate the loss of control inherent in the act of loving her. The strategy is a destructive one, and it is saddening to watch how he pursues it relentlessly until the end. When WOMAN finally says the words he wants to hear – "Thank you. / [...] And ... / Sorry" (71) – she admits both indebtedness and guilt, making herself vulnerable by exposing herself to his judgement and benevolence. Her expectation, however, that he might respond with a similar act of opening up and making himself vulnerable by admitting a mistake, is frustrated. "*She gestures for him to say something*" but the scene ends with "*Silence, for as long as it can be held*" (72). In a kind of vicious circle, the intensity of his love for WOMAN has made its precariousness unbearable for MAN. Incapable of enduring the loss of control inherent in love, he turns the anger about his powerlessness against his love object. As Simon May puts it, the fear caused by the uncontrollability and potential loss of the loved one "quickly leads to hate – unless the lover can genuinely accept his vulnerability to the loved one" (259). Unable to afford this acceptance, B's fear of losing control prevents him from adopting a position of vulnerability and inferior power by apologising. In refusing to accept the fundamental precariousness of love, however, he rejects one of its necessary conditions. In the concrete situation of MAN and WOMAN, what makes enduring love impossible is his denial of mutual openness and vulnerability.

Indeed, in Part Three, set a few years later, MAN is in a new relationship. His partner YOUNG WOMAN is *"the daughter of A+B,"* the couple of Part One, who *"is of legal age"* (2) by now but is much younger than MAN. Again, we see a single coherent scene during which the two characters unveil what they dislike about the other, but this time the tone is far less scathing and confrontational. When they declare what they would like to change about the other they display a healthy awareness of the imperfections of their partner, a reasonable scepticism towards the idea of unconditional love of the other in his or her full particularity, and an ability to talk about their likes and dislikes. In fact, even though their communication is far from perfect, the more serious communication problem lies outside their relationship. As MAN observes, YOUNG WOMAN suffers from the lack of communication with her father, which was already indicated at the end of Part One and which seems to have remained unimproved since then. It is hard to avoid the conclusion that YOUNG WOMAN seeks to compensate the silence between herself and her father, and also a proper paternal relationship in general, through the romantic relationship with the much older MAN. While she does not speak with her father, she drowns MAN in an incessant torrent of words, a habit he seems to find both loveable and unnerving. Her (off-stage) brother is quite unambiguous about his impression that she is projecting affectionate and romantic feelings on a surrogate father. As YOUNG WOMAN reports about her conversations with him, "He don't know what's wrong with me going out with someone as old as you. [...] he thinks I lost my mind (he) thinks I'm goin through a phase [...] thinks it's linked to Dad in some way (and) tries to psycho something about me says I got psychologicals" (80). But the uneasy atmosphere surrounding this seemingly incompatible and fragile couple notwithstanding, their love is no more improbable than the love between the previous couples. Conspicuously, their dialogue contains the only real declaration of love in the entire play:

MAN:	I love you.
Y. WOMAN:	I know.
MAN:	Love everything about you.
Y. WOMAN:	(I) know that too. (90)

Even if only seconds later he has to correct himself about the latter part of the declaration because he admits that "a slight amend on the endlessness" (91) of her talking is something he would change about her, and even if she lists quite a few amendable traits of his character, their mutual affection feels authentic and reliable right through to the end of the scene, which ends with a kiss. Their relationship is a battlefield of love just as much as those presented in Parts One and Two, and the peaceful tone on which the play ends might be no more than a temporary ceasefire, but the ending at least instils the faint hope that, despite everything, their love is still alive and stands a chance of surviving.

10.2 'Understanding' and 'Devotion'

tucker green's play addresses a wide range of love-related subjects and, with its episodic structure, singles out, repeats, contrasts, or re-emphasises various possible stumbling blocks lying at the heart of more or less severe disputes and power struggles between lovers. But, as I will argue in the following, in the plenitude of love topics covered by the play there are two issues that recur so frequently that they can justifiably be considered the play's major concerns. Moreover, the very title of the play seems to identify these two subjects as its core themes. On the one hand, the dictionary-style phrase gestures towards a problematisation of language and communication. Just as the title mimics attempts to understand and describe what love is, the characters in the play try hard to understand both the other and themselves, and to make themselves understood. On the other hand, the specific choice of words in the title contains an indication regarding its focus. The noun with which the term in question is defined, the word, that is, which comes closest to the conception of love underlying the play, is 'devotion.' 'Affectionate' and 'passionate' are only the determiners of this central element of the definition which, as I will demonstrate, provides the key for an understanding of the structures of need and desire presented in the play. 'Understanding' and 'devotion,' then, are what I take to be the overarching themes of the play. I will analyse them separately.

10.2.1 Love You More Than I Can Say: Epistemological and Terminological Limits of Understanding

How is 'love' translatable into (other) words? To what extent are we capable of communicating our intimate feelings and to what degree is this capability dependent on our willingness to speak out and listen to the other? These are the questions implied in tucker green's title and in the characters' often ineffective attempts at communication and self-expression. Most critics have stressed the problematisation of communication as a central concern of the play. Aleks Sierz, for instance, writes: "As you'd expect with love, the central issue is communication. The 80-minute piece is about what we say, and what we don't say. About how we listen, or don't. It's about our need to be heard, and our unwillingness to hear" (*Sierz.co*). For Fiona Mountford, "the through-line is the never-ending difficulty of intimate communication, of listening properly to what is being said to us" (*Standard*) and Natasha Tripney describes it as "a play about listening as much as [...] talking, about speaking and not being heard" (*The Stage*). Unsuccessful and refused communication are frequently made explicit in the secondary text in all three parts of the play. A "*is busy*"[61] while B is declaring his love (10), B "*is not listening*" when A complains that she is never allowed to watch a TV programme without being annoyed (29), MAN and WOMAN both frequently become and try to stay "*busy*" during their argument in order to evade the necessity of a direct reply or to appear unaffected by the other, and YOUNG WOMAN "*isn't particularly paying attention*" (74) to MAN during the first moments of Part Three, to name only a few examples. But intentional inattentiveness is only one form of miscommunication in the play. Often, the characters have the distressing experience that they have misunderstood or have been misunderstood by the other. A and B for example, who know each other very well and whose profound love is, despite their frequent squabbles, beyond doubt, have to learn that perfect mutual knowledge is impossible. B, who feels his entire personality misunderstood by A in Scene One, is later hit hard by the revelation that A "never liked" the "thing" he used to do with her (or "to" her, as she puts it). "You looked like you liked it" (13), he protests, but her reply reveals the fallibility of his capacity to 'read' her: "That look that looks like I'm enjoying that thing that something you do – did – done back then ... that look weren't that look that I liked it. It weren't that" (14). A similarly disillusioning insight into his imperfect knowledge of her comes with her assertion that "I've always

---

61 In the Royal Court production, 'busyness' was mostly performed by scrawling meaningless lines and forms onto the blackboard surrounding the stage.

hated getting undressed in front of you" (35). The idea that he makes her "feel uncomfortable" (36) is devastating to him as it shakes both his self-image and the image he had of their relationship. Whether what A says is true or rather a well-chosen strategy to gain the upper hand in their constant power struggle, the effect on B is the frustrating awareness that he can never have certain knowledge of A and her needs and desires. Roland Barthes describes this as an inevitable but no less irritating situation of a person in love:

> I am caught in this contradiction: on the one hand, I believe I know the other better than anyone and triumphantly assert my knowledge to the other ("*I* know you – I'm the only one who really knows you!"); and on the other hand, I am often struck by the obvious fact that the other is impenetrable, intractable, not to be found. I cannot open up the other, trace back the other's origins, solve the riddle. Where does the other come from? Who is the other? I wear myself out, I shall never know. (134)

The unknowability of the other is especially vexing if it concerns the other's needs and wants. After all, Barthes asks, "isn't *knowing someone* precisely that – knowing his desire?" (134). To learn that he is ignorant of A's sexual preferences – probably after years of marriage – is thus more than disillusioning for B, and the same is true for the two scenes in which first B and later A tell the other that sex has become boring for them. Again, the other is confronted with her or his lack of knowledge regarding their partner's sexual desire, a revelation that shakes the lover's self-image as the person who knows the beloved perfectly. For Barthes, the unknowability of the beloved turns love into a form of religion. As he argues, "to bestir oneself for an impenetrable object is pure religion. To make the other into an insoluble riddle on which my life depends is to consecrate the other as a god" (135). The analogy offers a striking image of love's precariousness. Like the faithful in a relationship to an impenetrable, unknowable god, the lover is dependent on 'grace' as there is no possibility of knowing the terms and conditions of the relationship. Ignorant of the beloved's desire, the lover can only hope but never be sure to be that which is desired.

The fundamental uncertainty about the other is not only the consequence of an inability or unwillingness to listen and comprehend. In fact, an even more basic factor is the insufficiency of the means of communication we have at our disposal and the resulting difficulties accompanying all efforts of self-expression. As Vesey-Byrne argues, tucker green "has created an insightful observation of how we [...] talk about talking, and what 'silence' means between lovers" (*Independent*), and I want to have a closer look now at two scenes which revolve around exactly these two points: the metalingual problematisation and the absence of verbal communication.

Scene Two starts with a declaration of love from B to A:

> I want you. Yeah.
> I want you to me. Want you with me want me us.
> I want you beyond what I can say
> beyond what I got words for
> beyond my vocabulary
> beyond any vocabulary
> beyond language
> beyond imagination,
> beyond what words can do beyond what my words can do. Beyond what words have been known to do, have yet to do, beyond what they will ever do.
> I wanted you beyond sentence in between syllables above vowels under consonants and after punctuation. (10)

B's preoccupation with linguistic terms bespeaks his effort to find a language that might express his emotions – an effort he has already accepted to be futile. The language he shares with A is an alien system, a language not of his own, inevitably inadequate for an expression of his most intimate feelings. Any such language is always necessarily insufficient for self-expression, and all he can try is to circumscribe what he feels with words that are not his own and the meaning of which, as understood by the other, is beyond his control. In a move that mirrors the play's title, B tries to explicate his love with synonyms – his primary choice is 'want' –, but senses immediately that he has won nothing by that since he is not able to express the specific nature and intensity of this 'want' with the words available to him. In an attempt to approximate it, he resorts to the use of an example: "But my days-our days seemed shorter with you in em, you shortened the hours, made minutes not matter and seconds seem shit, racing to build our old age together. Days couldn't be long enough with you in em and nights was like blinkin before daylight dawned quick –" (11). However, A's reaction is sobering: "(you) never said you wanted me that much – [...] never said none of this poetical ... *shit* – [...] Then. [...] That you seem to find so easy to say now" (11). Unimpressed by his present effort at self-expression, she resents his earlier silence on the topic. The information about A's death, given at a later point in the play, allows for the interpretation that her 'then' and 'now' refer to the time before and after her death. The scene would then probably be part of B's imagination, a sorrowful fantasy about his late wife expressing his regret at not having spoken about his feelings when she was still alive. However, A's reply castigates not only the lateness but also the form of B's declaration of love. When she calls his phrases 'poetical shit' – and when he is astounded by this verdict she repeats it: "Hmm. Yeh. *Shit*." (11) – she bluntly identifies the lack of authenticity that stains his declaration. Indeed, his words are at times highly reminiscent of the texts produced by the culture industry, in romantic films, novels, and songs. What is more, his very struggle to express himself, the insight into the insufficiency of language, his strained effort to 'say the unsayable,' is an age-old trope of literature, figuring most prominently in Romanticism. Hence both his struggle to find words and the words he finds are inauthentic; they are 'poetical shit,' neither capable of communicating what he wants to say nor appearing original and convincing to their addressee.

The impossibility to find authentic and appropriate verbal expressions for his love has probably caused B to refrain from declarations of love in the past. At the end of the scene, he tries to convince A, and himself just as much, that there was no need to use words.

> ... I didn't have to say ...
> [...]
> how much I ... Shouldn't have to say how much I ...
> (I) showed it instead.
> Didn't stop showing you.
> Did I?
> Couldn't stop showing you.
> Could I.
> Wouldn't stop. Would I.
> Thass how you did know – do know.
> You did know. You did know you did know, you knew how much ...? How much I...?
> Don't you?
> Didn't you? (12)

The question remains unanswered. Maybe she knew, maybe she did not, and again B is tormented by uncertainty, by the unknowability whether she knew how much he loved her. B's attempts at wordless communication were just as insecure as his verbal declaration, or even more so. To quote Barthes again in this context: "Whether he seeks to prove his love, or to discover if the other loves him, the amorous subject has no system of sure signs at his disposal" (214). Signs and their interpretation, Barthes argues, are even more ambiguous and uncertain than words so that, faced with their unreliability, "one falls back, paradoxically, on the omnipotence of language" (215). Compared to signs, the lover assumes, language is reliable and unambiguous.

> I shall receive every word from my other as a sign of truth; and when I speak, I shall not doubt that he, too, receives what I say as the truth. Whence the importance of *declarations*; I want to keep wrestling from the other the formula of his feeling, and I keep telling him, on

my side, that I love him: nothing is left to suggestion, to divination: for a thing to be known, it must be spoken. (215)

This is the position taken up by A. Instead of relying on signs to 'show' his love, B should have made a declaration of love, containing the 'formula of his feeling' and clarifying his emotions beyond doubt. However, what underlies this demand is the belief in the unambiguity of language and the 'sayability' of intimate feelings, the power of language to ensure perfect self-expression and perfect understanding. This is what the 'amorous subject,' in the desperate desire to communicate the inner self, wishes language to be, but it is not what language proves to be in reality. Hence, the recourse to language is 'paradoxical' for Barthes. Neither can language guarantee perfect understanding, nor can it serve as a means of authentic expression. After all, "every other night, on TV, someone says: *I love you*" (151), Barthes famously says, emphasising the unsuitability of the formula as an authentic expression of individual feelings. In the final analysis, B is trapped in a dilemma. Any attempt at verbal self-expression is necessarily insufficient and runs the risk of being denigrated as inauthentic 'poetical shit,' but remaining silent and hoping for the other's ability to interpret the 'signs' is no less precarious.

The latter option is the theme of the second scene I want to examine briefly. With reference to Vesey-Byrne's review, while Scene Two made its characters "talk about talking," Scene Six explores "what 'silence' means between lovers." In the end, however, they both address the same problem from two different angles. In Scene Six, A expresses her dissatisfaction with B's support during her depressive period after the death of her mother, while B takes her silence as an argument to defend himself.

A: I wasn't able to say anything.
B: I can't know –
A: was I? Wasn't able / to –
B: except I can't know what I don't know. I really can't –
A: you knew what was / happening
B: you didn't speak.
A: You saw what was happening.
B: You didn't speak
A: I didn't speak
B: you didn't speak to me, wouldn't speak to / me.
A: in all your poetical bollocks with how well you think you know me you coulda picked up on what I was goin through – I *couldn't* / speak. (24)

The roles of Scene Two are reversed in this episode. Where B used to be unable to 'speak' his love due to the lack of adequate means of expression, A went through a phase of psychosomatic speechlessness. Where B relied on A's ability to read the signs with which he thought to show his love, A hoped for his skill in deciphering the signs indicating what she needed during her illness. "A blind man could see what I was going through," she claims, "and you were no help" (24), whereas B holds that "[a] clue a hint a sign an indication about your non-verbals" (24) would have been necessary to show him *how* to help. In both scenes, silence appears as detrimental to their relationship.

It is a sign for the seriousness of the play, however, that it does not naively present 'better communication' as a solution. The communication failure between A and B, and between all the characters in the play, is not a consequence of their flawed individual personalities, something that 'better' characters might be able to avoid. Rather, it is a symptom of their humanity, a result of the insufficiency of language and the impossibility of perfect self-expression. This is not to say, however, that perfect communication and self-expression would be possible were it not for the deficiency of language. If one central argument of the play is the unknowability of the other, it is so not only due to the deficiency of language, which limits the possibility of self-expression, but also because the play denies the very idea of a coherent self that could be expressed and understood. Quite in line with Illouz, the play rejects the fashionable idea that proper introspection leads to knowledge of the self and that becoming aware of and communicating this 'essentialised self' will help in finding a 'compatible' partner. Neither is the self a stable, unchanging entity, nor do lovers usually arrive at a clear-cut and unambiguous definition of their own desires. Even before knowledge of the other and self-expression become a problem, knowledge of the self is a problem, as it is imperfect, tentative, and transitional. Consequently, lasting compatibility cannot be predetermined. It has to be constantly approximated through compromise and adaptation, while any hope to reach a permanent state of perfect compatibility is futile. Part Three insinuates as much. MAN has clearly developed since Part Two, where he refused to apologise or "say something just cos you want" (62) until the bitter end. Now, his entire behaviour is much more reconciliatory, and he offers to apologise if that is what YOUNG WOMAN needs to feel better (cf. 89). At the same time, he has still not grown out of many of the habits that strained WOMAN's tolerance in Part Two. Words of apology still don't come easy to him and, so far, he has "never apologised" to YOUNG WOMAN (89). He still has "that look thatchu do, that one where y' look like y' listenin, like you've heard" (89) while YOUNG WOMAN is unable to tell whether he really did or not. And there is a long list of other habits that she would change about him if she could:

you not lissenin I would change / that
[...]
Change your gaze of a look thatch do, would get rid of that.
Beat.
Change your hard-edged sighs when I haven't even finished a sentence.
[...]
Change your you kissin me to not solve nuthin –
[...]
Change how you snap at me –
[...]
Change you askin about you to / me
[...]
shit thass annoyin would change that would change your raised eyes of when you're bored of what / I'm sayin.
[...]
change your attitude and your tone generally
[...]
change your moods in a minute –
[...]
change all a them an' there's plenty a them / to change. (90–91)

MAN has brought down this litany of improvement suggestions on himself. The scene starts with his repeated assurance that he "[w]ouldn't change a thing" about her and his wish to know whether she would change anything about him (73). This wish to know, or rather the negative answer he hopes for, arises from his desire for recognition, and so does his request that she tell him "[a]nything about me ... ? That you (love) ..." (76). What he craves is recognition and affirmation of his self, his personality. He requests the validation of his self as loveable, a recognition of his self-worth through an intimate other. And he demonstrates the form this recognition is supposed to take when he reassures YOUNG WOMAN that he loves everything about her, for example her incessant talking: "Love how you talk – [...] love that it's non-stop. Love your lovely noise in my ears. All the time" (76). When she does not follow his invitation to name something she loves about him, he goes on: "(I) like how you wake up with words – your words. Bathe with a dialogue and breakfast with a chat. Like how you go to work missing me, call me up on your way. [...] like how you text me when you can't talk and – [...] and-and FaceTime me in your first break" (77). However, before long, it is divulged how unstable MAN's self-image, his 'essentialised self' for which he demands recognition and which determines what he loves and prefers, actually is. Any idea of a stable self that can be recognised once and for all in its full particularity is corroded by the very fact that MAN cannot even reach certainty about what he likes and loves. At the end of the play, he bluntly contradicts himself: "the endless phone calls of insecurity and texts of drama – all of that would change – FaceTime – what is the point? […] Your endless reportin of what your fucked-up family don't never say – would love to change that about you – you talkin *at* me […] *All* the time. That does need changing" (92). The point here is not to determine whether his first or second statement is true. The point is that he probably has no idea himself what is true; that he loves her talking sometimes and is annoyed by it at other times; that he is incapable of telling what it is exactly that he loves about her; and that perfect compatibility is an unlikely chimaera given the uncertainty and instability of selves and self-images. The scene leaves no doubt that the love between MAN and YOUNG WOMAN is a far cry from love of the other in his or her full particularity. For both, love means adaptation, compromise, and acceptance of unpleasant character traits in the other. They both know and accept that the other is anything but perfect – and this is a quite reasonable attitude since they seem to have no idea in the first place what the perfect partner, the perfectly compatible other, would have to be like.

It is helpful to remember Simon May's argument here, who considers the notion of loving everything about the other an absurdity (cf. 241–42). No person is flawless and it is unreasonable to assume that people in love start to love bad habits or unpleasant character traits that they would normally detest. Rather, he argues, people stay in love *despite* what is unlovable about the other, as long as the only fundamental condition of love is fulfilled: "We will love them in spite of almost any destructiveness, indifference, mean-spiritedness, and vindictiveness they might show us. […] Unless – and this is the only circumstance in which love can be killed – they stop being the sort of person who can arouse in us the hope of ontological rootedness" (244). In the following chapter, I will argue that the provision of the feeling of ontological rootedness is indeed a crucial topic of the play. As I will try to show, this rootedness can be understood as being tantamount to a recognition of the self – but not in the form of the recognition of loveable qualities or of a self-constructed 'essentialised self' that is the result of a process of introspection. Rather than affirming a predetermined 'essentialised self,' love in the play affirms the self independently of self-chosen categories. It does not so much confirm a self-image than create it. It is through love, not before it, that subjects have the chance to reach an 'essentialised self' through the confirmation of the meaningfulness and importance of their own existence. This longing for love as the longing for a validation of their own existence is implemented in the play as the characters' urgent desire to be needed by the other.

10.2.2 Devotion: I Want You to Want Me

The play mirrors its title with the conspicuous absence of the word 'love,' which is hardly ever used by the characters to denote or express their feelings. Instead, the verb used most frequently by the characters is 'want,' as for instance in B's declaration quoted above. In general, large parts of the play consist of discussions of the characters' wants: A wants to watch a TV programme in peace, B wants to have the bathroom to himself; WOMAN wants an apology and talks about the things "I could want that I trained myself not to. To stop. To not bother to want at all. Because of you" just as MAN regrets the "things I wanna do but don't cos I know you and know you won't" (63); MAN and YOUNG WOMAN name all the things they would want to change about the other. In this atmosphere of self-centred wants, love might be expected to appear as an ultimately selfish project aiming at what is good for the self. But it does not, and the reason why it does not is the inextricable link between selfishness and altruism that characterises erotic love. 'Need-love' and 'Gift-love,' to use C. S. Lewis' terms, are not clearly separable, neither in the philosophies of love outlined in Chapter 2.1 nor in tucker green's play. In the long tradition of philosophy that sees love as a form of compensation for a lack that affects the loving subject, and thus as inherently selfish, there has been no shortage of voices pointing out the acts of kindness, tenderness, and (seemingly) selfless concern for the other which love inspires. I have mentioned the positions of Spaemann and Pieper, for example, and their notion of a 'joy of giving,' the idea that 'selfless' care for the other is a source of happiness and meaning (and thus not entirely selfless) and that humans feel a need to bestow love on others. Lacan's dictum that "man's desire is the Other's desire [*le désir de l'homme est le désir de l'Autre*]" ("Subversion" 690) can be understood similarly as a reminder of the human desire to be desired or the need to be needed which in turn, I have argued, is a position much in line with Sartre's notion of being loved as affirmation of one's being and the sociological concept of the human demand for recognition. These approaches all indicate the deeply felt human need to have one's self validated and recognised through the other's desire, to be confirmed as desirable, useful, and needed. They indicate, as tucker green has it, "a profoundly affectionate, passionate devotion to someone."

It is hardly a coincidence that in tucker green's dictionary-like definition, 'devotion' takes the central position, whereas in other dictionary entries on love the most frequent surrogate noun is 'affection,' while 'devotion' is only of subordinate significance, if mentioned at all.⁶² Passion and affection are by no means neglected in the play, but it is indeed the 'devotion to someone' that carries most weight as the essential but also most precarious feature of love. Its central implication is the willingness to commit unreservedly to the beloved and to take care of his or her well-being.⁶³ Moreover, the religious connotation of the word 'devotion' in the sense of "religious worship or observance" denotes a kind of "love, loyalty, or enthusiasm for a person or activity" that resembles religious commitment (*Lexico*). With its combined meaning of altruistic bestowal and religious observance, 'devotion' expresses love's continual oscillation between 'Gift-love' and 'Need-love,' its simultaneous motivation by the sincere wish for the other's well-being and the pleasure that can be derived from providing this well-being as it engenders feelings of recognition and ontological rootedness. In tucker green's play, I will argue, the devotion demonstrated by the characters can be reconstructed precisely as this mixture of loving, altruistic commitment and a demand for recognition that manifests itself in the desire to be desired. It is a posture of heightened vulnerability and precariousness as the lovers make their feeling of self-worth and ontological rootedness dependent on something so absolutely uncontrollable as the beloved's desire. A series of episodes enquires into this tension area of desire.

Twice in the play, B has to make the bitter experience that his demand for recognition is frustrated, that he is made to feel useless where he thought he was needed and that his wish to be that which the other desires remains unfulfilled. The first moment comes in Scene Three, when A discloses that she "never liked" that "thing" (13) he used to do to (or with) her. Relying on the sounds she made and the looks she gave, B had always felt sure that he was giving her immense pleasure. When he now learns that he has completely misinterpreted her signs, his self-image as the man capable of giving her what she desires crumbles. His self-defensive reply cannot mask his disappointment: "*I* didn't particularly like doin it but done it cos I knew you liked it and wanted to-wanted you to, y'know … Something you liked that I could do you-give you – do with you doin it right. That you liked. Even though I didn't like doin it. Selfless. Generous" (14). A, in turn, doubts this claim of selflessness: "And what you was doin you were doin cos you-you liked, *really* looked like you were enjoying doin more'n me –" (14). Their debate about (probably) a certain sexual practice can easily

---

62 Cf. the entries on 'love' in the online editions of *OED*, *Merriam-Webster*, *Collins*, and *Longman*.
63 Cf. the entries in the online editions of *Merriam-Webster*, *Collins*, *Longman*, and *Lexico* for the verb 'devote,' which is circumscribed as the 'giving' of time, effort, or resources to the object of devotion.

be transferred onto a more universal level. In more abstract terms, what B is expressing is a lover's hope for "[s]omething you liked that I could do," a need or desire in the other that he – and, ideally, he alone – can satisfy. What A is questioning, on the other hand, is whether it is really her desire whose satisfaction is the primary objective in this business – whether B's devotion, in other words, is really 'selfless' and 'generous.' Their argument relates to the debate about the possibility of absolutely selfless love or 'disinterested concern' examined above at the beginning of Chapter 2. As I have argued, there is good reason to assume that even the most sacrificial and selfless acts of love are not entirely free of an element of self-love since the act of providing pleasure is at the same time a source of pleasure for the self. A's claim that B derived more pleasure from his effort to pleasure her, moreover, ties in with Lacan's reading of the *Symposium*, in which Alcibiades' wish to gratify Socrates' sexual needs is primarily motivated by Alcibiades' longing for a manifestation of Socrates' desire for him. Just as Alcibiades' primary motivation is not the gratification of Socrates but the satisfaction of his own desire to be desired, B's primary objective in doing the thing which he thinks A likes might be the satisfaction of his need to be needed. In fact, the palpable disappointment of B when he learns that he has misinterpreted her signs and has not satisfied her needs can only be explained based on the assumption that an important goal for him is the satisfaction of her needs.

The second and even more hurtful revelation of B's 'uselessness' comes in Scene Six, when A complains about his inadequate support during her depressive phase. B's argument that she denied him the chance to help with her refusal to speak and communicate her needs receives a sharp retort: "I wasn't there to give you chances. [...] I wasn't able. That was the point and you were no help are no help. Can't help. Couldn't help. Didn't help. Didju?" (25). But it takes the information conveyed some moments later to understand the full impact of A's accusation. B's long monologue that ends the scene reveals that, in fact, he cannot be blamed for a lack of support during A's illness. Rather, the destructive potential in A's accusation of uselessness lies in the implication that he failed *although* he tried, that among all the things he did to make her feel better none of them were what she really needed:

> Ten months of you sayin nuthin to me when I dried your tears and your tears.
> And your tears.
> When I fed you healthy
> when I watched you not sleep.
> When I rubbed your aches and set your baths
> when I dried your body and lay you down
> then watched you not sleep more.

> When I dressed you-undressed you, combed out your hair, oiled your hair creamed your skin and (I) didn't touch you when I wanted to and I wanted to – when I wanted you and I wanted you.
> When I sat with you
> when I excused you
> when I drove you – walked you, walked with you stood by you and when I tried to fill the gaps your gaps with sayin something with sayin anything tryin to fill you up with what little I had to offer with what I had left to give and now you're tellin me that was all wrong? That that was all no good that weren't what you wanted weren't what you needed and weren't a piece of it right or done right anyway – and that I was the 'let down'? Really?
> *Really?*
> *Beat.*
> Fuck. (28)

Like in Scene Three, B again is confronted with the information that what he did for A was not what she really desired. Again, apparently unable to satisfy her needs, his own need to be needed is frustrated. What is more, his inability to help her during her illness means not only that he did not *do* the right things but it also implies that he cannot *be* what he would like to be, namely, to use Sartre's words, "'the whole World' for the beloved" (229). Since A's depression is a reaction to the loss of her mother, B's inability to compensate this loss signals his insufficiency as A's only focus of need and desire. The fact that she mourns the loss of her mother despite his existence and presence is an unwelcomed reminder that he cannot be 'the whole World' for her and that her feelings are not all centred upon him. For Barthes, this amounts to the sobering insight that, however close a lover might feel to the beloved, there is an ineluctable boundary separating them into two distinct individuals:

> For at the same time that I 'sincerely' identify myself with the other's misery, what I read in this misery is that it occurs *without me*, and that by being miserable by himself, the other abandons me: if he suffers without my being the cause of his suffering, it is because I don't count for him: his suffering annuls me insofar as it constitutes him outside of myself. (57)

B is 'annulled' by A's suffering because it takes place independently from him. Neither did he cause it, nor can he alleviate it. No matter how close he feels to her, A's independent suffering emphasises her insurmountable alterity and uncontrollability – the very features she later proves again with her death.

The motif of caring for a sick partner is taken up again in Parts Two and Three, where it is MAN who nurses WOMAN. While in Part Two, in the heated argument with WOMAN, he depicts his caregiving as a selfless act of kindness for which she owes him gratitude, Part Three allows for the interpretation that his altruism, too, is shaped by his need to be needed. Now that he is in a new relationship with YOUNG WOMAN, he is still regularly visiting and caring for an unspecified old female friend, most probably his partner from Part Two. Understandably, YOUNG WOMAN displays signs of jealousy and doubts that the woman he is seeing is really sick.

Y. WOMAN:	you don't owe her anything.
MAN:	I go cos it's decent.
	I go cos she's not well.
Y. WOMAN:	Again.
MAN:	She's not been well
Y. WOMAN:	again.
MAN:	She's / not –
Y. WOMAN:	She's not your responsibility.
MAN:	She's not my responsibility.
	[...]
	I go and when I go I talk about you, that pisses her off
Y. WOMAN:	shouldn't go at / all
MAN:	I go to be decent, go to see she's alright, see she's alright talk about you, piss her off and leave.
Y. WOMAN:	She's not as sick as she says she / is. (83)

But whether she is really sick or not, whether MAN is "gullible" and "stupid" (85) because he believes her to need his care, is not the central question. The point is that he *wants* to believe it and that 'decency' is only little more than a pretext for MAN's frequent visits to a woman who is not his responsibility. YOUNG WOMAN's jealousy arises from her justified suspicion that his visits, which he describes as downright ordeals (cf. 83–84), are not self-sacrificial acts of altruism but have to be pleasurable in some way or another, that MAN needs and desires something else outside of their relationship. It is his need to be needed which is not fully satisfied by YOUNG WOMAN so that he eagerly believes in the neediness of his former partner in order to have his self recognised as useful and needed. Paradoxically, YOUNG WOMAN's insufficiency in satisfying MAN's need to be needed results from MAN's insufficiency to be all that YOUNG WOMAN needs. If MAN was able to be 'the whole world' for YOUNG WOMAN, his need for recognition would be satisfied. But her lack of well-being, which is thematised throughout Part Three, constantly reminds him of his insufficiency as her sole target of need and desire. Just as B could not compensate A's loss of her mother, MAN cannot compensate YOUNG WOMAN's loss of her's and the deficient relationship with her father. This experience of insufficiency – the incapability of making his beloved completely happy – threatens his self-image and urges him to seek additional recognition outside their relationship.

All these scenes, then, stress the precariousness of love as devotion. The wish to 'be there' for the other, to give the beloved something that he or she needs or desires, to *be* the other's centre of desire, to be 'the whole world' for the other – this wish is not only crucial in the project of grounding the self (or an image thereof) through recognition but it is also dependent on the utterly uncontrollable desire of the other. There is no guarantee for the lover that he can ever be what the other desires, let alone *all* she desires. At the same time, if the lover realises that he cannot be all the beloved needs, his self-image may begin to crack. The beloved, then, does not provide sufficient recognition to generate the sense of self-worth the lover expects from the relationship, making additional sources of recognition desirable. tucker green's play offers a subtle but eloquent description of this almost tragical vicious circle in which the insufficiency of the lover causes the insufficiency of the beloved and vice versa. With its scenes of tenderness and profound affection, the play is not a denial of love *per se,* but with its suggestion that lovers are necessarily insufficient and incapable of being 'the whole world' for the other, it is a denial of the possibility of a love where mutual compensation is perfect and precariousness is absent.

10.3 Conclusion

tucker green's fragmentary approach to love allows her to include a multitude of themes that could hardly be integrated into a coherent plot following the rules of linearity and finality. The inclusion of this multitude and their fragmentary presentation, in turn, are bound up with one of the key messages of the play: that love, as a term or concept, is elusive and impossible to pin down. However, this elusiveness does not prevent the play from confronting the audience with a number of highly recognisable glimpses at three fictional but realistic love relationships which, through repetition, contrast, and correspondence, accentuate a selection of topics that, above all, stress love's precariousness. In all three parts, love is a battlefield characterised by the couple's daily struggles for power. However, these power struggles are not so much symptoms of the characters' lack of love but rather arise from the experience of vulnerability and powerlessness in a relationship with an uncontrollable other. Their anxiety to gain power and control in their daily quarrels is, in fact, a redirection of their wish to compensate the loss of control, which is most drastically experienced in the loss of the other through death. What fuels the quarrels between the lovers is the lack of understanding that marks their communication. Apart from intentional inattentiveness or the refusal to communicate, what inhibits mutual understanding is the insufficiency and uncertainty of both the language and the signs of love. Neither self-expression nor understanding of the other can reach the precision that would be necessary to reach certainty about the other's feelings and desires. To make things even more adverse, the play also denies its characters any certainty about their own desire. Lacking secure and stable self-images, they neither know exactly what they want nor do they, if they demand the recognition of their selves through an intimate other, know what exactly it is they want to have recognised. What they *do* know, however, is that any such recognition means to be recognised as useful, important, or, ideally, indispensable by the other, and it is this recognition they are seeking in their devotion to the other, even though this makes them dependent on the beloved's acceptance of their devotion and hence makes their position precarious.

In spite of this uncompromising review of love and its language, the play does not deny the possibility of love as such. It argues quite eloquently that silence wreaks more damage to a relationship than even the most imperfect communication, and it applies the same logic to love and understanding: the impossibility of perfection in these projects does not mean that they are not worth striving for or that anyone should acquiesce to their total absence. The inevitable lack of perfection in love and its language does not relegate them to the realm of impossible ideals, it only makes them inevitably deficient and difficult and further increases love's inherent precariousness. The play's aesthetics repeats this thematic thread on a formal level. Its fragmentary episodic composition defies a secure reconstruction of its underlying story and the establishment of closure. Many informational gaps remain unfilled, producing an epistemological uncertainty that reinforces the impression of uncertainty and imperfection in love and its language. And yet, the play is anything but incomprehensible or meaningless. Its primary thematic concern and the subtle cohesion of its three parts shine through all its fragmentation, indicating coherent meaning as a definite possibility that is impossible only in the form of definite perfection. Thematically, the play decries any blue-eyed vision of love as based on or tantamount to perfect communication, total understanding, complete knowledge, and full compensation of one's own and the other's insufficiency, but this does not imply that it propagates the opposites of these goals and the impossibility of love. In fact, in all of its three relationships, love is put to the test and the partners wound each other with their miscommunication and struggle for power, but love, ultimately, survives in one form or another. A and B love each other beyond the grave, MAN pays visits to and cares for WOMAN although he is already in a new relationship, and MAN and YOUNG WOMAN end the play with a kiss despite all the things they would change about the other. None of these love relationships feels stable or secure – such is the inherent precariousness of love – but none of them is made impossible by the lack of certainty that characterises love and its language.

11 Coda

In "Word and Stage in Postdramatic Theatre" (2007) Hans-Thies Lehmann condenses and reemphasises some of the cornerstones of his pivotal *Postdramatic Theatre* (1999). He repeats his conviction that the dramatic form "with its tendency toward closure, logic, order (even in so-called 'open forms' of drama)" (44) is no longer appropriate in our modern world. Apart from the general problem of narrativisation or 'emplotment' which, as he claims, contradicts the "commonplace experience that even the individual life is lived less and less as a coherent drama or even story, but much more as a series of fragments, life-phases and episodes" (41),[64] his main argument for the inadequacy of the dramatic form is the suggestion that, as an artform which develops stories around momentous individual decisions, it lacks models in contemporary reality which it could imitate or 'dramatise':

> For theatre as a form of art which has its raison d'être in the capacity to problematize or even renew our perception and understanding of reality, it holds true that the pattern of drama somehow no longer seems to be able to grasp this reality. If we consider the principle of its form, drama is about conflict between protagonists and about decisions created in the sphere of dialogue. But it is beyond doubt that conflicts and decisions of relevance in contemporary society are less and less conflicts and decisions of personal protagonists. They result from tensions between anonymous blocks of power, economic interests, regional and global strategies, markets, shifts of balance of forces, and chaotic sudden implosions which are much less influenced by individual actors in political power than they themselves may imagine. Equally in everyday social life, conflicts have taken the form of the 'prose of civic life' (Hegel: 'Prosa des bürgerlichen Lebens'). Thus, in the absence of credible personal protagonists it becomes increasingly questionable to present socially and even individually relevant conflicts by dramatizing them. (40)

---

64 I will not discuss this argument any further, but I want to indicate that it is by no means self-evident. Firstly, while closed plays are by their nature not staging entire lives (coherent or not) but necessarily concentrate on a limited segment, open forms of drama are well suited for the depiction of fragmented, episodic lives. Secondly, it is simply questionable whether there is really a "commonplace experience" of a rupture between lived reality and its narrativisation. Hayden White may be right that "[w]e do not *live* stories" but only "give our lives meaning by retrospectively casting them in the form of stories" (1720), but this act of story-making, as White never doubts, is a necessary and quite natural way of human sense-making and understanding. Moreover, there are good reasons to consider the argument of philosopher David Carr that the proponents of what he calls the "discontinuity view," who claim "that life has no beginnings, middles, and ends," neglect the many "forms of closure and structure to be found" in human lives in between the definite beginning and end of birth and death – all the small-scale actions and projects that people start and finish in their lifetime (122). Stories, in other words, might not contradict but correspond to the human perception of the world.

In the face of the analysis of dramatic plays in this volume, I would like to disagree with Lehmann. In particular, I think the argument that individual conflicts and decisions in contemporary society are irrelevant and not worth dramatising cannot hold. Even if we accept for a moment the claim that individual decisions of personal protagonists play a minor role in political actions of far-reaching impact and social relevance, it still does not follow that there is no material for the modern dramatist. Drama was never restricted to the portrayal of large-scale heroic deeds of overarching political significance, and there is no reason to regard the 'prose of civic life' devoid of relevance for an audience who is interested in the vicissitudes of 'ordinary' life.

Quite to the contrary, I think, the plays I have discussed make plain how very 'dramatic' ordinary life can be. The hope for love's compensatory function – the longing for meaning, groundedness, security, and recognition – and the precariousness that always accompanies love are essential constituents of contemporary life. Love has the potential to generate moments of immense joy and deep crisis, it demands decisions of enormous personal relevance, and it takes place between individual protagonists. True, the decisions of the protagonists are hardly entirely individual in the sense that they are independent from the impact of external factors, and the plays find ways to reflect this influence. The episodic plots of Marber, Ravenhill, Kelly, Bartlett, and tucker green, for instance, enlarge the temporal and/or spatial scope of their plays so as to indicate the influence of urban isolation and an ideology of self-realisation (Marber), of the absence of meaning in a materialist world (Ravenhill, Kelly), of education and the prevailing *zeitgeist* (Bartlett), or of the attrition of long-term relationships and of 'hereditary' communication problems (tucker green) on the characters' romantic behaviour. Sarah Kane, in *Cleansed,* subjects her lovers to a system of oppression within a coercive institution, and Penelope Skinner evokes an atmosphere of social pressure through the cascades of words spoken by unseen voices in the dark. In short, in all these plays the protagonists are under the sway of external forces that impact on their actions and decisions – a situation that is further reinforced by the very fact which this volume has sought to demonstrate, namely the persistence of historical discourses of love which shape the modes of expression and experience of love. But despite the characters' subjection to impersonal and external influences and their embeddedness in society and its discursive formations, the plays still grant them sufficient space for individual decisions and responsibility. When Dan leaves Alice for Anna, when Mark fulfils Gary's suicidal desire, when Hippolytus drives Phaedra to commit suicide, when Rose degrades and mutilates herself in search of love, when David kills Jess for a car, or when Kenneth and Sandra decide to divorce in front of their children, they are all individually responsible for their actions, any external influence on their motivation notwithstanding, and the plays duly foreground their own awareness of this responsibility. When Lehmann rejects the dramatic form because it "has a tendency to create the less and less plausible impression that actions in this world essentially depend on the decisions and the psychology of individual human agents, while since Nietzsche, Marx and Freud the problem of 'agency' has turned out to be much more complex" (43), I think he overstresses the influence of social, materialist, and psychological determinism at the cost of individual responsibility. To be sure, the shaping influence of external or incontrollable forces is not to be denied and Nietzsche, Marx, and Freud have reminded us of their significance with their justified attacks on an idealised image of man governed entirely by individual will and reason, but neither did awareness of humans' exposure to external or incontrollable forces start with them nor is such awareness in conflict with the dramatic form. Just as Shakespeare's plays abound with examples of characters who cannot break free from the impact of their origins, social conventions, or irrepressible passions but are arguably nonetheless in harmony with the dramatic form, so the plays I have analysed present characters who are exposed to external forces and yet retain sufficient 'agency' to make their stories dramatic. They hence oppose the notion that the appeal which postdramatic theatre undoubtedly has results from the 'commonplace experience' that our world leaves no space for relevant individual decisions. In many spheres of 'ordinary' life, presumably, and in matters of love certainly, drama still has its place.

Love, in other words, has always been dramatic and the continuity of its dramatic potential reflects and is reflected in the continuity of the great 'matters' of love which, however, despite continuity, are also subject to change and modification. The simultaneity of continuity and change, of traditional and specifically contemporary concepts of love, has been the focal point of interest in this volume, and it is time now to pull a few threads together.

As the diachronic survey of philosophies of love in Chapter 2 suggests, erotic or romantic love has rarely been considered unconditional. Rather, it has been conceived of as a mode of compensation necessitated by the insufficiency of human beings. In the plays I have analysed we see three major and specifically modern or contemporary causes for the compensatory hope that is pinned on love. The first is a distinctly modern form of loneliness and isolation resulting from the dissolution of traditional forms of community and strong family bonds, which has not only liberated modern individuals but also stripped them of the security of 'belonging.' *Closer, Shopping and Fucking*, and *Eigengrau* offer the most obvious renditions of the theme, with protagonists who appear adrift in the big city, lacking close personal relations and expecting love to put an end to their solitary existence and feeling of incompleteness. The second is the post-traditional and post-religious absence of meaning and orientation for which love is supposed to recompense. For Gary in *Shopping and Fucking*, for Rose in *Eigengrau*, and for Jess in *Love and Money*, particularly, but to a lesser extent for many other figures as well, love acquires a quasi-religious status, endowing their lives with meaning and a form of teleology that makes up for the absence of parental guidance and the lack of orientation resulting from the end of all other master-narratives. Similarly, when Phaedra hopes that her love will rescue Hippolytus from self-hate and apathy, she implies the idea that love is a purveyor of meaning that will work as a redemptive force upon her stepson who ekes out a senseless existence full of luxuries but void of purpose. Thirdly, love fulfils the need for recognition through intimate others that has increased with the progress of individualism and the democratisation of society and that is now the main source of feelings of self-worth. Mark in *Shopping and Fucking*, Dan and Larry in *Closer*, Tim and Rose in *Eigengrau*, and the passionately 'devoted' characters in *a profoundly affectionate, passionate devotion to someone (-noun)* all display a pressing need to be needed, a desire for the other's desire, a longing for the recognition of their usefulness, lovability, and desirability and, by extension, an affirmation of the meaningfulness of their existence.

As throughout the history of love philosophy, so in contemporary British drama love's compensatory function comes into conflict with its many forms of precariousness. The desire for the other's desire, for instance, is directed at a notoriously problematic object. As the characters in Marber's and tucker green's plays have to realise, the desire of the other is fundamentally precarious as it is ultimately unknowable and uncontrollable for both agents of the dyad. For the self and for the other it resists fixation, knowledge, and communication – it alters, often remains subconscious, and lacks a universal language of its own – and hence also disavows the modern 'self-help ideals' of communication, commensuration, and compatibility. In *Shopping and Fucking*, Mark can neither understand nor control Gary's desire and inevitably fails to become its object. In *Eigengrau*, Tim is as powerless to win Rose's desire as she seems powerless to terminate its vain fixation on Mark, and her method of winning Tim's and controlling her own desire through self-blinding is marked as a fictional and certainly non-exemplary solution to this non-fictional problem. Of course, the most traditional form of love's precariousness, the unattainability or potential loss of the beloved, is also addressed regularly. Gary's desire is directed at an imaginary, otherworldly object and eventually emerges as a death wish. Phaedra's passionate, irresistible love for Hippolytus is brutally rejected, Mark is unattainable for Rose, and B is reminded of the precariousness of love by the death of his wife. The specifically contemporary aspect in this regard are the characters' futile efforts to oppose or immunise themselves against love's precariousness, which are symptomatic of a mindset that worships self-sufficiency and pathologises dependency and vulnerability. In his endeavour to avoid emotional dependency, Mark tries to turn his intimate relationships into 'transactions,' while Hippolytus shuns romantic commitment to prevent the pain of unrequited love he seems to have suffered once and which now tantalises his stepmother. Where these characters seek to prevent the vulnerability and lack of control inherent in love from taking hold in the first place, tucker green's characters engage in daily power struggles in order to win back at least some of the control they have lost. The same idle wish for control, independence, and invulnerability underlies the cases of commitment phobia in the plays of Bartlett and Marber, in which the protagonists prefer self-realisation to commitment or refuse to embrace love's indispensable openness and vulnerability for fear of appropriation and self-loss. Kane's plays are exceptional in this regard since for her characters, loss of self is not a looming threat but a Romantic fantasy of dissolving the boundaries that separate the demarcated, individuated self from the wholeness it desires. Lastly, what troubles some of the dramatic figures, especially in the plays of Marber and tucker green, is an uneasy feeling of inauthenticity in the experience and expression of romantic love. Moreover, both Marber's and Kelly's characters exhibit a postmodern incredulity towards love when Dan and Larry set off romantic passion against realist companionship or when Jess subverts her own belief in the power of pure love with her materialist desires.

The selection of plays in the analysis is inescapably arbitrary, apart from the fact that I have tried to make it representative of the broad variety of new writing within text-and story-based British drama of the past three decades. Many other plays would have been suitable for my approach and would have yielded comparable results, as I would like to indicate briefly. Philip Ridley's *Tender Napalm* (2011), for example, combines the problematisation of authenticity and imagination, the theme of love as a struggle for power, and the proximity of love, lust, hate, and violence. The play features two characters, Man and Woman, who fantasise about their love, creating their own myths full of passion, violent fights, and incredible events in a dialogue that blurs the boundaries between the couple's imaginary and real worlds. Above all, the characters' creative fabrication of their personal, intimate narrative(s) is a mine for investigation into the reoccurrence and adaptation of historically continuous discourses of love. David Eldridge's *Under The Blue Sky* (2000), in its three successive and subtly related pair scenes, thematises the difference between love and friendship, the uncontrollability of the other, the pain of unrequited love, and the nature of 'obstacles' to love in a time and place where they are no longer the effect of impersonal, external forces like wars or unbridgeable social boundaries but of self-chosen attitudes. Eldridge's 2017 two-hander *Beginning* starts when the party of the previous night has ended and it ends with the possible beginning of new love and even new life. In between, Laura and Danny turn from complete strangers interested in a one-night stand to potential lovers and parents. What they experience in these ninety minutes or so is the full range of love's precariousness: the vulnerability that follows the courageous move of opening up to a stranger, the fear of rejection or later desertion, the unknowability of the other's desire, and the difficulty of making a decision in a moment of crisis in which the baggage of their entire lives seems to accumulate and put them under pressure. Nick Payne's *Constellations* (2012), too, stresses the factor of decision-making with its dramatisation of the multiverse theory, in which each decision of Roland and Marianne opens up a different version of their story while they are at the same time subject to events in which they have no hand. In some versions they become a couple, in others not; in some they marry, in others not; and in some Roland agrees to support his terminally ill wife when she seeks assisted suicide, while in others he is unable to do so. Multiverse theory and the concomitant questions concerning the possibility of choice, control, free will, and right or wrong decisions are also discussed in the internal communication system between the cosmologist Marianne and the beekeeper Roland. When the latter, in his proposal speech to Marianne, envies honeybees for the simplicity of their existence and their imperturbable sense of purpose, he foregrounds the idea, implicit in the whole play, of love as a source of orientation, meaning and groundedness that promises to compensate for the absurdity, contingency, and uncontrollability of the universe. Laura Wade's *Other Hands* (2006), to give a last example, portrays the attrition of love in the long-term relationship of the seemingly incompatible Hayley and Steve, who both temporarily find new love interests in Greg and Lydia. Throughout the play, the decision between fixing or disposing of dysfunctional 'things' (computers, printers, employees, marriages) works as a leitmotif and in the end Hayley and Steve stay together, symbolically bound by their increasingly aching, bandaged hands. Although Steve quite clearly represents the disposition to mend what can be mended and to put effort into the survival of their love and marriage, what eventually draws him to his wife is arguably the same motivation that had temporarily drawn him to Lydia: the feeling of being needed and, consequently, the promise of ontological rootedness.

There is, to come to a conclusion, a direct link between the present and the past as far as the experience and expression of love are concerned. There is continuity and comparability in human emotions – just as the two otherwise diametrically opposed theoretical positions of (biological) universalism and (social/cultural) constructionism suggest, which both reject the idea of emotions as individually unique. While the universalist position stresses the continuity and comparability of emotions with reference to humanity's shared physiological design, constructionism points out synchronous differences between cultures or societies and diachronic differences due to the historic mutability of all societies but implies a degree of sameness for a given society at a given point in time. In both cases, emotions are shared phenomena experienced more or less comparably by members of a group. The discourse analytic approach of this study and the very conception of emotions as both biological and cultural make the universalist position untenable in its extreme form. As I hope to have demonstrated, however, the same is true for a position that denies any suprahistorical continuity of emotional experience. There are strong historical traditions regarding the expectations, hopes, and fears in love, and they are just as important as their (post-) modern transformations. In a sense, then, an examination of our contemporary discourses of love reveals both what is special about us, and what is not.

Works Cited

Abu-Lughod, Lila, and Catherine A. Lutz. "Emotion, Discourse, and the Politics of Everyday Life." *Emotions: A Cultural Studies Reader*, edited by Jennifer Harding and E. Deidre Pribram, Routledge, 2009, pp. 100–12.
Ahmed, Sara. *The Promise of Happiness*. Duke UP, 2010.
Akbar, Arifa. Review of *Love, Love, Love*, by Mike Bartlett. *Guardian*, 12 Mar. 2020, www.theguardian.com/stage/2020/mar/12/love-love-love-review-mike-bartlett-rachael-stirling-lyric-hammersmith.
Alderson, David. "Postgay Drama: Sexuality, Narration and History in the Plays of Mark Ravenhill." *Textual Practice*, vol. 24, no. 5, 2011, pp. 863–82.
Angelaki, Vicky. *Social and Political Theatre in 21st-Century Britain: Staging Crisis*. Bloomsbury, 2017.
Aragay, Mireia, Enric Monforte, and Pilar Zozaya. Introduction. *British Theatre of the 1990s: Interviews with Directors, Playwrights, Critics and Academics*, edited by Mireia Aragay et al., Palgrave Macmillan, 2007, pp. ix–xii.
Aragay, Mireia, and Martin Middeke. "Precariousness in Drama and Theatre: An Introduction." *Of Precariousness: Vulnerabilities, Responsibilities, Communities in 21st-Century British Drama and Theatre*, edited by Mireia Aragay and Martin Middeke, CDE Studies 28, De Gruyter, 2017, pp. 1–13.
Artaud, Antonin. "The Theatre of Cruelty: First Manifesto." 1932. *Selected Writings*, edited by Susan Sontag and translated by Helen Weaver, Farrar, Straus and Giroux, 1976, pp. 242–251. Reprinted in *Modern Theories of Drama: A Selection of Writings on Drama and Theatre, 1840–1990*, edited by George W. Brandt, Clarendon, 1998, pp. 188–94.
Badiou, Alain. "Happiness Is a Risk That We Must Be Ready to Take." Interview by Miri Davidson. *Verso*, 10 June 2015, www.versobooks.com/blogs/2032-alain-badiou-happiness-is-a-risk-that-we-must-be-ready-to-take.
Badiou, Alain. *In Praise of Love*. 2009. Translated by Peter Bush, New Press, 2012.
Baraniecka, Elżbieta. "Precariousness of Love and Shattered Subjects in Dennis Kelly's *Love and Money*." *Of Precariousness: Vulnerabilities, Responsibilities, Communities in 21st-Century British Drama and Theatre*, edited by Mireia Aragay and Martin Middeke, CDE Studies 28, De Gruyter, 2017, pp. 171–86.
Baraniecka, Elżbieta. *Sublime Drama: British Theatre of the 1990s*, CDE Studies 23, De Gruyter, 2013.
Barry, Elizabeth. "'Conscious Sin': Seneca, Sarah Kane and the Appraisal of Emotion." *Canadian Review of Comparative Literature*, vol. 40, no. 1, 2013, pp. 122–35.
Barry, Peter. *Beginning Theory: An Introduction to Literary and Cultural History*. 3rd ed., Manchester UP, 2009.
Barthes, Roland. *A Lover's Discourse: Fragments*. 1977. Translated by Richard Howard, Vintage, 2002.
Bartlett, Mike. *Love, Love, Love*. 2010. Methuen, 2015.
Bassett, Kate. Review of *Phaedra's Love*, by Sarah Kane. *Times*, 22 May 1996. *Theatre Record*, vol. 16, no. 11, p. 651.
Bataille, Georges. *Erotism: Death and Sensuality*. 1957. Translated by Mary Dalwood, City Lights, 1986.
Bauman, Zygmunt. *Liquid Love: On the Frailty of Human Bonds*. Polity, 2003.

Baxter, Leslie A., and Chitra Akkoor. "Aesthetic Love and Romantic Love in Close Relationships." *Communication Ethics: Between Cosmopolitanism and Provinciality*, edited by Kathleen Glenister Roberts and Ronald C. Arnett, Peter Lang, 2008, pp. 23–46.
Beck, Ulrich, and Elisabeth Beck-Gernsheim. *Das ganz normale Chaos der Liebe*. Suhrkamp, 1990.
Belsey, Catherine. *Desire: Love Stories in Western Culture*. Blackwell, 1994.
Belsey, Catherine. "Postmodern Love: Questioning the Metaphysics of Desire." *New Literary History*, vol. 25, no. 3, 1994, pp. 683–705.
Belsey, Catherine. *Poststructuralism: A Very Short Introduction*. Oxford UP, 2002.
Benardete, Seth, ed. *Plato's "Symposium": A Translation by Seth Benardete with Commentaries by Allan Bloom and Seth Benardete*. U of Chicago P, 2001.
Benardete, Seth. "On Plato's 'Symposium.'" *Plato's "Symposium": A Translation by Seth Benardete with Commentaries by Allan Bloom and Seth Benardete*. U of Chicago P, 2001, pp. 179–99.
Benedict, David. Review of *Cleansed*, by Sarah Kane. *Independent*, 9 May 1998, *Theatre Record*, vol. 18, no. 9, pp. 564–65.
Benedict, David. Review of *Closer*, by Patrick Marber. *Independent*, 31 May 1997. *Theatre Record*, vol. 17, no.11, p. 674.
Berlant, Lauren. *Desire/Love*. Punctum, 2012.
Billen, Andrew. Review of *Closer*, by Patrick Marber. *Observer*, 1 June 1997. *Theatre Record*, vol. 17, no. 11, p. 678.
Billington, Michael. Review of *a profoundly affectionate, passionate devotion to someone (-noun)*, by debbie tucker green. *Guardian*, 7 Mar. 2017, www.theguardian.com/stage/2017/mar/07/a-profoundly-affectionate-review-meera-syal-debbie-tucker-green.
Billington, Michael. Review of *Cleansed*, by Sarah Kane. *Guardian*, 7 May 1998. *Theatre Record*, vol. 18, no. 9, p. 566.
Billington, Michael. Review of *Closer*, by Patrick Marber. *Guardian*, 23 Feb. 2015, www.theguardian.com/stage/2015/feb/23/closer-review-donmar-warehouse-london.
Billington, Michael. Review of *Closer*, by Patrick Marber. *Guardian*, 31 May 1997. *Theatre Record*, vol. 17, no. 11, p. 678.
Billington, Michael. Review of *Love, Love, Love*, by Mike Bartlett. *Guardian*, 5 May 2012. *Theatre Record*, vol. 32, no. 9, p. 463.
Bloom, Allan. *Love and Friendship*. Simon and Schuster, 1993.
Brantley, Ben. Review of *Love, Love, Love*, by Mike Bartlett. *New York Times*, 19 Oct. 2016, www.nytimes.com/2016/10/20/theater/love-love-love-review.html.
Brown, Georgina. Review of *Closer*, by Patrick Marber. *Mail on Sunday*, 8 June 1997. *Theatre Record*, vol. 17, no. 11, p. 674.
Brown, Peter. Review of *Love, Love, Love*, by Mike Bartlett. *LondonTheatre.co.uk*, 3 May 2012, www.londontheatre.co.uk/reviews/love-love-love.
Brusberg-Kiermeier, Stefani. "Cruelty, Violence, and Rituals in Sarah Kane's Plays." *Sarah Kane in Context*, edited by Laurens De Vos and Graham Saunders, Manchester UP, 2010, pp. 80–87.
Brusberg-Kiermeier, Stefani. "Re-writing Seneca: Sarah Kane's *Phaedra's Love*." *CDE Vol. 8: Crossing Borders. Intercultural Drama and Theatre at the Turn of the Millennium*, edited by Bernhard Reitz and Alyce von Rothkirch, WVT, 2001, pp. 165–72.

Butler, Judith. *Precarious Life: The Powers of Mourning and Violence*. Verso, 2006.
Butler, Robert. Review of *Closer*, by Patrick Marber. *Independent on Sunday*, 1 June 1997. *Theatre Record*, vol. 17, no. 11, pp. 676–77.
Capitani, Maria Elena. "Dealing with Bodies: The Corporeal Dimension in Sarah Kane's *Cleansed* and Martin Crimp's *The Country*." *JCDE*, vol. 1, no. 1, 2013, pp. 137–48.
Carney, Sean. *The Politics and Poetics of Contemporary English Tragedy*. U of Toronto P, 2013.
Carr, David. "Narrative and the Real World: An Argument for Continuity." *History and Theory*, vol. 25, no. 2, 1986, pp. 117–31.
Cavendish, Dominic. Review of *a profoundly affectionate, passionate devotion to someone (-noun)*, by debbie tucker green. *Telegraph*, 7 Mar. 2017, www.telegraph.co.uk/theatre/what-to-see/profoundly-affectionate-passionate-devotion-royal-court-upstairs/.
Cavendish, Dominic. Review of *Love, Love, Love*, by Mike Bartlett. *Daily Telegraph*, 14 Oct. 2010. *Theatre Record*, vol. 30, no. 21, p. 1189.
Cavendish, Dominic. Review of *Love, Love, Love*, by Mike Bartlett. *Telegraph*, 13 Mar. 2020, www.telegraph.co.uk/culture/theatre/theatre-reviews/8061160/Love-Love-Love-Drum-Theatre-Royal-Plymouth-review.html.
Clapp, Susannah. Review of *Cleansed*, by Sarah Kane. *Observer*, 10 May 1998. *Theatre Record*, vol. 18, no. 9, p. 566.
Clapp, Susannah. Review of *Love, Love, Love*, by Mike Bartlett. *Observer*, 6 May 2012. *Theatre Record*, vol. 32, no. 9, pp. 463–64.
Coveney, Michael. Review of *Closer*, by Patrick Marber. *Daily Mail*, 30 May 1997. *Theatre Record*, vol. 17, no. 11, p. 676.
Curtis, Nick. Review of *Closer*, by Patrick Marber. *Evening Standard*, 30 May 1997. *Theatre Record*, vol. 17, no. 11, p. 677.
D'Monté, Rebecca, and Graham Saunders, eds. *Cool Britannia? British Political Drama in the 1990s*. Palgrave Macmillan, 2008.
De Beauvoir, Simone. "From *The Second Sex*." *The Philosophy of (Erotic) Love*, edited by Robert C. Solomon and Kathleen M. Higgins, UP of Kansas, 1991, pp. 233–40.
De Jongh, Nicholas. Review of *Cleansed*, by Sarah Kane. *Evening Standard*, 7 May 1998. *Theatre Record*, vol. 18, no. 9, p. 563.
De Rougemont, Denis. *Die Liebe und das Abendland*. 1956. Translated by Friedrich Scholz, Kiepenhauer & Witsch, 1966.
De Vos, Laurens. *Cruelty and Desire in the Modern Theatre: Antonin Artaud, Sarah Kane, and Samuel Beckett*. Fairleigh Dickinson UP, 2011.
"devote, v." *Collinsdictionary.com*, n.d., Collins, www.collinsdictionary.com/dictionary/english/devote.
"devote, v." *Lexico*, n.d., Oxford UP and Dictionary.com, www.lexico.com/definition/devote.
"devote, v." *Ldoceonline.com*, n.d., Longman Dictionary of Contemporary English. www.ldoceonline.com/dictionary/devote.
"devote, v." *Merriam-Webster.com*, n.d., Merriam-Webster. www.merriam-webster.com/dictionary/devote.
"devotion, n." *Lexico*, n.d., Oxford UP and Dictionary.com, www.lexico.com/definition/devotion.

Defraeye, Piet. "In-Yer-Face Theatre? Reflections on Provocation and Provoked Audiences in Contemporary Theatre." *Extending the Code: New Forms of Drama and Theatrical Expression*, edited by Hans-Ulrich Mohr and Kerstin Mälcher, CDE 11, WVT, 2004, pp. 79–97.

Deubner, Paula. *"Into the Light": Selbst und Transzendenz in den Dramen Sarah Kanes*. CDE Studies 22, WVT, 2012.

"Divorces in England and Wales: 2017." *Office for National Statistics*, 26 Sept. 2018, www.ons.gov.uk/peoplepopulationandcommunity/birthsdeathsandmarriages/divorce/bulletins/divorcesinenglandandwales/2017.

Dromgoole, Dominic. *The Full Room: An A–Z of Contemporary Playwriting*. Methuen, 2002.

Eco, Umberto. "Reflections on *The Name of the Rose*." *Encounter*, vol. 64, no. 4, 1985, pp. 7–19.

Edwardes, Jane. Review of *Cleansed*, by Sarah Kane. *Time Out*, 13 May 1998. *Theatre Record*, vol. 18, no. 9, p. 563.

Edwardes, Jane. Review of *Closer*, by Patrick Marber. *Time Out*, 4 June 1997. *Theatre Record*, vol. 17, no. 11, p. 678.

Enns, Diane. "Love's Limit." *Thinking About Love: Essays in Contemporary Continental Philosophy*, edited by Diane Enns and Antonio Calcagno, Pennsylvania State UP, 2015, pp. 31–45.

Esslin, Martin. *Artaud*. Fontana Modern Masters. Fontana/Collins, 1979.

Faulstich, Werner. "Die Entstehung von 'Liebe' als Kulturmedium im 18. Jahrhundert." *Liebe als Kulturmedium*, edited by Werner Faulstich and Jörn Glasenapp, Fink, 2002, pp. 23–56.

Firestone, Shulamith. "From *The Dialectic of Sex*." *The Philosophy of (Erotic) Love*, edited by Robert C. Solomon and Kathleen M. Higgins, UP of Kansas, 1991, pp. 247–56.

Fisher, Helen. *Why We Love: The Nature and Chemistry of Romantic Love*. Henry Holt, 2004.

Fordyce, Ehren. "The Voice of Kane." *Sarah Kane in Context*, edited by Laurens De Vos and Graham Saunders, Manchester UP, 2010, pp. 103–14.

Fragoku, Marissia. *Ecologies of Precarity in Twenty-First Century Theatre: Politics, Affect, Responsibility*. Bloomsbury, 2018.

Frankfurt, Harry G. *The Reasons of Love*. Princeton UP, 2004.

Freeland, Cynthia A. "Woman: Revealed or Reveiled?" *Hypatia*, vol. 1, no. 2, 1986, pp. 49–70.

Freud, Sigmund. "Concerning the Most Universal Debasement in the Erotic Life." *Sigmund Freud: The Psychology of Love*, edited by Jeri Johnson and translated by Shaun Whiteside, Penguin, 2006, pp. 250–61.

Freud, Sigmund. "Three Essays on Sexual Theory." *Sigmund Freud: The Psychology of Love*, edited by Jeri Johnson and translated by Shaun Whiteside, Penguin, 2006, pp. 111–237.

Freud, Sigmund. *Civilization and its Discontents*, edited and translated by James Strachey, Norton, 1962.

Fromm, Erich. *The Art of Loving*. 1956. Harper Perennial Modern Classics, 2006.

Gardner, Lyn. Review of *Eigengrau*, by Penelope Skinner. *Guardian*, 19 Mar. 2010. *Theatre Record*, vol. 30, no. 6, p. 286.

Gardner, Lyn. Review of *Love and Money*, by Dennis Kelly. *Guardian*, 6 Nov. 2006, www.theguardian.com/stage/2006/nov/06/theatre.

Geisenhanslüke, Achim. *Die Sprache der Liebe: Figurationen der Übertragung von Platon zu Lacan*. Wilhelm Fink, 2016.

Giddens, Anthony. *The Transformation of Intimacy*. Polity, 1992.

Gilliver, Peter. "precarious." *OED*, www.oed.com/public/precarious/loginpage.

Gore-Langton, Robert. Review of *Cleansed*, by Sarah Kane. *Express*, 10 May 1998. *Theatre Record*, vol. 18, no. 9, p. 563.

Grassi, Samuele. *Looking Through Gender: Post-1980 British and Irish Drama*. Cambridge Scholars, 2011.

Greer, Stephen. *Contemporary British Queer Performance*. Palgrave Macmillan, 2012.

Greig, David. Introduction. *Complete Plays: Blasted, Phaedra's Love, Cleansed, Crave, 4.48 Psychosis, Skin*, by Sarah Kane. Methuen, 2001, pp. ix–xviii.

Grieve, James. Introduction. *Love, Love, Love*, by Mike Bartlett. Methuen, 2015, pp. 3–20.

Gross, John. Review of *Cleansed*, by Sarah Kane. *Sunday Telegraph*, 18 May 1998. *Theatre Record*, vol. 18, no. 9, p. 567.

Hagerty, Bill. Review of *Closer*, by Patrick Marber. *News of the World*, 8 June 1997. *Theatre Record*, vol. 17, no. 11, p. 675.

Hähnel, Martin, Annika Schlitte, and René Torkler. Einleitung. *Was ist Liebe? Philosophische Texte von der Antike bis zur Gegenwart*, edited by Martin Hähnel et al., Reclam, 2015, pp. 9–35.

Harding, Jennifer, and E. Deidre Pribram, eds. Introduction. *Emotions: A Cultural Studies Reader*. Routledge, 2009, pp. 1–24.

Hatfield, Elaine, and Richard L. Rapson. "Passionate Love: New Directions in Research." *Advances in Personal Relationships*. Vol. 1, edited by W. H. Jones and D. Perlman, JAI, 1987, pp. 109–39.

Haydon, Andrew. "Theatre in the 2000s." *Modern British Playwriting: 2000–2009. Voices, Documents, New Interpretations*, edited by Dan Rebellato, Methuen, 2013, pp. 40–98.

Hegel, G. W. F. "A Fragment on Love." *The Philosophy of (Erotic) Love*, edited by Robert C. Solomon and Kathleen M. Higgins, UP of Kansas, 1991, pp. 117–20.

Hemming, Sarah. Review of *Love, Love, Love*, by Mike Bartlett. *Financial Times*, 4 May 2012. *Theatre Record*, vol. 32, no. 9, p. 464.

Hemming, Sarah. Review of *Phaedra's Love*, by Sarah Kane. *Financial Times*, 23 May 1996. *Theatre Record*, vol. 16, no. 11, p. 653.

Hendrick, Susan S., and Clyde Hendrick. *Romantic Love*. Sage, 1992.

Hitchings, Henry. Review of *Eigengrau*, by Penelope Skinner. *Evening Standard*, 16 Mar. 2010. *Theatre Record*, vol. 30, no. 6, p. 285.

Hitchings, Henry. Review of *Love, Love, Love*, by Mike Bartlett. *Evening Standard*, 4 May 2012. *Theatre Record*, vol. 32, no. 9, p. 463.

Holdsworth, Nadine, and Mary Luckhurst, eds. *A Concise Companion to Contemporary British and Irish Drama*. 2008. Wiley-Blackwell, 2013.

Hondrich, Karl Otto. *Liebe in den Zeiten der Weltgesellschaft*. Suhrkamp, 2004.

Howe Kritzer, Amelia. *Political Theatre in Post-Thatcher Britain: New Writing 1995–2005*. Palgrave Macmillan, 2008.

Illouz, Eva. "The Lost Innocence of Love: Romance as a Postmodern Condition." *Theory, Culture & Society*, vol. 15, no. 3, 1998, pp. 161–86.

Illouz, Eva. *Consuming the Romantic Utopia: Love and the Cultural Contradictions of Capitalism*. U of California P, 1997.

Illouz, Eva. "Einleitung: Gefühle als Waren." *Wa(h)re Gefühle: Authentizität im Konsumkapitalismus*, edited by Eva Illouz and translated by Michael Adrian, Suhrkamp, 2018, pp. 13–48.
Illouz, Eva. *Cold Intimacies: The Making of Emotional Capitalism*. Polity, 2007.
Illouz, Eva. *Hard-Core Romance: Fifty Shades of Grey, Best-Sellers, and Society*. U of Chicago P, 2014.
Illouz, Eva. *Why Love Hurts: A Sociological Explanation*. Polity, 2012.
Innes, Christopher. *Modern British Drama: The Twentieth Century*. Cambridge UP, 2002.
Jaggar, Alison M. "Love and Knowledge: Emotion in Feminist Epistemology." *Emotions: A Cultural Studies Reader*, edited by Jennifer Harding and E. Deidre Pribram, Routledge, 2009, pp. 50–68.
Jankowiak, William R., and Edward F. Fischer. "A Cross-Cultural Perspective on Romantic Love." *Ethnology*, vol. 31, no. 2, 1992, pp. 149–55.
Johnson, Samuel. "precarious." *A Dictionary of the English Language*. W. Strahan, 1755. Reprint G. Olms, 1968.
Johnston, Adrian, "Jacques Lacan." *The Stanford Encyclopedia of Philosophy* (Winter 2016 Edition), edited by Edward N. Zalta, 2 Apr. 2013, www.plato.stanford.edu/entries/lacan/.
Jung, Carl. "Marriage as Psychological Relationship." *The Philosophy of (Erotic) Love*, edited by Robert C. Solomon and Kathleen M. Higgins, UP of Kansas, 1991, pp. 177–89.
Kane, Sarah. *Cleansed. Complete Plays*. Methuen, 2001, pp. 105–51.
Kane, Sarah. Interview with Nils Tabert, 8 Feb. 1998. *Playspotting: Die Londoner Theaterszene der 90er*, edited by Nils Tabert, Rowohlt, 2001, pp. 8–21.
Kane, Sarah. *Phaedra's Love. Complete Plays*. Methuen, 2001, pp. 63–103.
Kaufmann, Jean-Claude. *The Curious History of Love*. 2009. Translated by David Macey, Polity, 2011.
Keats, John. "To Fanny Brawne." 13 Oct. 1819. Letter 160 of *The Letters (1819–1820)*. Vol. 8 of *The Poetical Works and Other Writings of John Keats*, edited by H. Buxton Forman, Phaeton, 1970, pp. 121–22.
Kelly, Dennis. *Love and Money. Plays One*. Oberon, 2008, pp. 205–87.
Kristeva, Julia. *Tales of Love*. Translated by Leon S. Roudiez, Columbia UP, 1987.
Kuhn, Helmut. *Liebe: Geschichte eines Begriffs*. Kösel, 1975.
Kuti, Elizabeth. "Tragic Plots from Bootle to Baghdad." *Contemporary Theatre Review*, vol. 18, no. 4, 2008, pp. 457–69.
Lacan, Jacques. "On Freud's 'Trieb' and the Psychoanalyst's Desire." *Écrits*. Translated by Bruce Fink, Norton, 2005, pp. 722–25.
Lacan, Jacques. "The Subversion of the Subject and the Dialectic of Desire in the Freudian Unconscious." *Écrits*. Translated by Bruce Fink, Norton, 2005, pp. 671–702.
Lacan, Jacques. *The Seminar of Jacques Lacan Book VIII: Transference*, edited by Jacques-Alain Miller and translated by Bruce Fink, Polity, 2015.
Lacan, Jacques. *The Seminar of Jacques Lacan Book XI: The Four Fundamental Concepts of Psychoanalysis*, edited by Jacques-Alain Miller and translated by Alan Sheridan, Norton, 1998.
Lehmann, Hans-Thies. "Word and Stage in Postdramatic Theatre." *Drama and/after Postmodernism*, edited by Martin Middeke and Christoph Henke, CDE 14, WVT, 2007, pp. 37–54.

Letts, Quentin. Review of *Eigengrau*, by Penelope Skinner. *Daily Mail*, 16 Mar. 2010. *Theatre Record*, vol. 30, no. 6, p. 285.
Letts, Quentin. Review of *Love, Love, Love*, by Mike Bartlett. *Daily Mail*, 4 May 2012. *Theatre Record*, vol. 32, no. 9, p. 463.
Levinas, Emmanuel. "Time and the Other." *The Levinas Reader*, edited by Séan Hand and translated by Richard A. Cohen, Basil Blackwell, 1989, pp. 37–58.
Lewis, C. S. *The Four Loves*. 1960. Harper Collins, 2002.
Leys, Ruth. "The Turn to Affect: A Critique." *Critical Inquiry*, vol. 37, no. 3, 2011, pp. 434–72.
Liebermann, Matthew D. *Social: Why Our Brains Are Wired to Connect*, Crown, 2013.
"love." *Collinsdictionary.com*, n.d., Collins, www.collinsdictionary.com/dictionary/english/love.
"love, n." *Ldoceonline.com*, n.d., Longman Dictionary of Contemporary English, www.ldoceonline.com/dictionary/love.
"love, n." *Merriam-Webster.com*, n.d., Merriam-Webster, www.merriam-webster.com/dictionary/love.
"love, n.1." *OED*, Oxford UP, June 2018, www.oed.com/view/Entry/110566?rskey=9Ofqhw&result=1#eid38905025.
Lublin, Robert I. "'I love you now': Time and Desire in the Plays of Sarah Kane." *Sarah Kane in Context*, edited by Laurens De Vos and Graham Saunders, Manchester UP, 2010, pp. 115–25.
Luhmann, Niklas. *Liebe als Passion: Zur Codierung von Intimität*. Suhrkamp, 1982.
Luhmann, Niklas. *Liebe: Eine Übung*. 1969. Edited by André Kieserling. Suhrkamp, 2014.
Lukowski, Andrzej. Review of *Love, Love, Love*, by Mike Bartlett. *Time Out London*, 25 Feb. 2020, www.timeout.com/london/theatre/love-love-love-review-cancelled.
Macaulay, Alastair. Review of *Cleansed*, by Sarah Kane. *Financial Times*, 8 May 1998. *Theatre Record*, vol. 18, no. 9, p. 567.
Macaulay, Alastair. Review of *Closer*, by Patrick Marber. *Financial Times*, 31 May 1997. *Theatre Record*, vol. 17, no. 11, p. 675.
Mahler, Margaret. "The Psychological Birth of the Human Infant." *Literary Theory: An Anthology*, edited by Julie Rivkin and Michael Ryan, Blackwell, 1998, pp. 215–22.
Malinowska, Anna, and Michael Gratzke, eds. "Introduction: Love Matters." *The Materiality of Love: Essays on Affection and Cultural Practice*. Routledge, 2018, pp. 1–9.
Marber, Patrick. *Closer*. Methuen Modern Plays. Methuen, 1997.
Marlowe, Sam. Review of *Cleansed*, by Sarah Kane. *What's On*, 13 May 1998. *Theatre Record*, vol. 18, no. 9, pp. 566–67.
Marlowe, Sam. Review of *Eigengrau*, by Penelope Skinner. *Time Out London*, 25 Mar. 2010. *Theatre Record*, vol. 30, no. 6, p. 287.
May, Simon. *Love: A History*. Yale UP, 2011.
McGinn, Caroline. Review of *Love, Love, Love*, by Mike Bartlett. *Time Out London*, 10 May 2012. *Theatre Record*, vol. 32, no. 9, p. 465.
McMahon, Darrin M. *Happiness: A History*. Grove, 2006.
Middeke, Martin, Peter Paul Schnierer, and Aleks Sierz. Introduction. *The Methuen Drama Guide to Contemporary British Playwrights*, Methuen, 2011, pp. vii–xxiv.
Mitchell, W. J. T. "Representation." *Critical Terms for Literary Studies*, edited by Frank Lentricchia and Thomas McLaughlin. 2nd ed., U of Chicago P, 1995, pp. 11–22.
"Modern Love." *New York Times*, www.nytimes.com/column/modern-love.

Morley, Sheridan. Review of *Cleansed*, by Sarah Kane. *Spectator*, 16 May 1998. *Theatre Record*, vol. 18, no. 9, p. 568.
Mountford, Fiona. Review of *a profoundly affectionate, passionate devotion to someone (-noun)*, by debbie tucker green. *Standard*, 7 Mar. 2017, www.standard.co.uk/go/london/theatre/a-profoundly-affectionate-passionate-devotion-to-someone-noun-theatre-review-enervating-and-a3680206.html.
Müller, Klaus Peter. "British Theatre in the 1980s and 1990s: Forms of Hope and Despair, Violence and Love." *What Revels Are in Hand? Assessments of Contemporary Drama in English in Honour of Wolfgang Lippke*, edited by Bernhard Reitz and Heiko Stahl, CDE Studies 8, WVT, 2001, pp. 81–107.
Munteanu, Dana LaCourse. "Empathy and Love: Types of Textuality and Degrees of Affectivity." *The Palgrave Handbook of Affects Studies and Textual Criticism*, edited by Donald R. Wehrs and Thomas Blake, Palgrave MacMillan, 2017, pp. 325–43.
Nancy, Jean-Luc. "Shattered Love." 1986. *The Inoperative Community*, edited by Peter Connor and translated by Lisa Garbus and Simona Sawhney, U of Minnesota P, 1991, pp. 82–109.
Nancy, Jean-Luc. *Die Liebe, übermorgen*. Translated by Ignaz Knips and Joan Catharine-Ritter, Salon, 2010.
Neu, Jerome. "Plato's Homoerotic *Symposium*." *The Philosophy of (Erotic) Love*, edited by Robert C. Solomon and Kathleen M. Higgins, UP of Kansas, 1991 pp. 317–35.
Nietzsche, Friedrich. *Die Geburt der Tragödie. Unzeitgemäße Betrachtungen I–IV. Nachgelassene Schriften 1870–1873*. Vol. 1 of *Sämtliche Werke: Kritische Studienausgabe*, 15 Vols., edited by Giorgio Colli and Mazzino Montinari, dtv, 1999.
Nietzsche, Friedrich. "Selections." *The Philosophy of (Erotic) Love*, edited by Robert C. Solomon and Kathleen M. Higgins, UP of Kansas, 1991, pp. 140–150.
Nightingale, Benedict. Review of *Cleansed*, by Sarah Kane. *Times*, 8 May 1998. *Theatre Record*, vol. 18, no. 9, p. 565.
Nightingale, Benedict. Review of *Eigengrau*, by Penelope Skinner. *Times*, 17 Mar. 2010. *Theatre Record*, vol. 30, no. 6, p. 285.
Nikcevic, Sanja. "British Brutalism, the 'New European Drama,' and the Role of the Director." *NTQ*, vol. 21, no. 3, 2005, pp. 255–72.
Norman, Neil. Review of *Love, Love, Love*, by Mike Bartlett. *Daily Express*, 11 May 2012. *Theatre Record*, vol. 32, no. 9, p. 465.
Nussbaum, Martha C. "The Speech of Alcibiades: A Reading of Plato's *Symposium*." *The Philosophy of (Erotic) Love*, edited by Robert C. Solomon and Kathleen M. Higgins, UP of Kansas, 1991, pp. 279–316.
Nussbaum, Martha C. *Upheavals of Thought: The Intelligence of Emotions*. Cambridge UP, 2001.
Ortega y Gasset, José. *On Love: Aspects of a Single Theme*. 1939. Translated by Tony Talbot, Meridian, 1957.
Pascal, Blaise. *The Thoughts of Blaise Pascal*. Greenwood, 1978. Trans. of *Pensées*, 1670.
Pattie, David. "Theatre Since 1968." *A Companion to Modern British and Irish Drama: 1880–2005*, edited by Mary Luckhurst, Blackwell, 2006, pp. 385–97.
Peter, John. Review of *Cleansed*, by Sarah Kane. *Sunday Times*, 10 May 1998. *Theatre Record*, vol. 18, no. 9, p. 564.

Peter, John. Review of *Closer*, by Patrick Marber. *Sunday Times*, 8 June 1997. *Theatre Record*, vol. 17, no. 11, p. 679.

Pettman, Dominic. "Love Materialism: Technologies of Feeling in the 'Post-Material' World." *The Materiality of Love: Essays on Affection and Cultural Practice*, edited by Anna Malinowska and Michael Gratzke, Routledge, 2018, pp. 13–24.

Pettman, Dominic. *Love and Other Technologies: Retrofitting Eros for the Information Age*. Fordham UP, 2006.

Pewny, Kaharina. *Das Drama des Prekären: Über die Wiederkehr der Ethik in Theater und Performance*. Transcript, 2011.

Pfister, Manfred. *The Theory and Analysis of Drama*. Translated by John Halliday, Cambridge UP, 2000.

Pieper, Josef. *Über die Liebe*. Kösel, 1972.

"precarious, adj." *OED*, Oxford UP, 2017, www.oed.com/view/Entry/149548.

Quay, Christine. *Mythopoiesis vor dem Ende? Formen des Mythischen im zeitgenössischen britischen und irischen Drama*. CDE Studies 16, WVT, 2007.

Rabey, Ian David. *English Drama Since 1940*. Longman, 2003.

Rathmayr, Bernhard. *Geschichte der Liebe: Wandlungen der Geschlechterbeziehungen in der abendländischen Kultur*. Wilhelm Fink, 2016.

Ravenhill, Mark. "A Tear in the Fabric." *New Theatre Quarterly*, vol. 20, no. 4, 2004, pp. 305–14.

Ravenhill, Mark. Interview with Nils Tabert, 17 Jan. 1998. *Playspotting: Die Londoner Theaterszene der 90er*, edited by Nils Tabert, Rowohlt, 2001, pp. 66–78.

Ravenhill, Mark. *Shopping and Fucking. Plays: 1*. Methuen, 2001, pp. 1–91.

Rebellato, Dan. "Love and Information." *danrebellato.co.uk*. 11 Sept. 2012, www.danrebellato.co.uk/spilledink/2013/3/11/love-and-information.

Rebellato, Dan. "New Theatre Writing: Dennis Kelly." *Contemporary Theatre Review*, vol. 17, no. 4, 2008, pp. 603–08.

Rebellato, Dan. "Sarah Kane: An Appreciation." *New Theatre Quarterly*, vol. 15, no. 3, 1999, pp. 280–81.

Rebellato, Dan. Introduction. *Plays 1: Shopping and Fucking, Faust is Dead, Handbag, Some Explicit Polaroids*, by Mark Ravenhill, Bloomsbury, 2013, pp. ix–xx.

Rees, Catherine. "Sarah Kane." *Modern British Playwriting: The 1990s: Voices, Documents, New Interpretations*, edited by Aleks Sierz, Methuen, 2012, pp. 112–37.

Rivkin, Julie, and Michael Ryan. "Strangers to Ourselves: Psychoanalysis." *Literary Theory: An Anthology*, edited by Julie Rivkin and Michael Ryan, Blackwell, 1998, pp. 119–27.

Rubik, Margarete. "Saying the Unspeakable: Realism and Metaphor in the Depiction of Torture in Modern Drama." *What Revels are in Hand? Assessments of Contemporary Drama in English in Honour of Wolfgang Lippke*, edited by Bernhard Reitz and Heiko Stahl, CDE Studies 8, WVT, 2001, pp. 121–38.

Sartre, Jean-Paul. "From *Being and Nothingness*." *The Philosophy of (Erotic) Love*, edited by Robert C. Solomon and Kathleen M. Higgins, UP of Kansas, 1991, pp. 227–32.

Saunders, Graham. *'Love Me or Kill Me': Sarah Kane and the Theatre of Extremes*. Manchester UP, 2002.

Saunders, Graham. "Mark Ravenhill." *Modern British Playwriting: The 1990s. Voices, Documents, New Interpretations*, edited by Aleks Sierz, Methuen, 2012, pp. 163–88.

Saunders, Graham. "The Persistence of the 'Well-Made-Play' in British Theatre of the 1990s." *Non-Standard Forms of Contemporary Drama and Theatre*, edited by Ellen Redling and Peter Paul Schnierer, CDE 15, WVT, 2008, pp. 227–39.
Saunders, Graham. *About Kane: The Playwright and the Work*. Faber, 2009.
Saunders, Graham. *Patrick Marber's* Closer: *Modern Theatre Guides*. Continuum, 2008.
Scheer, Edward, ed. *Antonin Artaud: A Critical Reader*. Routledge, 2004.
Schopenhauer, Arthur. "Metaphysik der Geschlechtsliebe." *Die Welt als Wille und Vorstellung*. Zweiter Band. Zweiter Teilband. Vol. 4 of *Züricher Ausgabe: Werke in zehn Bänden*, Diogenes, 1977, pp. 621–56.
Secomb, Linnell. *Philosophy and Love: From Plato to Popular Culture*. Edinburgh UP, 2007.
Sharma, Devika, and Frederik Tygstrup, eds. Introduction. *Structures of Feeling: Affectivity and the Study of Culture*. De Gruyter, 2015, pp. 1–19.
Shore, Robert. Review of *Eigengrau*, by Penelope Skinner. *Metro*, 18 Mar. 2010. *Theatre Record*, vol. 30, no. 6, p. 286.
Sierz, Aleks. *In-Yer-Face Theatre: British Drama Today*. Faber and Faber, 2001.
Sierz, Aleks. *Modern British Playwriting: The 1990s. Voices, Documents, New Interpretations*. Methuen Drama, 2012.
Sierz, Aleks. Review of *a profoundly affectionate, passionate devotion to someone (-noun)*, by debbie tucker green. *Sierz.co*, 6 Mar. 2017, www.sierz.co.uk/reviews/a-profoundly-affectionate-passionate-devotion-to-someone-noun-royal-court/.
Sierz, Aleks. Review of *Cleansed*, by Sarah Kane. *Tribune*, 15 May 1998. *Theatre Record*, vol. 18, no. 9, p. 568.
Sierz, Aleks. Review of *Phaedra's Love*, by Sarah Kane. *Tribune*, 31 May 1996. *Theatre Record*, vol. 16, no. 11, p. 651.
Sierz, Aleks. *Rewriting the Nation: British Theatre Today*. Methuen Drama, 2011.
Singer, Irving. *Philosophy of Love. A Partial Summing-Up*. MIT Press, 2011.
Singer, Irving. *The Nature of Love 1: Plato to Luther*. 1966. 2nd ed. U of Chicago P, 1984.
Singer, Irving. *The Nature of Love 2: Courtly and Romantic*. 1984. MIT Press, 2009.
Singer, Irving. *The Nature of Love 3: The Modern World*. U of Chicago P, 1987.
Singer, Irving. *The Pursuit of Love*. Johns Hopkins UP, 1995.
Skinner, Penelope. *Eigengrau*. Faber and Faber, 2010.
Solomon, Robert C. "The Virtue of (Erotic) Love." *The Philosophy of (Erotic) Love*, edited by Robert C. Solomon and Kathleen M. Higgins, UP of Kansas, 1991, pp. 492–518.
Soloski, Alexis. Review of *Love, Love, Love*, by Mike Bartlett. *Guardian*, 20 Oct. 2016, www.theguardian.com/stage/2016/oct/19/love-love-love-review-richard-armitage-baby-boomers.
Spaemann, Robert. "Antinomien der Liebe." *Was ist Liebe? Philosophische Texte von der Antike bis zur Gegenwart*, edited by Martin Hähnel, Annika Schlitte, and René Torkler, Reclam, 2015, pp. 102–16.
Spencer, Charles. Review of *Cleansed*, by Sarah Kane. *Daily Telegraph*, 7 May 1998. *Theatre Record*, vol. 18, no. 9, p. 565.
Spencer, Charles. Review of *Eigengrau*, by Penelope Skinner. *Daily Telegraph*, 17 Mar. 2010. *Theatre Record*, vol. 30, no. 6, p. 286.
Spencer, Charles. Review of *Love and Money*, by Dennis Kelly. *Telegraph*, 23 Nov. 2006, www.telegraph.co.uk/culture/theatre/drama/3656708/Vivid-portrait-of-our-times.html.

Spencer, Charles. Review of *Love, Love, Love*, by Mike Bartlett. *Daily Telegraph*, 7 May 2012. *Theatre Record*, vol. 32, no. 9, p. 464.

Stöckl, Korbinian. "'Experiential, not speculative': Love In and After In-Yer-Face." *After In-Yer-Face: Remnants of a Theatrical Revolution*, edited by William C. Boles, Palgrave Macmillan, 2020, pp. 217–30.

Surya, Michel. *Georges Bataille: An Intellectual Biography*. 1992. Translated by Krzysztof Fjalkowski and Michael Richardson, Verso, 2002.

Svich, Caridad. "Mark Ravenhill." *The Methuen Drama Guide to Contemporary British Playwrights*, edited by Martin Middeke, Peter Paul Schnierer, and Aleks Sierz, Methuen, 2012, pp. 403–24.

Taylor, Paul. Review of *Love and Money*, by Dennis Kelly. *Independent*, 23 Nov. 2006, www.independent.co.uk/arts-entertainment/theatre-dance/reviews/love-and-money-young-vic-london-425463.html.

Teachout, Terry. Review of *Love, Love, Love*, by Mike Bartlett. *Wallstreet Journal*, 20 Oct. 2016, www.wsj.com/articles/love-love-love-review-not-since-stoppard-1476997177.

Tönnies, Merle. "Das zeitgenössische englische Drama: Kategorien und Schreibweisen." *Das englische Drama der Gegenwart: Kategorien – Entwicklungen – Modellinterpretationen*, WVT, 2010, pp. 1–12.

Tripney, Natasha. Review of *a profoundly affectionate, passionate devotion to someone (-noun)*, by debbie tucker green. *The Stage*, 7 Mar. 2017, www.thestage.co.uk/reviews/a-profoundly-affectionate-passionate-devotion-to-someone–noun-review-at-royal-court-london.

tucker green, debbie. *a profoundly affectionate, passionate devotion to someone (-noun)*. Nick Hern, 2017.

Urban, Ken. "An Ethics of Catastrophe: The Theatre of Sarah Kane." *PAJ*, vol. 23, 2001, pp. 36–46.

Vesey-Byrne, Joe. Review of *a profoundly affectionate, passionate devotion to someone (-noun)*, by debbie tucker green. *Independent*, 7 Mar. 2017, www.independent.co.uk/arts-entertainment/theatre-dance/debbie-tucker-green-a-profoundly-affectionate-passionate-devotion-to-someone-noun-review-royal-court-a7615776.html.

Vire, Kris. Review of *Love and Money*, by Dennis Kelly. *Time Out*, 24 Jan. 2012, www.timeout.com/chicago/theater/love-and-money-at-steep-theatre-theater-review.

Vlastos, Gregory, ed. "The Individual as an Object of Love in Plato." *Platonic Studies*. Princeton UP, 1973, pp. 3–42.

Wade, Leslie A. "Postmodern Violence and Human Solidarity: Sex and Forks in *Shopping and Fucking*." *Theatre Symposium*, vol. 7, 1999, pp. 109–15.

Wald, Christina. *Hysteria, Trauma and Melancholia: Performative Maladies in Contemporary Anglophone Drama*. Palgrave Macmillan, 2007.

Wallace, Clare. "Sarah Kane, Experiential Theatre and the Revenant Avant-garde." *Sarah Kane in Context*, edited by Laurens De Vos and Graham Saunders, Manchester UP, 2010, pp. 88–99.

Wallace, Clare. *Suspect Cultures: Narrative, Identity and Citation in 1990s New Drama*. Litteraria Pragensia, 2006.

Wandor, Michelene. *Post-War British Drama: Looking Back in Gender*. Routledge, 2001.

Waugh, Rosemary. Review of *Love, Love, Love*, by Mike Bartlett. *The Stage*, 12 Mar. 2020, www.thestage.co.uk/reviews/love-love-love-review-at-the-lyric-hammersmith-london–resonant-and-perceptive-revival.

Wehrs, Donald R. "Introduction: Affect and Texts: Contemporary Inquiry in Historical Context." *The Palgrave Handbook of Affect Studies and Textual Criticism*, edited by Donald R. Wehrs and Thomas Blake, Palgrave Macmillan, 2017, pp. 1–93.

Wetherell, Margaret. *Affect and Emotion: A New Social Science Understanding*. Sage, 2012.

White, Hayden. "The Historical Fact as Literary Artifact." *The Norton Anthology of Theory and Criticism*, edited by Peter Simon, Norton, 2001, pp. 1712–29.

Wiechens, Peter. *Bataille zur Einführung*. Junius, 1995.

Williams, Raymond. "Structures of Feeling." *Structures of Feeling: Affectivity and the Study of Culture*, edited by Devika Sharma and Frederik Tygstrup, De Gruyter, 2015, pp. 20–25. Excerpt from "Structures of Feeling," *Marxism and Literature*, Oxford UP, 1977, pp. 128–35.

Zimmermann, Heiner. "Theatrical Transgression in Totalitarian and Democratic Societies: Shakespeare as a Trojan Horse and the Scandal of Sarah Kane." *Crossing Borders: Intercultural Drama and Theatre at the Turn of the Millennium*, edited by Bernhard Reitz and Alyce von Rothkirch, CDE 8, WVT, 2001, pp. 173–82.

Index of subjects

absorption   59 f., 118, 120, 142, 178, 238, 247
adventure   43, 59, 75, 79, 112, 132, 231, 234 – 238
advertising   71 – 73, 75, 89, 199, 214 f.
affair   9, 67, 76, 96 – 98, 106 f., 109, 117, 209 f., 216, 223, 234, 237
affect   4 – 7, 74, 149, 195
– affective community   195 f., 200
– affective turn   2
agape   3, 18, 21, 80, 236
agency   8, 41, 269
alienation   13, 35 f., 38, 73, 141, 170, 178 f., 183, 189, 195, 201
alterity   59 – 61, 120 f., 123, 246 f., 263
altruism   3, 19, 133, 212, 219, 260 f., 263 f.
ambivalence   2, 30, 51, 63, 84, 123, 167, 170, 177, 208
apathy   154 – 157, 159, 162, 180, 223, 238, 242, 270
appraisal theory   4 f., 63
appropriation   33, 47, 56, 62, 118 f., 123, 139 f., 170, 271
ascent   27, 30 – 32, 48, 50, 63
atmosphere   47, 85, 90, 95, 116, 124, 131, 164, 168, 227, 230, 232, 241, 250, 260, 268
attraction   2 f., 27, 32, 50, 77, 82 f., 96, 99, 101, 104, 172, 204 f., 217 f., 228, 234, 237
authenticity   73 f., 76, 89, 113 – 116, 123, 132, 161, 201, 206, 210, 217, 251, 254 – 256, 271
autonomy   13, 34, 37, 61 f., 64, 80, 82 – 84, 87 f., 140, 181, 207, 227, 237, 240, 242

belonging   88, 135 f., 195, 197 f., 269
biology   1 f., 5 f., 11, 13, 18, 84, 151, 272 f.
body   2, 4 f., 11 f., 23, 27, 29 f., 32, 44, 46, 50, 55, 60, 87, 109, 150 – 153, 161 – 166, 173 f., 176, 178, 180 f., 217, 234, 247

capitalism   13, 17, 66 – 68, 72 – 74, 86, 88 f., 95, 124, 126 f., 129, 132, 134, 140 f., 184, 189, 191, 197 – 199, 206, 227, 240
choice of partner   19, 57, 65, 80 f., 83 f., 99, 101, 105, 107 f., 116, 136, 139, 204, 215, 220
– abundance of choice   13, 81, 83, 90, 101, 157, 215
citationality   74 – 76, 114 f., 141, 143
cliché   2, 74 f., 113 – 116, 184, 211
code   7, 9, 66, 71, 74, 115, 151
cognition   4 f., 11, 39, 94, 151
comedy   52, 93 f., 100, 123, 204
– comic   26, 49, 153, 245
– comical   52, 95 f., 203, 206
commitment   2, 9, 13, 19, 34, 75, 77 – 80, 82, 85, 89 f., 93, 99 – 101, 103, 106, 119, 129, 132, 156 – 159, 171 f., 176, 183, 192 f., 195, 197, 200, 205, 213, 228 f., 236, 261, 271
– commitment phobia   13, 81, 100, 106, 123, 171, 193, 271
communication   4, 9, 11, 36, 68, 71, 74, 77, 79 f., 89, 123, 144, 149 f., 166, 172, 193, 208, 215, 244 f., 248, 250 – 253, 255 – 257, 262, 265 f., 268, 270
compatibility   36, 75, 77, 79 – 83, 89, 101 f., 104, 132, 206, 210, 214, 217, 229 f., 238, 257, 259, 270
compensation   11 – 13, 18, 21 – 23, 26, 29, 31 – 34, 36 – 41, 45 f., 64 f., 67 f., 86, 88, 90, 98 f., 116 f., 123 f., 132 f., 135, 137, 140, 149, 155, 158 f., 178, 180, 182, 188, 190, 201, 203, 212 – 214, 220, 223, 247, 249 f., 260, 263 – 266, 269 – 271, 273
completeness   25 f., 34 f., 63, 139, 173 f., 181,
– completion   22, 31, 62, 178, 186
– incompleteness   24, 26, 34 f., 62, 180 – 182, 269
compromise   82, 102 – 104, 106, 183, 237, 257, 259

Index of subjects

condition 39, 47, 58f., 62, 72, 75, 114f., 121, 123, 128f., 131, 133, 157–159, 170, 180, 188, 231, 253
- condition for love 1, 21, 45, 59, 62–64, 77, 243, 250, 259
- human condition 11f., 21, 26, 46, 64, 140, 158, 220

connection 2, 95, 104f., 127–130, 183, 186, 188–192, 196f., 204, 237
constructionism 5f., 11, 272f.
consumerism 72f., 118, 124, 127, 130, 132, 134, 139f., 183f., 190, 195–200, 202, 227
- consumption 73, 130, 139
contingency 59, 68f., 78, 214, 218, 272
control 31, 40, 60, 63, 79f., 84, 87, 90, 100, 104–106, 120–123, 132, 136, 175f., 214–216, 218, 221, 223, 246–250, 254, 265, 270–272
- uncontrollability 11f., 21, 30f., 47f., 52f., 59–61, 63f., 69, 78, 90, 104–108, 110f., 132, 134f., 155–157, 164, 168, 171f., 175, 203, 214f., 220f., 243, 247–250, 261, 263, 265, 269–272
courtly love 7, 22f., 32f., 53, 86, 103f., 108, 113–115, 135, 213
- *fin' amors* 53
cruelty 44, 136f., 140, 151, 171, 191, 193
- theatre of cruelty 151, 162
cultural emotion 1–7, 11, 13, 17f., 32, 61, 65f., 68, 71–74, 80–88, 180, 208, 213, 226, 272f.

death 21, 35, 41–45, 48, 53f., 95f., 98, 106, 111, 116, 135–138, 148, 152, 156, 160–162, 164, 166–168, 172–174, 178, 180–182, 184f., 190, 212, 227, 245–248, 254, 256, 263, 265, 268, 270
- death wish 127, 137, 175, 270
- love-death/*Liebestod* 53f., 178
decision 68, 75, 81, 90, 102f., 107f., 113, 115, 119f., 129, 139, 157, 172f., 201, 205, 209, 215, 226f., 235, 267–269, 272
delusion 38, 44, 155, 157f., 218
dependence 6, 12, 20, 31, 47f., 52, 56, 64, 78, 87f., 125, 129, 131f., 134, 136, 139– 141, 156f., 175, 180, 191, 193, 201, 207, 219, 221, 242, 253 265f., 271
- co-dependence 64, 90, 138–141, 193
- independence 31, 34f., 39, 63, 78, 80–82, 87, 131, 193, 214, 240, 242, 247, 263, 271
- interdependence 12, 63, 72, 80, 90, 125
difference (alterity) 40, 59f., 117f., 136, 163, 179
disappointment 38, 58, 66, 71, 74, 78, 105, 115, 198, 223, 225f., 261f.
discourse 1f., 4, 6–13, 18, 23, 36, 47, 76, 79, 82, 87, 94f., 114, 124, 131–133, 138, 141, 143f., 149, 162, 170, 188, 201, 203, 212, 218, 220, 243, 268, 271, 273
- discourse analysis 1, 13, 144, 273
- discursive field 9, 18, 23, 46, 64, 104, 124, 130, 135, 143, 155, 203, 211, 217
disenchantment 83–85, 89
disinterestedness 20, 28, 61, 65, 89, 190, 195, 219
- disinterested concern 19f., 100, 262
dissolution 34, 37f., 43f., 54, 57, 90, 137, 142, 152, 161f., 164, 176–178, 180–182, 240, 269, 271
divorce 9, 97f., 100, 106, 222–224, 234–238, 240, 268
dramatic form 8f., 143–148, 152f., 185, 201, 203, 241, 243, 268f.
drive 3, 32, 38f., 44f., 181f.
durability 69–71, 78, 83, 95, 99, 104f., 171–173, 183, 192, 227

egoism 54, 98f., 106, 124, 132f., 139, 193, 212, 223
emotion 4–8, 10f., 17, 39, 63, 68, 72, 74, 79, 83–86, 94, 108, 132, 142, 145f., 148, 150f., 157, 168f., 174f., 208, 210, 236, 241, 254, 256, 272f.
episodic play 8, 131, 153, 165, 184, 243f., 251, 256, 261, 266, 268
equality 58, 84, 207–209, 247
- inequality 57f., 207f., 214
eros 3, 18, 21–24, 26–31, 34, 43, 48f., 51
eroticism 6, 32, 36, 43–45, 84, 162, 208

Index of subjects

erotic love   2f., 18, 22, 26–32, 40, 45f., 48, 50, 59f., 63f., 130f., 141, 165, 170, 173, 180, 182, 260, 269
essentialised self   81f., 90, 117, 123, 132, 193, 257–259
eternality   10, 27, 29, 31, 48, 54, 65, 67, 70, 78, 83, 89, 104f., 165f., 170–172, 235
ethics   11f., 18f., 51, 59–61, 102, 107f., 125, 138, 232
evolutionary origins of emotions   5f., 11, 45, 84
excess   18, 30f., 69f., 129, 162
exclusivity   3, 60, 65, 70
existentialism   21, 168
– existential   2, 20, 26, 37, 59, 65, 159, 200, 214, 220
expectation   6, 13, 71f., 74–76, 78–82, 85, 89, 104f., 118, 184, 188, 198, 200f., 210, 249, 265, 269, 273
experientiality   8, 83, 94f., 124, 142, 144, 146, 149, 151, 183
experimentalism   8, 145, 183
exposure   12, 47f., 62, 86, 122, 132, 170f., 181, 189, 200, 220f., 249, 269
expressionism   153, 164, 183
expression of love   5f., 8–10, 17, 73f., 104, 107, 113–115, 148–151, 153, 166, 171, 178, 183, 186, 243, 245, 252–257, 260–262, 268, 271f.
extreme situations   21, 48, 128f., 135, 158, 169–171, 183f., 188, 243

falling in love   17f., 74, 81f., 107, 113, 126, 132, 157, 159, 175, 206, 209f., 216f., 222
family   66f., 69, 71, 88, 98f., 102, 107, 128, 155, 162, 185, 187, 199, 213, 222–224, 227, 230f., 233–235, 237–240, 242, 259, 269
figure characterisation   95, 143, 146–148, 152f., 183, 201–203, 205, 213, 216
figure conception   129, 133, 144–148, 202, 204, 211, 213, 220, 227, 241
– stock character   100, 106
– type-figure   100, 123, 146, 183, 202–205, 213, 216, 227, 241

freedom   13, 38, 55–58, 66, 69, 77–82, 102, 107, 118, 120, 136, 138–140, 161, 222f., 225–240, 242, 246f.
– liberation   13, 29, 32, 45, 66, 69, 73, 81, 86, 88f., 99, 108, 227f., 269
frustration   49, 52f., 85, 89, 101, 118, 121, 184, 202f., 223, 226, 234, 240, 249, 261, 263
fusion   25, 33, 36, 44–46, 54, 57, 59f., 62, 83f., 170, 176–178, 180, 182

gender   10, 17, 24, 58, 84, 207–209, 214, 246
gift-love   20, 195, 219, 260f.
good, the (idea of)   20, 27–31, 46, 48, 63, 139
grand narrative/*grand récit* see master narratives
gratification   3, 20f., 49–51, 78, 132, 216, 262
groundedness   21, 46, 184, 197, 214, 268, 272
guidance   10, 127f., 240, 242, 270

happiness   37–39, 47, 53, 58f., 66, 72, 87, 99–101, 106, 108, 115, 123, 132, 134, 158, 180, 184, 191, 193, 195–200, 222f., 227, 232, 234–238, 242, 249, 260, 264
heart   19, 35, 62, 65, 106, 135, 147f., 154, 186, 198f., 228
hope   13, 21, 46, 64, 72, 90, 103, 120f., 125, 135, 142f., 159, 170f., 198, 202, 212, 214, 220f., 226, 251, 253, 259, 262, 268–270, 273

identity   9, 17, 53f., 57, 64, 67, 77, 82, 86, 120–122, 127, 131f., 136, 159, 169f., 176, 178f., 181, 206, 208, 236
illusion   20, 38, 53, 75, 83, 87, 90, 155, 158, 201, 213, 216, 220
image repertoire   4, 6–8, 84
incredulity towards love   72, 89, 103–116, 123, 170f., 271
indeterminacy   29, 118–123
individualism   17, 22, 54, 71, 76, 95, 222, 227, 240–242, 270
– individualisation   66, 77, 81, 90, 99, 129

individuality   6, 17, 22, 29f., 34f., 39, 43, 47, 54, 57, 64, 70, 77, 79, 117, 164, 177–181, 193, 203, 211, 227, 241, 246f.
individuation   35, 38f., 44, 64, 137
infatuation   3, 20, 32, 75, 80, 105, 107, 166, 206, 216–218, 221, 228, 230
infinity   65
insufficiency   20–22, 29, 31f., 34, 40, 46, 64, 90, 123, 133, 135, 140f., 193, 221, 263–266, 269
insufficiency of language   253–257, 265
integrity   54, 56, 58f., 87, 118, 160, 177, 180f.
intentionality   4f.
intertextuality   9, 178, 213, 217
intimacy   2, 6f., 10, 60, 66f., 77, 102, 121f., 125, 140, 192
– intimate relationship   10, 42, 69f., 79, 84, 89f., 106, 116f., 122f., 132, 190, 193, 271
introspection   8, 80f., 89, 131, 257, 259
in-yer-face theatre   8f., 94f., 124, 142, 145f., 183, 201
ironic attitude to love   75f., 84f., 89f., 103f., 112f., 186
irrationality   4, 30, 65f., 68, 75, 79f., 83, 89, 108, 156, 158f., 230, 235, 237
isolation   35, 38, 43, 98f., 116, 123f., 128, 130, 137, 141, 161–164, 178, 181, 189–191, 193, 195, 213f., 220, 268f.

jealousy   31, 48, 63, 126, 244f., 247, 264

lack   10, 21f., 29, 32, 41, 46, 51, 90, 98, 116, 122, 127, 133, 159, 163f., 180f., 219, 247–250, 260
ladder of love   27–30, 48
language   2, 11, 25, 40f., 71, 76, 84, 113–115, 131, 163, 244, 251, 254–257, 265–267, 270
– dramatic language   8, 94, 132f., 142–144, 146, 148–154, 171, 183, 201f., 244, 243f.
loneliness   22, 66, 79, 99, 116, 128, 183, 190, 203, 212f., 269
loss   10, 29, 37, 40–43, 46f., 99, 127, 135, 137, 153, 163f., 175f., 179, 200f., 203, 205, 212–214, 244, 246, 249f., 263–265, 270
– self-loss   22, 47, 54, 57f., 61, 64, 77, 90, 102, 118, 120, 143f., 150, 155, 168–170, 176–179, 246, 271
love at first sight   65, 75f., 83, 113, 136, 204, 207, 228
lovesickness   87, 142, 168
lust   3, 6, 20, 109, 271

market   72f., 80f., 88, 101, 127, 130, 140f., 183, 189, 191, 196, 215, 240, 268
– marketing   89, 204–206, 213–215
marriage   9, 17, 52f., 67–71, 74–78, 82, 88, 101, 184, 187f., 192–196, 200f., 222f., 227–230, 234f., 238, 240, 253, 272
master narrative   1, 26, 66, 72, 89, 128f., 131, 141, 161, 201, 270
materialism   7, 34, 189f., 192, 194, 196–200, 202, 204, 227, 268f., 271
meaning   2, 12, 19–21, 58, 65f., 68, 71f., 88, 90, 112, 124, 127f., 130, 158–162, 171, 183f., 186, 192, 197f., 201, 203, 212–214, 219f., 259f., 267f., 270, 272
– meaninglessness   68f., 81, 157–159, 197f.
media images   71–74, 76f., 89, 114, 116, 199, 202, 207, 209
medium of communication   68f., 74, 88, 151
merging   22, 25, 31, 33, 35f., 38, 46, 54, 59f., 163, 170, 172–174, 177–182
metaphor   9, 41, 119, 124, 129f., 132, 134, 143f., 147, 149, 153–155, 159, 165, 168, 177, 181–183, 201, 217f.
metaphysics   21, 33, 35f., 42, 54, 72, 84, 105, 149–151, 159, 171, 186, 197, 212f.
mimesis   7–9, 141f., 146, 148, 153
mind-body dualism   4, 161–164, 169, 173f., 178
money   68, 102, 116, 126, 128, 130f., 134, 138f., 183–187, 190–194, 196–198, 200f., 209, 223, 231–233, 237, 239
mutuality   36, 52, 56, 58f., 70, 78, 133, 136f., 139–141, 186, 247, 250, 265
myth   22, 25–28, 31, 36, 42, 98, 124, 132, 136, 145, 156f., 173f., 179, 213, 271

Index of subjects

narcissism   41, 59, 63, 225
narratives   6, 9–11, 18, 47, 63, 66 f., 73–77, 83, 89, 103 f., 107 f., 113, 120, 129, 136, 141, 184, 199, 202, 227, 271
naturalism   8, 94, 146 f., 152, 164, 183, 203, 211, 244
nature vs. nurture   4, 6
neediness   18, 20, 27, 31 f., 34, 46, 110 f., 125, 127, 132–134, 139–141, 180 f., 198, 210 f., 213, 219, 221, 242, 247, 264
need-love   20, 260 f.
need to be needed   20, 108–110, 130, 133, 219, 244, 259 f., 262–264, 270
Neoplatonism   23, 31–33, 53
new writing   8, 146, 271

objectification   47, 51, 55 f., 79, 118–123, 216
object relations theory   39 f., 63
objet petit a   41 f., 52
obsessiveness   63, 156, 204, 211, 213
obstacle   10, 47, 52, 54, 59, 70, 75, 81, 102, 166 f., 170, 173, 181, 203, 238, 272
oceanic feeling   37–39, 135
omnipotence   46, 63, 140, 248
oneness   22, 31–36, 38–40, 45, 59 f., 135, 178, 182
ontological rootedness   21, 45 f., 135 f., 201, 213, 259, 261, 272
openness   49, 58, 61–64, 117, 120–123, 162, 174, 176, 181 f., 189 f., 193, 196, 216, 220 f., 247, 249 f., 271 f.
optimism   65 f., 126, 130, 138, 140, 143 f., 170, 174, 176 f., 180
orientation   13, 65, 124, 128, 197, 213, 220, 236, 270, 272
– disorientation   124, 127 f., 201, 220
other's desire, the   47, 51, 56, 108, 110, 133, 174, 203, 210, 216, 219, 221, 260, 265, 270, 272
overvaluation   33, 58

paradox   2, 42, 44, 54 f., 60, 65, 69, 72, 77, 89 f., 114, 221, 226, 230, 240, 242, 255 f., 264
parents   67, 77, 81, 98, 102, 106, 120, 127 f., 199, 222–242, 270, 272

passion   3 f., 6, 50, 52, 66, 69–71, 75 f., 83, 89, 104–108, 132, 136, 147, 155–157, 172, 207, 214, 227–230, 234–236, 244, 261, 269, 271
– passionate love   18, 23, 32, 52, 61, 67, 69–71, 75, 89, 104 f., 107 f., 121, 136, 156, 214, 217 f., 227, 230, 251, 260, 270
pathologisation   37, 82, 87 f., 140, 157, 173, 248, 271
pederasty   48 f.
permanence   27, 31, 52, 66, 69 f., 88 f., 174, 180 f., 214, 231, 234, 257
philia   3, 18 f.
possession   18, 20, 27 f., 31, 51, 54–56, 60–63, 70, 118–121, 135–140, 186, 195, 200, 216, 221, 225, 246 f.
postdramatic   8, 11 f., 146, 267, 269
postmodernity   65, 72, 74–77, 89, 95, 113–115, 124, 128 f., 131, 141, 227
– postmodern love   2, 7, 72, 74–77, 83, 85, 89, 95, 97, 103, 106, 110–115, 122–124, 184, 199, 202, 271
power   10, 61 f., 68, 84, 90, 97, 120, 134–136, 175–176, 208
– powerlessness   175, 214, 249 f., 265, 270
– power of love   36, 59, 64, 66, 72, 85, 136, 143, 154, 158 f., 171, 174, 176 f., 184, 199, 218, 271
– power struggle   52, 54 f., 57, 97, 117 f., 120, 214, 243 f., 246–249, 251, 253, 265 f., 271
– unequal power   57, 106, 140, 170, 246
precariousness   10–13, 18, 21, 23, 31, 36, 38, 46–48, 52–55, 58–61, 63–65, 77, 80, 83, 86–88, 90, 110 f., 116–118, 123–125, 132, 134 f., 141, 143, 149, 154, 157, 164, 167–170, 175–177, 180, 183, 188–193, 198, 201, 203, 212, 214, 216, 218, 220–222, 230, 238, 244, 246–250, 253, 256, 261, 265 f., 268, 270, 272
projection   120, 127, 165 f., 168, 170, 173, 175, 250
psychoanalysis   1, 13, 22, 26, 29, 35 f., 39 f., 46, 52, 62–64, 71, 79, 84, 89, 135, 159, 170

Index of subjects — 291

psychology 1, 3, 17, 37, 39, 41, 65, 79, 81 f., 84, 87, 89, 115, 123, 131, 138, 146, 148, 175, 180, 248, 250, 269
pure relationship 78 f., 82, 101, 235

rationality 67 f., 77, 80 f., 83, 88 f., 200, 227–230, 235 f.
– romantic rationality 76 f., 79, 89, 230
realism (formal) 8 f., 93, 123, 129 f., 133, 141, 143–149, 152 f., 183, 201–203, 217, 220, 241, 265
realism vs. idealism 22, 29 f., 50, 63, 65 f., 74–77, 87, 89, 100, 104, 113, 121, 158 f., 170–172, 184, 192, 194, 198–202, 217 f., 232, 271
real, the 40 f., 163
reason 4, 19, 34, 45, 66, 77 f., 107–108, 167, 169, 216, 269
reciprocity 3, 47, 53, 56, 72, 87, 228
recognisability 4, 7–9, 13, 94, 136, 143, 146 f., 149, 211, 243 f., 265
recognition (affirmation) 51, 55 f., 61, 68 f., 85–89, 110, 116–119, 121–123, 136 f., 177, 196, 200, 207 f., 214, 232, 240, 258–261, 264–266, 268, 270
rejection 51, 86–88, 117, 125, 155–157, 159, 166, 189–191, 270, 272
religion 1, 21, 38, 69, 81, 88, 127, 134, 178, 184, 198, 235, 253
– religious 20–22, 31–33, 37, 65, 86, 122, 135, 160, 178, 198, 201, 204, 261, 271
representation (in art/drama) 7 f., 10, 76, 140, 142, 146, 153, 170, 190, 197, 212, 220, 272
– representative (figures) 100, 104, 106, 117, 122–124, 128 f., 133, 135, 170 f., 183, 202 f., 212 f., 216, 222, 227, 239, 241
representation of love 7, 42, 71–74, 76, 142, 170, 201, 220
responsibility 31, 41, 60, 69, 79, 107 f., 125, 129, 194, 200, 219, 223, 226–228, 230, 235, 240–242, 264, 268 f.
romance 6, 73, 75, 95, 110, 112 f., 224, 228
– commodification of romance 72 f., 89, 131, 139, 199, 231
romance (genre) 10, 52, 71, 83, 112 f., 167

Romanticism 26, 31–36, 53, 86, 105 f., 178, 255
romcom 114, 203, 206, 220

sacrifice 42, 57, 78 f., 84, 100–102, 106, 135, 160 f., 164, 172 f., 178, 180 f., 187, 192–196, 200, 216 f., 221, 225, 232 f., 237, 264
satisfaction 19, 21, 25, 37 f., 40, 49, 60, 78 f., 82, 108, 120, 134, 137, 158 f., 161, 178–181, 193, 195 f., 201, 208, 213, 262–264
scepticism 32, 66, 72, 75 f., 89 f., 105, 112 f., 150, 177, 198, 250
scholasticism 31 f.
security 3, 21, 39, 66 f., 78, 81, 85 f., 90, 98 f., 106, 113, 116, 120, 124, 128, 135–137, 141, 183 f., 213, 220, 227, 232 f., 240, 242, 268 f.
self-destruction 54, 86, 131, 137, 143, 160 f., 164, 168, 170, 181
self-expression 252–254, 256 f., 266
self-help 1, 64 f., 76, 79, 82, 131, 271
selfishness 30, 56, 59–61, 99 f., 133, 222, 224, 227 f., 230, 235, 240, 248, 260
– self-centred 100, 129, 195, 223, 227, 260
selflessness 20 f., 60, 65, 67, 80, 195, 230, 241, 260–263
self-realisation 13, 77, 79 f., 82, 89 f., 98–101, 103, 106, 123, 222, 227, 230, 235–238, 241 f., 268, 271
self-renunciation 102 f., 169
– self-abandonment 83, 106, 136
– self-abdication 58
– self-abnegation 156
self-sufficiency 20, 50, 62, 87, 90, 125, 132 f., 138–141, 156–158, 161, 175, 180 f., 191, 193, 219, 271
self-worth 85–87, 89, 109, 116 f., 123, 168, 200, 214, 216, 218 f., 240, 258, 261, 265, 270
separation 22, 24, 26, 33–41, 43–45, 54, 59, 63, 163 f., 173 f., 178–182, 240, 247, 263, 271
sex (intercourse) 24 f., 27, 29, 36, 41 f., 44 f., 52 f., 72, 79, 94, 109 f., 130 f., 134,

137, 142, 154, 156 f., 162, 174, 207 f., 210, 215, 233, 245, 253, 261
- sexuality   6, 10, 27, 29, 31, 33, 37, 39 46, 137, 175, 208 f.
sexual desire   2 f., 25, 27, 29 f., 32, 37, 49 – 51, 53, 58, 84, 109 f., 125 f., 129, 132, 137, 168 f., 173, 207 – 209, 215 – 217, 246, 253, 262
sociology   1, 9, 12 f., 17, 19, 65 f., 71, 74, 76 f., 81 f., 88, 97 f., 100 f., 122 – 124, 184, 222, 224, 226 f., 241 f., 260
sovereignty   57, 64, 118, 122, 175, 207
structure of feeling   1, 9, 98, 241
sublimation   38, 48, 53, 150
submission   56 f., 88, 113, 135 f., 207, 216, 221
substitute   1, 25 – 27, 41 f., 52, 66, 71, 85, 88, 158 f., 175, 184
subsystem   17, 65, 88
suffering   44, 46 – 48, 66, 80 f., 85 – 88, 105, 107, 126, 151, 153, 163, 172, 193 f., 200, 203, 214, 218, 247, 263
surrealism   153, 162, 164, 179
surrender   35, 55, 57, 64, 90, 135 f., 178
surrogate   37, 64, 128, 135, 137, 159, 163 f., 198, 250, 260
symbiotic   38 – 40
symbol   129 f., 151, 166
- symbolic   142, 150, 153, 165, 170, 203, 220
- symbolism   147, 203, 211, 216
symbolic order   40 f., 163 f., 205

therapy   1, 64, 76, 79, 82, 87, 89, 131, 133, 167
tradition   13, 18, 21 f., 30 f., 33, 36, 45 f., 64, 104, 108, 113, 137, 171, 177, 208, 236, 260, 269, 273
- post-traditional society   13, 66, 69, 77, 81, 86, 88, 97 – 99, 122, 227 – 231, 235, 239 f., 269 f.

tragedy   52, 101, 122, 137, 139, 167, 179, 184 – 186, 188, 200 – 202

unattainability   10, 38, 41, 47, 52 – 54, 64, 87, 135, 137, 164 f., 170, 175 f., 203, 214, 270
uncertainty   11 f., 47, 52, 58, 61, 64, 69, 78, 84 – 86, 103, 207 – 209, 231, 244, 253, 255, 259, 265 f.
unconditionality   19 – 21, 45, 65, 67 f., 77 f., 83, 89, 156, 158, 235, 240 f., 247, 250, 269
understanding/knowledge (of the other)   60 f., 100, 104 – 106, 108, 111, 120 – 122, 171, 197, 244 f., 247 f., 251 – 253, 256 f., 265 f., 270
uniqueness   2 f., 11, 50, 65, 70 f., 78, 83, 85 f., 104, 108, 235, 272
unity/union   3, 26, 29, 32 – 36, 38 f., 41, 45 f., 53 f., 105, 161, 163 f., 171, 173 f., 178 f., 182
- reunion   25, 34 f., 37, 39, 161, 163 f., 172, 178
unrequited love   52, 62, 87, 156, 160, 165, 175, 271

values   19, 29 f., 32, 53, 63, 66, 68, 80, 127, 129 f., 140, 183 f., 186, 188 – 190, 195 – 199, 224, 241 f.
violence   44, 49, 62, 94, 116, 119, 130, 137 f., 141 – 143, 153 f., 161 f., 164, 167, 180 f., 205, 271
vulnerability   11 f., 21, 31, 46 f., 58, 62 – 64, 81, 83, 85, 90, 122, 134, 175, 183, 189 f., 193 f., 211, 214, 216, 220 f., 242, 245, 249 f., 261, 265, 271 f.

wholeness   22, 25 f., 28, 31, 34 – 37, 40 f., 46, 90, 137, 139, 163 f., 170, 173 f., 176, 178 – 180, 271

Index of authors

Ahmed, Sara 195
Alcibiades 23, 28, 47–52, 135, 216, 262
Aristophanes 23–29, 31, 34, 36, 42, 46, 132, 135, 173f., 179
Artaud, Antonin 142, 149–151, 161–164, 169, 178f.

Badiou, Alain 58–61, 118, 121
Barthes, Roland 4, 6, 8, 47, 107, 113f., 136, 168, 243, 253, 255f., 263
Bataille, Georges 35, 42–45, 54, 137, 161f., 164, 178
Bauman, Zygmunt 78, 82, 88, 100f., 122, 171, 195, 227, 240, 242
Beauvoir, Simone de 57–59, 64, 118, 216, 246
Beck, Ulrich 66f., 71f., 74, 77–79, 88, 99, 122, 184, 227
Beck-Gernsheim, Elisabeth 66, 71, 74, 76–79, 88, 99, 105, 122, 184, 227
Belsey, Catherine 2, 10, 40f., 52, 72, 74, 89, 103, 184, 199, 202
Berlant, Lauren 10, 140
Bloom, Allan 24, 26f., 48, 50, 57
Butler, Judith 11f., 21, 119

Coleridge, Samuel Taylor 34

Diotima 27, 49f.

Firestone, Shulamith 58f., 64, 118, 121f., 216
Fisher, Helen 2f., 5, 157
Frankfurt, Harry G. 19f., 100, 219
Freud, Sigmund 36–40, 43, 45–48, 51, 134f., 137, 151, 269
Fromm, Erich 35, 45

Giddens, Anthony 78f., 82, 88, 101, 122, 235

Hegel, G. W. F. 34f., 38f., 54, 137, 178, 267

Illouz, Eva 1f., 10, 65, 71–77, 79–89, 95, 99–101, 103f., 112f., 115, 117, 122, 132, 171, 180, 184, 199, 202, 207f., 218, 227, 230f., 240, 257

Jaggar, Alison M. 4f., 17
Johnson, Samuel 47
Jung, Carl 39

Kristeva, Julia 7

Lacan, Jacques 40–42, 44f., 50–52, 56, 108, 133, 137, 162–164, 216, 219, 221, 260, 262
Lehmann, Hans-Thies 8, 267–269
Levinas, Emmanuel 11f., 58–61, 118, 121, 246f.
Lewis, Clive Staples 20, 219, 260
Lucretius 52f.
Luhmann, Niklas 68–71, 74, 77, 83, 88, 99, 227

Mahler, Margaret 39
May, Simon 1, 17, 20f., 23f., 26f., 29f., 32f., 35, 45f., 52–54, 65f., 135, 179, 250, 259

Nancy, Jean-Luc 22, 58, 61f., 121f., 181f., 193, 247
Nietzsche, Friedrich 54f., 57, 136, 151, 162, 198, 216, 246, 269
Nussbaum, Martha 18, 30f., 39f., 48–50, 62f., 121, 140, 247f.

Ortega y Gasset, José 216f.
Ovid 52f., 74, 214, 246

Plato 12, 22–24, 27–34, 36, 48–50, 52, 63, 132, 214
Plotinus 32

Rochefoucauld, François de la 74f.

Sartre, Jean-Paul 55–57, 118, 246, 260, 263
Schopenhauer, Arthur 44, 57, 162
Socrates 22f., 26–32, 34, 46, 48–52, 216, 262

Spaemann, Robert 20, 219, 260
St. Augustine 32

Wagner, Richard 53f.
Williams, Raymond 1

www.ingramcontent.com/pod-product-compliance
Lightning Source LLC
Chambersburg PA
CBHW020932180426
43192CB00036B/606